GOING 'GREEN'

Forests, fire and a flawed
conservation culture

GOING 'GREEN'

Forests, fire and a flawed conservation culture

Mark Poynter

Connor Court Publishing

Connor Court Publishing Pty Ltd

Copyright © Mark Poynter 2018

ALL RIGHTS RESERVED. This book contains material protected under International and Federal Copyright Laws and Treaties. Any unauthorised reprint or use of this material is prohibited. No part of this book may be reproduced or transmitted in any form or by any means, electronic or mechanical, including photocopying, recording, or by any information storage and retrieval system without express written permission from the publisher.

PO Box 7257
Redland Bay QLD 4165
sales@connorcourt.com
www.connorcourt.com

ISBN: 9781935501926

Cover design by Maria Giordano

Printed in Australia

Cover photo: Burnt ash forest, Howmans Gap, Victoria, October 2006, taken by David Bruce, Bushfire and Natural Hazards CRC.

CONTENTS

Acknowledgements vii

Preface viii

Prologue ix

1: Forestry and the emergence of a 'conservation culture' 1

2: Environmental activism: Building the culture 35

3: Media: Entrenching the culture 98

4: Conservation science: Advancing the culture 128

5: Politics and the bureaucracy: Implementing
   a cultural agenda 183

6: Politics and bureaucracy 1:
   Reserving Victoria's river red gum forests (2005-08) 220

7: Politics and bureaucracy 2:
   Reserving Tasmanian forests (2010-14) 262

8: Active management or benign neglect?
   The burning question for forest biodiversity 322

9: Going 'Green': Conservation or preservation? 367

Index 380

Acknowledgements

There are many who have supported this book either by reading draft chapters and providing feedback, or indirectly by (perhaps unknowingly) providing advice and encouragement over many years. This includes in alphabetical order: Alan Ashbarry for his depth of knowledge and insights into Tasmanian forestry; Barrie Dexter for his impeccable knowledge of Victorian forestry and especially its river red gum forests; Andrew Denman, for his detailed knowledge of the Tasmanian 'forests peace' process and his determination to examine the political and bureaucratic processes associated with it; Alan Eddy for his encouragement and review of several draft chapters; Peter Fagg for reviewing the draft and providing useful insights; Mike Leonard for his encouragement and thought-provoking analysis of early draft chapters; Peter Rutherford for providing insights into Victorian and NSW forestry over many years; Mike Ryan for reviewing a draft chapter and providing other advice and insights into Victorian forestry over many years; Roger Underwood for reviewing early draft chapters and his historical insights of Western Australian forestry over many years; and Peter Volker for his knowledge of Tasmanian forestry and insights into the so-called 'forest peace deal' process.

Preface

This book is primarily about the forests and woodlands of south-eastern (Victoria, NSW, Tasmania) and south-western (the south-west corner of WA) Australia. In large-part due to their proximity to our major population centres, they have been subject to around 50-years of dispute regarding their management and use.

For those who may be new to the public discourse surrounding this long-running conflict, the term 'Going Green' refers to the past 15 to 20 years of progressive aquiescence of Australian forest policy to the ideological aims of environmental extremism and hard-left politics, which are collectively and commonly described as 'green'.

This has happened as the nation's 'conservation culture' has matured to make it politically attractive to 'save' forests through ever-more national park declarations. Undoubtedly a substantial proportion of Australians believe that this can only be a good thing, and it is they who are most in need of reading this book.

Prologue

In late February 1975, I was a callow youth travelling on the late evening train bound for the historic town of Creswick, in west-central Victoria. Just a few months out of high school, I'd been awarded a three-year bursary by the Forests Commission to reside and study for a diploma at their Victorian School of Forestry.

The school had been training foresters since 1910 in Creswick's former hospital and doctor's residence built decades earlier at the height of the gold-rush era. Other buildings had since been added to serve as additional classrooms and student accommodation. All was set amongst expansive, sloping gardens, including an arboretum of mature trees of many Australian and overseas varieties.

At the bottom of the gardens, pedestrian entry to the school was gained through a heavy, 10-foot high wrought iron gate which, with the application of some strength, would open to the accompaniment of a grating metallic screech. Wide, white-gravelled pathways led uphill towards the oldest buildings. At night, the whole scene took on a Gothic aura that might not have been out-of-place in a Sherlock Holmes' mystery.

A photo of that gate had featured prominently for some years on a careers promotion pamphlet under the now politically-incorrect heading – "*A Gateway to a Man's Career*". But this nod to a long tradition of masculinity would change just a year later when four female students were accepted into the school. By the 1990s, around 50% of those studying forestry around Australia would be young women.

For decades, ten or fewer new students had started at the school each year, but in anticipation of a growing need for forestry expertise, the Forests Commission had increased their 1975 intake to nineteen. This was the first year of a new era of larger intakes that would eventually require the building of additional student accommodation and other infrastructure in the picturesque school grounds.

Unbeknownst to me, several other boys of the 1975 intake were bound for Creswick on the same train. This became apparent when we all alighted at the town's dimly-lit railway station at around midnight. Traditionally at this point, the assembled new boys would apparently navigate the several kilometres from the railway station to the school on foot. According to the school's folklore, they'd sometimes be met by a posse of the town's toughest lads on the bridge over Creswick Creek, within sight of the school gate. This was no friendly welcome to the country, but a stern marking of territory – a threat-laden warning to the 'new-kids-in-town' to keep their distance from the local 'sheilas'. It was a warning which would invariably fail to deter the progression of youthful romance.

Fortunately times had changed, and we were spared from any potential for an ugly confrontation when the school principal met us at the railway station with the school bus which promptly transported us to our new 'digs' and the beginnings of a career.

Writing this book has given some cause for reflection on what has changed since I began studying and working in southern Australia's forests all those years ago. In hindsight, I was probably fortunate to at least begin my career at a time when forestry was a widely valued and well regarded profession that provided important community services. However, it would be only a few short years before increasingly vocal critics would become politically influential and start to drive changes that were not always warranted or beneficial to the forests, rural communities, or indeed to foresters or forestry and timber workers.

Arguably, the greatest changes have been a dramatic reduction in the extent to which we now produce wood from our own native forests and a corresponding, but irrationally increased, belief that our forests are so threatened that they can only be 'saved' if they become national parks that exclude human uses.

In the late 1970s, our native forests were the nation's primary source of wood products while foresters were simultaneously establishing a substantial softwood plantation estate to help meet projected future needs. The Victorian Forests Commission's 1979 Annual Report noted that around 25,000 hectares of native forest was harvested that year right around the state. Nowadays, less than one-fifth of that area is annually harvested and regenerated and this is largely restricted to forests in central and eastern Victoria. Similarly substantial reductions of native forest use have also occurred in all Australian states. Yet despite this reduction, there continues to be extensively publicised claims that today's timber production is responsible for flora/fauna extinctions and supposedly all manner of dire environmental crises.

My earliest Forests Commission postings were to Cann River and Orbost in Victoria's East Gippsland region. These towns were then substantial hubs of native hardwood timber production. During 1978, when I was a junior Assistant Forester at Cann River, the town of just 400 people boasted five sawmills and most employment was either at those mills or in the bush supplying logs to them.

Passing through Cann River several years ago, I noted just one remaining sawmill employing 17 people. Further to this, a local service station attendant informed me that there were no longer any timber harvesting or log haulage contractors or their employees living in the town. During the writing of this book, the last Cann River sawmill closed and, according to a recent government report into social welfare, the town now has only 113 adult residents of which one-third are on disability pensions – the highest rate of social welfare of any town in the state.

Reflecting the times, Cann River now has three cafes operating along a 100 metre stretch of the Princes Highway – two of which were 'for sale' during my visit – eking out what must be a precarious existence as they compete with each other to service the needs of passing travellers.

The bigger town of Orbost, some 75 km to the west, has also seen a drastic decline of the local timber industry, but the surrounding region has always supported various agricultural enterprises and a stronger tourism sector based on closer proximity to the coast. A friend and his family who've run the town's clothing shop since the 1980s speak of how a strong majority of their sales were once male work clothes, but not so today.

Orbost now boasts five cafes compared to the single one that operated during my time living there in the early 1980s, but the town's population is diminished and of a different make-up compared to what it was back then. Today's adult residents are more likely to be either long-term unemployed or retirees lured from the city by the cheap real estate, pleasant lifestyle, and presence of a district hospital.

To a large extent, the reduction in Australia's native forest harvesting is related to the maturing of the nation's softwood plantation estate which, after the mid-1990s, enabled radiata pine to almost completely replace native hardwood in traditional high volume/low value uses, such as house framing. Indeed, plantation pine now meets around 80% of Australia's domestic sawn timber needs. This has understandably dislocated the timber industry workforce. Many traditional timber towns like Cann River and Orbost are set amongst extensive public native forests and are distant from the plantations estate. Consequently there was never scope for the local industry to transition to plantations.

As plantations have progessively reduced the imperative for native forests to solely meet the nation's timber needs, it has been only right and proper that a greater proportion could be devoted to conservation. However, the extent to which this has happened is problematic, particularly given that there is still strong demand for high value native forest hardwood that cannot be met by plantation pine. Consequently, we are now importing considerable volumes of tropical forest hardwoods from south-east Asia to make up for a shortfall in local sawn hardwood created in-part by needlessly over-reserving our own forests.

Arguably the most disturbing change has been a nationwide loss of community respect for forestry, foresters, and forest industries and their workers. Like any other resource use that has evolved over such a long period, forestry has had its share of errors, trials and tribulations as it negotiated a way between society's pragmatic socio-economic demands and the needs of the forests themselves. This has understandably created a cadre of critics whose concerns have been amplified by decades of 'save-the-forest' environmental campaigning and mostly one-sided media reporting. Ultimately this has equated the broad science-based discipline of forestry with only one activity – wood production. Foresters, who are in fact academically-trained scientists conversant with all aspects

Forests, fire and a flawed conservation culture

of forests, are now commonly regarded as 'loggers' supposedly driven by greed and entangled in systemic corruption. Indeed, disclosing that you work as a forester during polite dinner party conversation in the gentrified inner suburbs of our largest cities has at times engendered the sort of under-the-breath contempt normally reserved for criminals.

Wood production is now commonly discussed as though it is simply a "red-neck" recreational pursuit rather than a legitimate and gainful resource use. There is little public acknowledgement of its societal benefit in supplying essential materials, or consideration for the lives and livelihoods of those who work to produce them. They don't seem to matter to the inner urban demographic which is both the most remote from the forests, and yet the most vocally disaffected critic of what occurs within them. This is despite this same demographic harbouring arguably the greatest appreciation for the durability and beauty of native hardwood floors, furniture, and other decorative and practical uses.

Mostly missing from the public discourse has been the scientific justification for forest management and use. Forest science as practiced by foresters is integral to the hows and whys of forest management. But because it sits somewhere between the polarised protagonists – environmental activists and timber industries – it has rarely been afforded a voice. Accordingly, the average person in the street, armed with knowledge gleaned only from skewed mainstream media coverage, can be forgiven for believing that forests are being logged without any foresight or regulation simply because a few rural workers need a job.

In fact, Australia is a highly developed country in which forest-based resource use is more carefully controlled than ever before, and where information about forest management practices on state-owned land is freely available and open to community view, consultation and participation. While timber production is the lightning rod for environmental campaigns and 'green' politicking, only somewhere between 5 – 8% of Australia's forests and woodlands (on the combined area of public and private land) are being managed for this purpose on a cycle of harvest and regrowth.

The fact that such a relatively minor, and also renewable, use continues to be so contentious raises serious questions about environmental ideology and the way its messages have evolved into a superficially appealing and politically-correct 'conservation culture'. This equates environmental protection to a simple change in land tenure (usually from State forest to national park) which, on its own, does nothing to address real environmental concerns such as unnatural fire regimes and invasive pest species.

From the rarely heard perspective of a forestry practitioner, this book sets out to examine the basis of this 'conservation culture' and the ways in which it has and is being promoted, supported, and presented to the community specifically to influence political outcomes. While I am under no illusion that the revelations in this book can quickly turn around this culture and transform it into a more realistic appreciation of effective forest management, recording what has happened so far is surely the first step to informing a better future.

Unsavoury aspects of this story are eminently worthy of investigative journalism by the mainstream media which would inevitably reach a far greater audience. However, hard experience shows that our city-centric media is predisposed to uncritically accept and support the simplistic messages of environmental activism, rather than challenge their veracity as such an investigation would have to do. Indeed, it is primarily because of the uncritical publicity given to environmental activism, that we have become a society in which 'green' urban myths are now largely accepted as absolute truths, while rural realities are dismissed as self-serving myths.

Mark Poynter, Melbourne, March 2018

1

Forestry and the emergence of a 'conservation culture'

"Forestry is the wise and sustained fostering, production and use by people of the many values, benefits and products of forests"

Dr Leslie Carron, 1985[1]

Human use of forests and trees for shelter and other needs is as instinctive as hunting, fishing or gathering food. Furthermore, the manipulation of forests for particular needs must be recognised as a part of human evolution given the tens of thousands of years of careful forest and woodland burning by Australia's indigenous peoples to facilitate improved hunting and easier travel.

It is impossible to know exactly when such practices shifted from being instinctive and needs-driven to being deliberately 'managed' through restraint in deference to future needs. However, there are records from as early as the 5th century of monks on the Adriatic coast successfully establishing pine plantations to meet future fuel and food requirements; and Visigoths in the 7th Century instituting a code to ensure that oak and pine forests would be conserved for future generations.[2]

In Western Europe, the recognition of forest management as requiring dedicated practitioners to protect and husband valuable resources, stretches at least as far back as 1215 when the Magna Carta famously addressed King John and his 'archbishops, bishops, abbots, earls, barons, justices, foresters, sheriffs, stewards, servants, and to all his officials and loyal subjects'.

Whilst the earliest 'foresters' may have been largely preoccupied with protecting game animals in the royal estates, Europe's medieval forests were being variously used to supply wood for building and fuel, pitch for boat building, and stock grazing. The extent to which this was regulated is unknown although historical records suggest that systematic management of some forests to supply timber was occurring in Germany as early as the 14th century.[3]

By the 16th century, timber was assuming even greater importance due to the need for ships to facilitate a growth in maritime trade, a boom in housing construction, and the evolution of mining which required timber stays and fuel for ore smelting. At that time, the critical value of forests as sources of wood was eloquently articulated by John Evelyn in his 1664 forestry text, *Sylva*:

> Since it is clear and demonstrable that all arts and artisans whatsoever must fail and cease if there were no timber and wood in a nation ... I say when this shall be well considered it will appear that we had better be without gold than without timber.

The concept of creating a sustainable timber supply was reportedly first formally canvassed by German mine administrator, Hans Carl von Carlowitz, in his book, *Sylvicultura oeconomica, oder haußwirthliche Nachricht und Naturmäßige Anweisung zur wilden Baum-Zucht,* published in 1713. His treatise was prompted by severe wood shortages stemming from the depletion of forests for mining timbers and the resultant threat this posed to the livelihoods of thousands of miners working under his charge. His assertion that only so much wood should be cut as could be regrown would eventually become a guiding principle of forestry.[4]

Throughout the history of western civilisation, the trajectory of societal development has been typified by an initial period of uncontrolled resource exploitation to meet immediate needs, followed by growing concerns for the future as society stabilises. This has invariably created an imperative to, if possible, conserve and restore resources and establish sustainable supplies. It is a natural order which supports the notion of societal affluence being an essential pre-requisite for effective environmental conservation.

The use of Australia's forests since the beginnings of European settlement has largely followed the same script. For the first 100 or so years after the arrival of the First Fleet in 1788, there was largely uncontrolled exploitation of the most easily accessible (and seemingly inexhaustible) forests to serve an evolving society in need of sawn lumber, fuel, and a host of other commodities such as fence posts, poles, and mine stays. By the 1890s, the severe impacts of this demand were becoming apparent and this instigated a period of reflection, including several Royal Commissions which investigated the damage and recommended new ways forward.

Arising from this, starting around 1900, was a 10 to 20 year period during which the colony's various States gradually put in place the legislative, bureaucratic, and administrative structures needed to define a permanent public forest estate and manage its protection and use. During this time, European-trained foresters were recruited and several forestry training schools were established, including the Victorian School of Forestry at Creswick in 1910, in order to ensure a future supply of 'home-grown' forest managers more attuned to local conditions and issues.

Protracted efforts to enshrine a permanent public forest estate were to continue into the 1920s in the face of stiff opposition from some self-interested land developers and their political backers who were intent on continuing settlement programs. These programs had already resulted in substantial areas of forest and woodland being lost to agriculture during the preceding 50-years. [5]

Pioneering foresters such as Lane-Poole, Jolly, Swain and Kessell eventually convinced State governments of the need to permanently reserve large tracts of the surviving forest for future generations. Their efforts have been integral to the conservation of Australia's forest-based biodiversity.

From the 1920s to the 1960s, there ensued a period of evolving control over the use of forests as knowledge of the dominant eucalypt associations grew. This was occurring against a backdrop of critical demands for wood associated with the war effort in the early 1940s, and the subsequent post-war building boom to house a growing population augmented by increased rates of immigration. Ultimately, this unceasing demand for wood compelled the nation's foresters and the government institutions which they served, to embark on an extensive program of plantation establishment from the mid-1960s. This was based upon the realisation that the nation's native forests would, on their own, be incapable of fully meeting Australia's projected future needs.

The late-1960s arguably marked the start of the modern era of Australian forestry – a period in which the greater level of industrialisation associated with the plantation expansion program and the beginnings of export woodchipping, fuelled growing community concerns that challenged the evolving approach to forest management. From the early 1970s, State-based forestry authorities moved to address these concerns by directing greater efforts towards catering for non-wood values such as recreation, aesthetics, and conservation, in accordance with the key forestry concept of 'multiple-use management'.[6]

The history of Australian forestry is, of course, far more complex than the above very brief outline. Whole books have been written detailing its evolution both nationally and on a State-basis, so it cannot be afforded sufficient justice in just a few paragraphs. Suffice to say that, like any other natural resource use which has emerged over such a lengthy period, forestry has been shaped by pragmatic societal demands and has endured its share of mistakes, trials and tribulations which have provided the lessons for its progress and future successes.

Multiple-use forestry and its recent decline

The concept of multiple-use forestry stems from the 1920s teachings and writings of Sir William Schlich. A decorated Professor of Forestry at Oxford University, Schlich had been trained in the conservative German tradition and had for several years served as India's Inspector General of Forests.[7]

In his *Manual of Forestry*, first published in 1921, Schlich defined manageable forest values in terms of 'indirect effects' (such as landscape beauty, preservation of wildlife, prevention of erosion, and regulation of water), and 'produce' with an economic value (such as timber, honey, tannin, game hunting, or firewood). He reasoned that:

> It rests with the owner of the forest ...... to determine what the object of management shall be, and it then becomes the duty of the forester to see that these objects are realised to the fullest extent and in the most economic manner. In some cases the realisation of indirect effects requires a special and distinct management, but in the majority of cases they can be produced with economic working.[8]

Multiple-use forestry was subsequently widely embraced internationally. It is a landscape-scale concept which dictates that, with careful management, a range of values may be obtained from a forest in a complementary manner that doesn't unduly compromise the whole of other values.

It was never meant to entail every hectare of forest being managed to achieve multiple objectives to an equal extent at all times. Instead, it dictates that, for example, producing a sustained yield of timber from a portion of a large forest may not significantly affect the biodiversity values of the forest as a whole. Further to this, it can confer an overall conservation benefit by raising funds and employing workforces that can then help to manage whole-of-forest threats, such as damaging bushfires or pest plants and feral animals.

From the earliest days of organised government forestry agencies up until the 1960s, Australia's State-owned native forests were the nation's primary source of wood products (and would remain so until the mid-1990s) with their management necessarily focussed on meeting what were regarded as critical economic needs. The forests also contained a greater area that was unsuited to wood production and this provided for a range of non-wood uses and protected a suite of critically important environmental values at the landscape-scale. However, it is fair to say that up until this point, the deliberate management of Australian forests for 'multiple-uses' had been somewhat constrained by societal demands for wood commodities above all else.

By the 1970s, most forests in Australia's southern and eastern states were classed as multiple-use State forests managed by foresters still primarily focussed on wood production, but increasingly attempting to address the community's growing demand for recreation and interest in conservation. Ironically, this increased adherence to the multiple-use doctrine was occurring at around the same time as organised environmental activism began to publicly villify forestry as a destructive force rather than a critical community service. Arguably this began to sow the seeds for ending multiple-use forest management just as it was beginning to find its feet.

State forests being managed for multiple-use allow a broad range of economic, recreational (e.g. hunting, horse riding, dog walking) and public service (e.g. personal firewood collection, free access) activities that are either not permitted or are somewhat restricted in Australian national parks and conservation reserves overwhelmingly focussed on ecological protection and controlled visitor use. Despite its detractors, multiple-use management offers the greatest potential to optimise the societal benefit of forests.

The multiple-use forestry landscape of the early 1970s is somewhat exemplified by the Victorian experience. In response to growing community demands, Victoria's Forests Commission established a Forest Environment and Recreation Branch in 1971. According to the 1973 Australian Year Book, Victoria at that time had 2.64 million hectares of 'forest reserves' (combined State forests and national parks). Of this, approximately 61% was classified as 'productive' for timber, with 20% being classed as 'unproductive' and the remaining 19% being contained in parks, other reserves or vacant Crown Land.[9]

The defined area of parks and reserves included 107 areas totalling 26,000 hectares which had been set aside within State forests by the Forests Commission.[10] Outside of the State forests, there was a national park estate of around 150,000 ha contained in 20 parks, of which only three were of significant size: Wilsons Promontory, Wyperfield, and Mount Buffalo.[11] Aside from these dedicated formal reserves, recreation and conservation values were also incidental within the productive and unproductive parts of the 'forest reserves'.[12]

Unfortunately this doesn't give a full picture of the forest estate of the time as the Year Book's nominated 'forest reserves' didn't include the several million hectares of uncommitted Crown Lands. These were then also being managed by the Forests Commission and were concurrently protecting environmental values and providing further recreational opportunities. For all intents and purposes these were being treated at the time as additional multiple-use forests.[13]

Similarly, the notion that a significant majority of the forest was dedicated to timber production at that time needs to be considered in the context of whether those forests were capable of actually being used for that purpose. Later efforts to distinguish potential forest productivity for timber from actual capability to produce it have shown that the majority of forests were never usable for that purpose. In 1986, Victoria's Timber Industry Strategy would note that "although 63% of the (Victoria's) public land area is designated for wood production along with other uses, only about 31% is currently considered to be available *[i.e. suitable]* for wood production."[14]

It is difficult to directly compare the 1973 situation with today due to changed definitions of public land tenure as well as revised definitions of what constitutes forest and woodland. However, the extent of profound change which has occurred in the management of Victoria's forests over the past 40 years can be gauged by the growth of forested national parks and other formal conservation reserves from an estimated 7% of the State-owned forest area in 1972,[15] to 51% by 2012.[16] During this period, the number of Victorian terrestrial national parks increased from 20 to 45 (plus 13 marine national parks) alongside a host of other formal reserve categories.[17]

In addition, in keeping with the multiple-use management concept, there are now also significantly greater areas of informal reserves within Victoria's State forests that bring the total reservation of public forest primarily dedicated to biodiversity conservation to over 75% of the combined all-tenures area of public forest. In addition, much of the State forest outside the reserve system is unsuitable for timber production due to topographic and productivity constraints, and when these areas are also added to the formal and informal reservations, around 87% of Victoria's forests are now either reserved for conservation, or are effectively acting primarily as conservation reserves by dint of not being usable for commercial purposes.[18]

By 2013, the net proportion of Victorian public forest that was regarded as both available and suitable for long-term timber supply had fallen to around 9% after deducting inaccessible and unproductive areas and Code of Practice operational and management reserves; with a further 4% classed as potentially suitable subject to future market demand.[19] In addition to this, some of the ostensibly available State forest occurs in areas where there is no longer any timber industry presence.

Overall, there was a more than two-thirds reduction in the proportion of Victorian forest being managed on a cycle of harvest and regeneration for long term timber supply in the 27 years since 1986. This dramatic decline coincided with a sharp increase in forested national parks and other conservation reserves over the same period.

Similar trends have occurred in Australia's other southern states. In NSW, Labor Governments from the mid-1990s routinely promised new national parks to attract 'green' voters in the lead-up to state elections. Under Premier Bob Carr, the area of NSW national parks was doubled by the addition of over two million hectares from 1995 to 2005, most notably substantial new parks declared in the Brigalow cypress and ironbark forests and in the north coast forests. Carr's successor, Morris Iemma, followed the trend by declaring 21 new national parks during February 2006[20] and his successor, Nathan Rees, turned most of the Riverina red gum State forests into national park on his last day in office in December 2009.

This was said to have completed Labor's 'forest preservation program' which had been initiated by Premier Carr some 14 years earlier,[21] and has left just 9% of the state's area of public forest still classed as multiple-use State forest.[22] For his efforts, Carr was awarded Life Membership of the Wilderness Society and received an International Parks Merit Award from the World Conservation Union.[23]

In Western Australia, the outcome of the 2001 State election was said to have turned on forestry issues, with the newly-elected Labor government immediately ending timber production in those 'old growth' forests in the state's south west corner that weren't already reserved, and creating 30 new national parks covering 400,000 hectares. This included the 12 new parks that had already been agreed to under the Regional Forest Agreement (RFA) process of the late 1990s.[24] The WA Government's decision to override the RFA so soon after its signing was a political response to environmental activism that had continued to reject the RFA's attempt to broker an enduring compromise between forest conservation and use. The resultant increase in national parks and other conservation reserves drastically reduced the combined annual production of high value jarrah and karri sawlogs from 704,000 m³ under the 1994-2003 Forest Management Plan, to just 185,000 m³ under the 2004-13 Plan.[25] Despite this being touted as a solution to decades of conflict, environmental campaigns against forestry activities have continued.

In Queensland in 1999, the Beattie Labor Government began a 25-year phased-closure of its south-eastern native forest hardwood industry after striking an agreement with local environmental groups to end anti-logging campaigns in return for a complete transference of multiple-use State forests to the national parks estate by 2024. Under the terms of this SE Queensland Forests Agreement, half of the State forests were immediately reserved while the other half were to be intensively harvested to meet wood supply contracts until 2024. Concurrently a hardwood plantation expansion program was funded to enable the industry to transition to plantation-grown resource beyond 2024. Unfortunately, the plantation program failed comprehensively creating a problem for future wood supply that was further exacerbated in 2010 when Labor Premier, Anna Bligh, sold off the state's softwood and hardwood plantations to a large corporate entity which reportedly has little interest in producing future hardwood.[26]

In 2013, newly-elected Liberal Premier, Campbell Newman, abandoned the 1999 SE Queensland Forests Agreement, thereby stopping the complete transference of State forests to the conservation reserves estate as had been planned. However, his Government maintained the intensive harvesting regime in the remaining State forests and was thereby fostering the same end result of completely exhausting the available timber by 2024. With the replacement of the Newman Government by

Anna Palaszczuk's Labor Government in 2015, there was an expectation that a substantial increase in forested national parks would be promised at the next State election.[27] However this failed to eventuate when Labor retained Government in a close election contest in November 2017 during which both major parties committed to ongoing support for the forestry sector.[28]

In Tasmania, around 50% of the State-owned public forest area resides in national parks and other conservation reserves.[29] The net area of multiple-use State forest being managed on a renewable cycle of timber harvest and regeneration – and in which almost all anti-logging campaigns are focussed – comprises just 25% of the State-owned forests, and about 15% of Tasmania's total (public and private) forest area.[30] This reality is starkly at odds with the widely-promoted perception that Tasmania's forests must be 'saved' by stopping all wood production.

At a national level, just 11.5% of Australia's public (State-owned) native forests are now State forests being managed for multiple-use, with only about half of this being actually used for long term timber supply due to constraints such as topography, species suitability, and environmental management restrictions.

Conversely, 24.3% of the nation's public native forests are in formal conservation reserves including national parks.[31] This may not seem much, but when added to the huge and extensive areas of leasehold and privately-owned woodlands and forests with often little or relatively minimal usage, it can be assumed that the majority of Australia's forests are effectively acting primarily as conservation reserves. Albeit that most of this area harbors a range of environmental problems such as weeds, feral animal infestations, and unnatural fire regimes.

Despite this favourable weighting towards conservation-focussed land tenures, campaigns for huge new forested national parks continue on the premise that so-called 'unprotected forests' are being lost to wood production. The latest are calls for a 350,000 hectare so-called 'Great Forest National Park' in Victoria's Central Highlands which was proposed by the Greens and their environmental group associates prior to the 2014 State Election campaign and still remains under consideration by the Andrews Labor Government at the time of writing in late 2017; and a proposed 315,000 hectare 'koala preserve' that was promised by the Labor Opposition in the lead-up to the 2015 NSW State Election.

The waning importance of multiple-use forestry is exemplified by the declining use of State-owned native forests for wood production. According to *Australia's State of the Forests Report 2013*:

- Public native forest net harvestable area – has declined by 45% since 1995.
- Public native forest sawlog sustainable yield – has declined by 47% since 1992.
- Annually harvested area of public native forest – has declined by 32% since 2006.[32]

Apart from the dramatic increase in national parks and related conservation reserves, there are a range of other reasons for these changes, including:

- the maturing of Australia's softwood plantation estate displacing native forests as the nation's major source of wood from the mid-1990s;
- the associated displacement of native hardwood from its traditional low grade construction markets by plantation softwood;
- increased imports of tropical hardwood sawn timber displacing the use of Australian hardwoods for high value uses; and
- declining market demand for native forest woodchips due to the increased availability of higher grade plantation-grown hardwood chips.

However, while these factors have progressively reduced the imperative to harvest Australia's native forests, this needn't have detracted from the concept of multiple-use which could have easily accommodated a changed balance between conservation and human use. But, seemingly to Australia's environmental movement, multiple-use forestry is synonymous with 'destruction by logging', and their determination to replace it with a conservation-only land tenure that would (supposedly) forever guarantee 'forest protection' has underpinned the largely politically-expedient expansion of national parks over the past 20-years.

Over time, this has seriously damaged the multiple-use forestry concept. Bit-by-bit, forests with both wood and significant non-wood values have been appropriated into new national parks or other conservation-focussed reserves. In NSW, this has left only a small core of supposedly multiple-use forest designated primarily for only one use – wood production. Tasmania was also heading in that direction following the 2013 Tasmanian Forests Agreement until a new Liberal Government was elected in 2014 with a mandate to defer and likely abandon the transfer of 400,000 hectares of State forest into the national parks estate as had been planned under the previous Labor-Greens Government.[33] With the Liberals re-elected in March 2018, these forests are now likely to revert back to multiple-use State forest.

In Victoria and WA, although larger areas of multiple-use forests still remain, commercial wood production has been separated from broader forest management by the creation of profit-focussed commercial agencies – VicForests and the Forest Products Commission respectively – which are charged only with a narrow focus on planning and managing timber harvesting and regeneration. This is fostering a polarised model of forest management whereby forests are either classed as conservation reserves or for productive use which is somewhat at odds with the multiple-use concept of using and conserving within the same forest unit.

Accordingly, foresters who had traditionally managed a range of economic, conservation, and recreational activities in native forests have been increasingly shut-out of the latter two areas of expertise and most now only manage wood production and/or fire, despite the profession's considerable past expertise in non-wood related values and activities.

Emergence of a simplistic 'conservation culture' that prioritises land tenure over management

The enviro-political campaigns that have in large-part driven this change reflect the emergence of a pervasive community belief which equates guaranteed environmental protection with national park declaration. This effectively represents a false premise that nature can only be truly protected if humans are largely separated from it.

The high level of community acceptance of such a myopic notion is seemingly rooted in the environmental movement's landmark victories of the late 1970s and 80s, particularly against logging in the rainforests of north coastal NSW, and in defeating the proposal to dam Tasmania's Franklin River for hydro-electric power. As Bob Brown articulated in an interview in 1983, those victories set the template for the mass marketing strategy used in environmental campaigns ever since:

> We have grabbed ideas from wherever we could. We looked at the way other people sell cheese and paper tissues, how they do it, and thought if that sells an idea than how much more important that (it) be grafted by us into saving wilderness.[34]

The repetitive and emotionally-charged 'save-the-environment' sloganeering that has ensued over the subsequent three decades has meant that the substantial body of Australians with concerns about the environment but little inclination to take a deeper interest in it, have been regaled with errant assertions about forests being 1) in grave danger; 2) needing to be saved from greedy human exploitation; and 3) only able to be protected in national parks quarantined from human interference.

Mostly missing from this narrative has been any reference to the scale and proportion of continuing human resource use and the adverse social and economic consequences of stopping it. In the absence of such important context – in-part due to media reticence to publicise more complex alternative views – the slogans gradually strengthened into the powerful force which has fostered the cultural belief in the need for more national parks – from here on referred to as the 'conservation culture'.

The fact that more nuanced explanations of the true situation have made little or no in-roads to the acceptance of this 'conservation culture' is arguably indicative of human nature. It is reasonably well established that we are pre-disposed to allow negative events or portrayals (such as environmental threats) to shape our cultural beliefs to a greater extent than positive or beneficial activities (such as supplying wood). This is apparently most pronounced amongst those who believe the world to be dangerous, who are often also key figures in so-called 'cultural transmission networks'.[35]

The now prevailing 'conservation culture' is constantly being reinforced whenever influential celebrities or public commentators express their support for each new campaign to save the next forest or its cute animal/bird inhabitants; whenever opportunities arise for less-than-perfect forestry practices to be publicly vilified; or for relatively benign practices to be misportrayed as having far greater impacts than they actually do.

Hoffman (2001) contended that the evolution of many environmental beliefs has been built around milestone events that periodically reinvigorate and reinforce the core belief.[36] With regard to Australian forests, such milestones could include the wider introduction of the contentious clearfall-burn-sow silvicultural approach in wet eucalypt forests from the late 1950s; the start of the plantation expansion program in the mid-1960s; the introduction of export woodchipping in the late 1960s; the Terania Creek rainforest logging protest in 1978; periodic admissions of unsustainable timber harvesting rates in some regions; and occasional fuel reduction burn escapes.

Forestry may be somewhat different to most other issues opposed by environmentalists in that many of its perceived negative milestones are not discrete events that occur at one time, but enduring activities that have a starting point and either no known endpoint, or an endpoint many years in advance. A significant observation is that the environmental movement's response to these milestones is rarely to call for improved forestry practices, but mostly to demand their cessation.

As well as the force of public opinion associated with this 'conservation culture', its political influence is derived from the extent to which it has infiltrated the mindset of journalists who often set or heavily influence the political agenda, as well as the bureaucrats and politicians who devise and implement policies to address these agendas. There is much to suggest that amongst those groups, there is a very high level of acceptance of this cultural belief to such an extent that it has now become politically-incorrect to publicly defend or advocate the societal benefits of natural resource uses such as forestry, fisheries or mining.

Unfortunately, a culture based on a 'dumbed-down' version of what are mostly very complex issues begets overly simplistic solutions. The notion that 'saving' forests and their flora and fauna is as easy as evicting human use and declaring a new national park either ignores the need to actually manage the conservation threats, or affords it a minimal priority on an irrational expectation that once an area becomes a park it will somehow revert to its 'natural' condition. Such idealistic optimism forgets that threats such as feral animals, weeds, or unnatural fire regimes don't respect land tenure boundaries and won't just magically disappear of their own volition.

Looking at the environmental movement's self-proclaimed history of 'successes' suggests that it's never actually given much thought to active conservation/land management. The Australian Conservation Foundation (ACF) proclaims that it "stands for ecological sustainability", but its list of 'success stories' makes little mention of improving the management of ecological threats. Instead, its past 'conservation gains' have been based upon 'saving' forests from human use, or stopping proposed resource use developments, such as two Tasmanian pulp mills (including the most recent Gunns' proposal that was to be based entirely on plantation-grown wood).[37]

Similarly, The Wilderness Society (TWS) lauds its past success in terms of evicting human uses or stopping proposed developments in forests rather than specifically addressing critical ecological threats.[38] That TWS is overtly focussed on area-based conservation gains is also exemplified by its close ties to the US-based Wildlands Network (formerly the Wildlands Project)[39] which has advocated huge reservations of rural lands for biodiversity conservation – some critics have claimed it to be proposing the reservation of as much as 50% of the land mass of the continental USA.[40]

One of the founding scientists of the US Wildlands Project, Dr Reed Noss, articulated its underlying philosophy in a World Wildlife Fund (WWF) Discussion Paper in 1995:

> The native ecosystems and the collective needs of non-human species must take precedence over the needs and desires of humans ... Many ecologists (myself included) would just as soon see huge areas of land kept off-limits to human activities of any kind.[41]

The current Wildlands Network website lists the 'Australian Wilderness Society' as one of its conservation partners, and the Network's founder and current Board member, Michael E. Soulé – Emeritus Professor of Environmental Studies and Father of Conservation Biology at the University of California – was an inaugural member of The Wilderness Society's WildCountry Science Council,[42] and helped to broker a Cooperative Agreement between TWS and the US Wildlands Project.

The Wilderness Society also has a WildCountry Vision which is described on its website as:

> ... an Australia-wide program designed not only to protect our remaining wild places and wildlife, but also to help define the path towards restoration ... It is inspired by a project in the United States - the Wildlands project, which has set a new agenda for the US conservation movement.[43]

The core thrust of the WildCountry Vision for south-eastern Australia is to end timber harvesting and expand forested national parks by reserving new areas and joining existing parks together to create the:

> ... Melbourne to Brisbane Conservation Link [which] would establish a visionary and world-class national Park system for south-eastern Australia that would provide for the long-term biological and recreation needs for Australia's most densely populated region.[44]

While this would supposedly foster 'environmental restoration', the WildCountry Vision contains no specific strategies for improving fire management or reducing the impacts of feral animals and weeds, which conservation scientists generally acknowledge as critical to restoring ecological integrity. Instead, it claims that hugely expanded conservation links "will buffer against large-scale landscape-wide impacts such as climate change, bushfires and widespread habitat destruction".

Accordingly, it appears that implementing the WildCountry Vision would involve little more than re-badging the land tenure of huge areas of public forests, including a majority that, although not currently national park, is either already informally reserved or effectively already acting as a conservation reserve by dint of not being usable for any commercial purpose. In reality, this would add little if anything to the existing state of biodiversity conservation – apart from creating an illusion of environmental protection.

The unmistakable implication from all this is that Australia's mainstream environmental movement has seriously skewed priorities. It regards human use as the primary threat to forests despite the reality that modern-day resource use in this country is limited in extent, well planned, and highly regulated. Whereas it inexplicably affords almost no priority to non-human threats such as severe bushfires and feral pests that are far less controllable with almost unlimited potential to cause catastrophic, landscape-wide ecological impacts.

Environmental threats and the irrelevance of land tenure

Returning unnatural fire regimes back to something like what they were prior to European settlement is a major challenge in southern and eastern Australia's forests. While there is continuing debate about what the natural, pre-European fire frequencies actually were across the range of ecosystem types, it is now pretty well accepted that, apart from wet schleophyll forests and rainforests, the great majority of Australian forests and woodlands have been regularly burnt by a combination of aboriginal and lightning-ignited fires for tens of thousands of years.[45]

It follows from this that it is unnatural for the vast majority of forests to be left unburnt for many decades as has happened to varying degrees since European settlement, particularly in the more settled regions of southern and eastern Australia. Doing less burning and actively extinguishing bushfires as soon as possible after they start, eventually allows huge build-ups of flammable material that inevitably fuels unnaturally severe and uncontrollable bushfires. Now commonly referred to as 'mega fires' they have become more prevalent over the past 15-years with catastrophic impacts on soil, water, and wildlife.[46]

Despite it being such a major ecological threat, Australia's environmental groups have traditionally said almost nothing about forest fire, and if anything, have campaigned (often subtly) against broadscale prescribed burning that is one of the few tools which can mitigate the severity of bushfires. A recent search of The Wilderness Society's website for 'forest fire management' found only two substantive references, both of which were reports produced in the aftermath of recent severe Victorian bushfires.[47] Neither report completely dismisses the practice of fuel reduction burning, but both only grudgingly accept it as a protective tool that (they believe) should be overwhelmingly restricted to narrow strips of land adjoining private property, with its broader-scale use in more remote forests strongly discouraged.

A similar search of the Australian Conservation Foundation website also found only two short articles, both of which claimed to be supportive of fuel reduction burning but then dismissed it as having little benefit. The second of these reports claimed that:

> … low intensity high frequency fuel reduction burning, in a conservation sense, is … nothing like the 'natural' regime of high intensity low frequency fires that many woodland and forest ecosystems are adapted to. It too often results in a loss of species diversity and habitat quality.[48]

This view contains a serious misconception about the frequency of fuel reduction burning which, in reality, is most often of the order of 15 to 20 years between treatments and so is generally not what could be regarded as 'high frequency'. This also somewhat exemplifies opposition to the practice which is all too often – including amongst parts of academia – predicated on misunderstandings of the limited logistical capability of fire management agencies and the narrow window of opportunity they have to operate in.[49]

Worse still is that it also misrepresents the 'natural' pre-European fire regime as being akin to the present era of low frequency but very hot fires burning in heavy fuels, whereas – as historian Bill Gammage (and others) has pointed out – there is a "tsunami of evidence" that the pre-European fire regime of most Australian forests and woodlands was characterised by more frequent low intensity fires burning in light fuels kept that way by the regularity at which they were burnt.[50]

The refusal of the environmental movement and some conservation scientists to fully support the use of fuel reduction burning effectively amounts to advocacy of a minimal or 'no-burn' strategy for the vast majority of forests. This would ultimately be ecologically catastrophic based on previous experiences of severe fires burning in heavy fuels around Australia, and is astonishingly at odds with the claimed intent of environmental activists to restore and protect the environment.

Similar searching of the TWS and ACF websites for 'control of feral animals and weeds' at least found that there was recognition of this as a critical problem and an acknowledgement of the need for action, despite there being no suggestions about how this could be achieved. In stark contrast to its lack of any suggested strategy to improve environmental management, the TWS website outlines various strategies to attack the legitimacy or integrity of resource uses in order to convince State governments to declare new national parks.

Given the practical limitations of protesting, it may well be understandable that the environmental movement focusses primarily on stopping human uses by creating ever more parks and reserves. It is relatively easy to campaign against timber harvesting, for example, by blockading coupes and chaining protesters to equipment, or even by promoting market boycotts of wood products. But there is no easy template for drawing attention to poor fire management or unsatisfactory efforts to control pests and weeds. In any case, the type of management interventions needed to address these problems – such as controlled burning programs, or poison baiting and/or shooting – are largely at odds with the movement's philosophical belief that Australian ecosystems are exceedingly fragile and must be left undisturbed.

Effective fire management relies on human forest use to generate income, employ workforces, and create imperatives to undertake activities such as maintaining forest road access and fuel reduction burning. Accordingly, campaigning for new national parks as a means of ending human forest use is tantamount to reducing the capacity to control and manage fire, thereby fostering worse environmental outcomes.

Devaluing the national parks concept?

While the emergence of a 'conservation culture' has substantially damaged the concept of multiple-use forestry, it has arguably also devalued the national parks concept. Instead of being an instrument for protecting outstanding natural features, the parks concept has in Australia been misappropriated by the environmental movement as a political tool specifically to end human resource uses and guarantee that they can never return. The over-use of the concept for this purpose has fostered many inappropriate declarations of huge new national parks in highly modified or degraded landscapes with few special features, which are unworthy of such high level conservation status.

At its 10th General Assembly in New Delhi in 1969, the International Union for the Conservation of Nature (IUCN) recommended that areas declared as a national park should be relatively large with "one or several ecosystems not materially altered by human exploitation and occupation, where plant and animal species, geomorphological sites and habitats are of special scientific, educational, and recreational interest or which contain a natural landscape of great beauty."[51]

Many of the forested national parks declared in Australia over the past 20-years would struggle to meet such a definition either due to extensive human use that was occurring right up to the time of their declaration, or an absence of sufficiently significant features. Invariably, the areas with truly significant landscapes, flora and fauna, or geomorphology had already been declared as national parks years before (e.g. Wilsons Promontory, Blue Mountains, Tasmanian highlands), or were contained in some lesser form of reserve category more appropriate to their level of significance. Indeed, many of the recent forested national park declarations have in-part embraced substantial areas of already existing reserves of lesser conservation status and – without sufficient justification – rebranded them to a higher status.

A good example has been the recent expansion of national parks in the extensive floodplain river red gum forests and woodlands on both the Victorian and NSW sides of the Murray River. These are amongst the most modified landscapes on the continent chiefly due to natural flooding regimes being over-turned by 80 years of river regulation for irrigated agriculture. In addition, most parts have had up to 150-years of widespread cattle grazing, selective timber harvesting, and exclusion of low intensity fire. Some have contended that this has turned formerly extensive reed beds, grasslands and low density woodlands into dense forests that are now regarded as 'natural' despite not even being present prior to European settlement.[52]

The building of large impoundments on the Murray River itself (Lake Hume) and on various major tributaries has substantially curtailed the natural frequency and duration of winter-spring flooding as water is stored for later release to irrigate crops during drier summer months. It was this winter-spring flooding which, for example, in the region's largest Barmah and Millewa red gum forests, naturally occurred on average 8 out of every 10 years and had for thousands of years sustained the trees and a wide range of habitats supporting specially-adapted flora and fauna.[53]

Clearly, declaring most of these river red gum forests as new national parks has done nothing to relieve the summer drought stress they now routinely suffer due to the altered flooding regime. By comparison, renewable selective timber production and cattle grazing were relatively minor factors, but environmental campaigning for the new parks was focussed on removing them under the false premise that this would 'save' the forests. These forests will continue to rank amongst Australia's most stressed environments despite their new national park status.

This episode also exemplified how pre-existing regimes of mixed multiple-use land management offering adequate and appropriate levels of environmental protection, have been overridden by political acquiescence to the ingrained cultural belief that only national park tenure can truly protect the environment. On the Victorian side of the Murray River prior to the declaration of new parks in 2008, over two-thirds of the region's river red gum forests and woodlands were already primarily conserving biodiversity in existing national parks, State and regional parks, nature conservation/natural features reserves, other reserves, and State forest reserves. Following the Victorian Government's park expansion exercise, the proportion of multiple-use State forests plummeted from 43% to just 5%.[54]

A similar pattern is evident in other recent national park declarations in forests such as those in Victoria's Otways Ranges, the NSW cypress pine and ironbark forests and woodlands, and in parts of Tasmania's Southern Forests which all had extensive histories of human use and were still being used right up until they were declared as national parks – or added to a World Heritage Area in the Tasmanian instance.

As many, if not most, of these new forested national parks declared since the mid-1990s have stemmed from election-eve commitments to appease environmentally-concerned, so-called 'green' voters, they've been largely emblematic of a simplistic political land grab. This has now largely overridden former conservation reserve networks which had been based on detailed science-based considerations of natural values, impacts and needs which recognised that effective environmental protection is more complex than simply reserving more land.

Both the National Reserve System and the Regional Forest Agreement (RFA) processes of the mid to late 1990s incorporated the need for a rationally-based comprehensive, adequate, and representative (CAR) network of conservation reserves. This was founded upon the nationally-agreed JANIS criteria developed from 1993-96 by an inter-governmental working group comprising representatives from state forestry and conservation agencies, and the CSIRO.

The JANIS criteria specified minimum levels of representation for ecosystems in the CAR reserve system. These minimum reservation targets are:

- 15% of the pre-1750 distribution of each vegetation type (higher than the internationally accepted 10% target);

- at least 60% of the remaining extent of defined 'vulnerable' ecosystems; and
- 100% of the remaining extent of defined 'rare' and 'endangered' ecosystems.[55]

These aims exceeded those of most overseas countries, regions or provinces with economically important forest use industries. For example, in 2001, Canada had just 7% of its forests in reserves.[56] Canada's British Columbia province was aiming to include just 12% of its forests in protected areas, whereas only about 2% of Sweden's productive forests, and about 1% of Finland's productive southern forests were contained in protected areas in 2003.[57] Most Swedish and Finnish productive forests are privately-owned, whereas like Australia, most of British Columbia's wood production forests occupy public land.

Overall by 2000, the RFA process had boosted the total area of native forests in formal and informal conservation reserves in Australia's ten RFA regions by some 36%, with the total reserved area within those regions comprising 28% of the pre-1750 area – significantly exceeding the nationally-agreed 15% target.

Only in vegetation classes that had been substantially reduced by agricultural development was the 15% pre-1750 area target unable to be met.[58] For example, 34 of Tasmania's 50 forest communities had at least 15% of their pre-1750 extent protected in reserves, but 10 communities that occur mainly on private land had less than 7.5% protected.[59] Over 40% of Tasmania's native forests (both public and privately owned) are in formal conservation reserves – reportedly one of the highest rates of formal forest reservation in the world.

In large part, the commitments made in the RFAs have now been overridden by national park expansions over the past 15-years. Nationally, 18% of Australia's native forests (i.e. public and privately-owned) are now in formal conservation reserve categories thereby significantly exceeding the minimum 10% target nominated by the International Union for the Conservation of Nature.[60] In stark contrast to forests, ten Australian bioregions located in non-forested, arid central and northern areas of the country have less than 1% of their area formally protected in reserves.[61]

With regards to national parks, European countries often encourage economic use as a means to fund park management and maintenance. However, Australia has adopted the American philosophy of park management whereby the land is sacrosanct and extracting natural resources, such as timber or minerals, is an anathema.

In Australia, only tourism offers potential for significant revenue-raising in national parks. Eco-tourism is a growing industry sector mostly revolving around activities in parks or other high value conservation reserves. Some areas such as Tasmania's high country, Victoria's Grampians region or its alpine areas, or the Blue Mountains in NSW, have exceptional and marketable natural features that can underpin commercially viable eco-tourism. However, most of our forested national parks lack such qualities and receive only low levels of tourist visitation.

This was exemplified by the NSW Government's 2005 declaration of 348,000 hectares of new national parks and reserves in the cypress pine/ironbark forests and woodlands of the state's north central-west. Prior to this, the centerpiece of the region's conservation reserves had been the small but iconic 21,000 hectares Warrumbungle National Park, which had been reserved since 1952. Based on 1996-97 economic data for the Warrumbungle National Park, scaled forward with CPI to match 2005 forest industry data, it was estimated that 211,000 additional visitors to the region's expanded parks and reserves system would be required to replace the lost jobs and reduced regional gross output caused by substantially paring back the local timber industry to create the new parks.

Such dramatically increased visitation was at the time regarded as highly unlikely given that reportedly only 90,000 people were visiting the far more spectacular Warrumbungles National Park each year, while the much larger new parks and reserves mostly contain flat and comparatively featureless country, most of it formerly known as the Pillaga Scrub.[62]

The adverse socio-economic outcome of national park declaration has certainly been evident in the southern NSW town of Mathoura. It sits on the western edge of a substantial river red gum forest which was declared as the Murray Valley National Park in 2010 after generations as a State forest supporting a vibrant local timber industry. By 2014, the town had reportedly suffered a 28% reduction in visitors and the loss of timber industry jobs four years earlier had since been compounded by the closure of the IGA supermarket, the bakery, and a local bed and breakfast, while the pub was up for sale.[63] As former State forest activities such as free and unregulated camping, hunting, firewood collection, and horse-riding were now either prohibited or restricted in the new national park, it appeared that tourist visitation had effectively been discouraged.

Even where tourist visitation rates are relatively high, the economic yield that tourism generates can be very low if there is a lack of appropriate infrastructure and experiences available to visitors.[64] However, attempts to redress this problem by encouraging high-end eco-tourism developments in those national parks where it could well be viable, are typically opposed by environmentalists – usually the same ones who have hypocritically campaigned for new national parks by extolling the supposedly superior socio-economic credentials of eco-tourism compared to resource use industries.[65]

The refusal of most environmental activists to countenance any economic use in national parks or other conservation reserves lends credence to the view that most see park expansion as a step in the re-creation of a natural wilderness where humans can visit only on foot. There is nothing particularly wrong with such a vision for appropriate parts of the landscape, but to campaign for it by claiming that it will provide a huge boost to local jobs and regional incomes is simply dishonest when it comes with a hidden intent to oppose any tourism development beyond free-of-charge bushwalking.

Without any significant revenue-raising capability, most of Australia's national parks are chronically underfunded. This is unlikely to change given that they must compete for government budgetary appropriations against higher-order social priorities such as health, law and order, and education. Accordingly, the prospect of removing economic activities from further large areas of forest to add them to an already under-resourced parks and reserves network makes little sense.

By international standards, our conservation reserve network has traditionally been ranked amongst those with the lowest budgets and staffing levels per hectare in the developed world. In 1999, Australia was spending about double that being spent per hectare in developing countries, but only about one-sixth of the average expenditure per hectare in developed countries such as the USA and Canada. In 2006, the CSIRO was concerned enough to call for increased funding to avoid Australia's protected areas becoming 'paper parks' that meet no conservation objectives and become havens for feral animals, weeds, and sources of fire.[66]

There is little to suggest that per-hectare park funding has increased since then. Governments still love to declare new national parks, but allocating the requisite funding to manage their growing area is far less 'politically sexy'.[67] For example in Victoria in early 2016, it was reported that national park funding had declined by 37% over the previous three years creating significant difficulties for the management of environmental threats and maintenance of park infrastructure.[68] A year later, the Andrews Labor Government was investigating the alleged misuse of expense accounts by Parks Victoria's senior executives amidst claims of insufficient funding to maintain and adequately service even basic in-park visitor facilities.[69]

This is not just an Australian problem. In late 2015, it was reported that the US National Parks Service budget had declined by 12% over the previous five years creating a steadily growing backlog of infrastructure maintenance estimated at $11.5 billion.[70] As government funding has declined, US park managers have been urged to compensate through renewed campaigns for private donations and by offering corporate sponsorships that, for some, have raised the unpalatable prospect of, for example, the "Starbucks' Yosemite National Park".[71]

Questioning the value of national parks

The dramatic extent of conversion of multiple-use State forests to national Parks in Australia over the past 20-years has done little to address the threat posed to biodiversity by feral animals, weeds, and unnaturally damaging bushfires. This view was strongly articulated in a range of stakeholder submissions to the 2013 NSW Legislative Council inquiry into the management of that State's public lands.

> Beyond questioning the quality of management that is being delivered in national parks under the current approach, some Inquiry participants challenged the very premise that national parks are the best and most appropriate means of conserving biodiversity and protecting the environment. They asserted that the change in tenure to the national park estate has had a detrimental impact on not only the environmental value of the land, but on all other values that make the land significant.[72]

In addition, Australia's political rejection of the multiple-use forest management philosophy in favour of conservation-focussed national parks appears to be at odds with international trends. For example, the United Nation's (UNESCO) Biosphere reserves scheme has been widely acknowledged as a successful means of integrating conservation and agriculture benefits through a 'whole-of-landscape approach to land management'. While this approach is pursued enthusiastically elsewhere around the world, it is being demonstrably rejected in Australia as the landscape-scale multiple-use forestry concept is lost with each national park declaration.[73]

Similarly, the NSW Legislative Council inquiry heard evidence of successful multiple-use land management approaches in Zimbabwe, France, and an acclaimed Australian example – the Innamincka Regional Reserve in South Australia where the protection of significant wetlands occurs alongside controlled stock grazing, mining, petroleum exploration, and managed recreational use.[74]

Arguably, most State forests could lay claim to the successful integration of conservation with use – be it commercial or recreational – and this is exemplified by the ongoing determination of environmentalists to turn State forests into national parks because they contain high conservation values. One of the best examples of successful multiple-use land management was the riverine red gum forests of northern Victoria, which since the early 1980s had protected internationally-recognised Ramsar-listed wetlands alongside managed stock grazing and selective timber production which had occurred for over 100-years. This was overturned in 2008 when the Victorian Government converted 90% of the multiple-use State forests to a mix of national parks and other reserves.

The effectiveness of national parks and other 'protected areas' in maintaining biodiversity is also being increasingly questioned in academic circles. Given the rarity of historical data to enable comparisons of conservation parameters before and after park or 'protected area' declarations, assessments of park value over the past 30-years have usually involved small local studies comparing conservation values within parks against those in adjacent non-park areas.[75] Clearly there is considerable scope for bias in such comparisons subject to the nature of these non-park areas, such as cleared farmland.

While these studies have unearthed examples of worrying declines in biodiversity within some 'protected areas' in Africa and the tropics, a recent analysis by Coetzee et al (2014) found that generally across the globe, 'protected areas' have higher conservation values than outside areas. However in Australia, they could find no evidence that 'protected areas' had significantly better conservation values than outside areas.[76] While this was tentatively attributed to a lack of local studies, it may also reflect the fact that a significant portion of Australia's national parks have only been relatively recently declared, or that the prevalence of landscape-scale fire regularly equalises the conservation state of reserved and non-reserved areas.

While Coetzee et al is generally positive (although somewhat uncertain) about the conservation value of 'protected areas', others aren't so sure. In 2012, noted conservation scientists Peter Kareiva, Michelle Marvier, and Robert Lalasz observed that the 'war to protect nature' is being lost despite the hard-fought battle which has led to a 10-fold increase in national parks and other 'protected areas' since 1950. Even though around 13% of the world's land mass is now 'protected', biodiversity continues to decline.[77]

In agreeing with this prognosis, Australian conservation scientists, Professor Bob Pressey and Euan Ritchie, recently asserted that this was due to the world's national parks and other conservation reserves being largely located in the wrong places – mainly in high, cold, arid, steep, or infertile areas where they supposedly least interfere with extractive uses such as agriculture, mining, or forestry.[78]

The notion that national parks and other reserves are not protecting the most productive and biodiverse ecosystems may well be true in the international context and may also hold true for Australia across all terrestrial ecosystems. However, it is a misconception that Australia's most productive forests are not adequately 'protected' from extractive uses after the past 20-years of expansion of forested parks and reserves in NSW, Victoria, south west WA, south east Queensland and Tasmania where timber industries were always most concentrated. This is glaringly obvious from the extent to which Australian native timber production has fallen as the area of forested parks and other conservation reserves has grown.

Pressey and Richie also warned that assessing conservation progress only in terms of the number and proportional extent of protected areas can provide an illusion of environmental improvement that may be completely unjustified in the absence of any greater efforts directed specifically towards addressing key environmental threats. This has particular resonance for Australia's forests where the wide celebration of each new national park as a 'conservation gain' ignores the fact that this alone does nothing to address the most critical ecological threats.

Australia is a vast place, and it would be almost impossible to effectively manage critical environmental threats to an equal extent across the whole landscape. Threats that face Australian forests and woodlands know no boundaries and exist across all land tenures. In reality, pest management is probably no better or worse in national parks compared to adjacent State forests or other 'unprotected' public lands. Fire is somewhat different because evidence and logic dictate that the capability to manage unnatural fire frequency and severity declines once an area becomes a national park, particularly if it was formerly State forest, notwithstanding that even in these so-called 'unprotected' forests, fire management is still far from satisfactory and is no longer being as actively addressed as it once was. These realities are at odds with the key assurance of the 'conservation culture' that once an area becomes a national park it has been saved.

The environmental impacts associated with our prevailing 'conservation culture' are not just limited to Australia. A recent analysis of forest products imports over the period from 1994-2010 noted that declining hardwood production from our own native forests had corresponded with a doubling of wood product imports from Asia-Pacific countries over that period. This had reportedly required an estimated 2.7 million hectares of tropical rainforests to be harvested in those countries to meet Australia's continuing strong demand for decorative and durable hardwoods. As the management of timber harvesting in those countries can be problematic due to poor planning and endemic corruption, this is likely to have had far greater environmental impacts than would have occurred if Australia had continued to produce significant wood volumes from its own forests.[79]

In 2010, Australia was ranked sixth in the world for per capita forest cover with our average of 7.1 ha per person far exceeding the world average of just 0.6 ha per person.[80] Coupled with this is that Australians have traditionally been amongst the top five per capita consumers of natural resources on the planet.[81] Under these circumstances it could be argued that it is morally reprehensible for Australia to quarantine its relatively abundant native forests from human use while out-sourcing its demand for forest products to developing countries beset with far more serious environmental problems.

Chapter 1 Endnotes

1 L.T. Carron, 1985, *A History of Forestry in Australia*, Australian National University Press.

2 N. Mirov and J. Hasbrouck, 1976, *The Story of Pines*, Indiana University Press.

3 S. Buttinger, 2013, "Idee der Nachhaltigkeit [The Idea of Sustainability]", *Damals* (in German) 45 (4):8.

4 Environment and Society Portal 2015: *Hans Carl von Carlowitz and 'Sustainability'*.

5 J. Dargavel, 1995, *Fashioning Australia's Forests*, Oxford University Press, Chapter 1

6 F.R. Moulds, 1991, *The Dynamic Forest: A history of forestry and forest industries in Victoria*, Lyndoch Publications, Chapter 11.

7 Wikipedia 2015, *Sir William Schlich* – accessed 6 February 2015.

8 W. Schlich, 1925, *Schlich's Manual of Forestry: Volume 3 Forest Management, 5th Edition*. Bradbury, Agnew and Co. London, p. 1

9 Australian Bureau of Statistics 1974, *Australian Year Book 1973*, Chapter 24, p. 879.

10 F.R. Moulds, op. cit., pp. 146-148.

11 D. Garden, 2012, *Conservation Journeys – A Short History of the Victorian National Parks Association*. Accessible from the VNPA website.

12 F.R. Moulds, op. cit., pp. 146-148.

13 Ibid.

14 Government of Victoria 1986, *Victoria's Timber Industry Strategy 1986*.

15 This figure is derived by assuming the current day total public forest area of 6.538 million hectares and then determining the proportion of this occupied by the reserved areas in State forests and in national parks specified in the *Australian Year Book 1973* (see endnote 5).

16 Australian Government, Department of Agriculture/ABARES 2013, *Australia's State of the Forests Report 2013*. Criterion 1, Table 1.6, p. 41.

17 D. Garden, op. cit.

18 State of Victoria, Department of Environment and Primary Industries 2014, *Victoria's State of the Forests Report 2013*. Indicator 1.1a Table 3, p. 54; and Indicator 1.1c Table 1, p. 61.

19 State of Victoria, Department of Environment and Primary Industries, op. cit., Indicator 2.1, p. 92.

20 Elise Eggleton, 2006, "Iemma follows Carr down national parks trail", *The Australian*, 21 February 2006.

21 Wikipedia 2015, *Nathan Rees, former NSW Premier* – accessed 7 February 2015.

22 Australian Government, Department of Agriculture/ABARES, op. cit.

23 Wikipedia, Bob Carr.

24 P. Attiwill, P. et al, 2001, *The Environmental Credentials of Production, Manufacture and Re-Use of Wood Fibre in Australia*. Prepared for the Australian Government Department of Agriculture Fisheries & Forestry by the University of Melbourne, p. 128.

25 D.S. Houghton, 2012, "Protecting Western Australia's old growth forests: the impact of the 2001 policy changes", *Australian Forestry* Vol 75:2 (June 2012).

26 Sean Ryan, 2016, "Innovation is the only answer left for the Queensland Hardwood Timber Industry", *Australian Forest Grower*, Vol 39: Spring 2016.

27 Ibid.

28 *State Election: We do things differently in Queensland, but timber industry mounted effective campaign*, by Jim Bowden, *Timber and Forestry E-News*, Issue No. 491, 1 December 2017

29 Australian Government Department of Agriculture/ABARES, op. cit.

30 Forestry Tasmania 2014, *Stewardship Report 2013-14*, Appendix 2, Table 2.1, p. 3.

31 Australian Government Department of Agriculture/ABARES, op. cit.

32 Australian Government Department of Agriculture/ABARES, op. cit., Criterion 2, pp. 118-172

33 Forestry Tasmania 2014, op. cit., p. 9.

34 Bob Brown, 1983, In *The Rest of the World is Watching*, Eds. R. Flanagan, C. Pybus. Pan Books, 1990.

35 D.M. Fessler, A.C. Pisor and C.D. Navarrete, 2014, "Negatively-biased credulity and the cultural evolution of beliefs", *PLOS One* 9(4): e95167. Doi:10.1371/journal.pone.0095167

36 A.J. Hoffman, 2001, *From Heresy to Dogma: An institutional history of corporate environmentalism*, Stanford University Press.

37 Australian Conservation Foundation, 2015, ACF website: About Us – Our Successes.

38 The Wilderness Society, 2015, TWS website: About Us – Our Successes.

39 The Wildlands Network (formerly Project).

40 Wildlands Project, April 2013. Agenda 21 Course: Lesson 3: This online lecture claims that the Wildlands Project aims to put 50% of the USA's rural lands under reservation for conservation purposes.

41 Reed Noss, 1995, In *Maintaining Ecological Integrity in Representative Reserve Networks*. World Wildlife Fund Canada, Discussion Paper, p. 12.

42 The Wilderness Society 2003 *Renowned conservation scientist leads Adelaide Wild Country Science Council meeting*. From the TWS website.

43 The Wilderness Society 2013 website: *WildCountry – Victoria*.

44 Ibid.

45 B. Gammage, 2011, *The biggest estate on earth – how aborigines made Australia*. Allen and Unwin, Chapter 6, pp. 157-186; Appendix 1, pp. 325-342.

46 V. Jurskis, 2015, *Firestick ecology – fair dinkum science in plain English*, Connor Court, Chapter 8, pp. 160-193.

47 Chris Taylor, 2009, *Victoria's February 2009 Bushfires – A Report on Driving Influences and Land Tenures Affected*. Report prepared jointly for the Wilderness Society, the Australian Conservation Foundation, and the Victorian National Parks Association; and The Wilderness Society 2009 *A Bushfire Action Plan which protects people, property, and nature*.

48 Mark Stockdale, 2011, *Bushfire management*. This article had first appeared in the Australian Conservation Foundation's *Habitat* magazine.

49 Australian Fire and Emergency Services Council 2015, *Overview of prescribed burning in Australasia*. National Burning Project, Sub-Project No.1, Summary, p.ii.

50 B. Gammage, op. cit., pp.

51 S. Gulez, 1992 "A method for evaluating areas for national park status". *Environmental Management*, 16: 6, 811-818.

52 V. Jurskis, 2015, *Firestick Ecology: Fairdinkum science in plain English*, Connor Court. Chapter 11: *Saving forests that never were* (pp. 225-238).

53 B.D. Dexter and M.W. Poynter, 2005, *Water, Wood and Wildlife: Opportunities for the Riverain Red Gum Forests of the Central Murray*. Submission to the Victorian Environmental Assessment Council's River Red Gum Forests Investigation prepared on behalf of the National Association of Forest Products, the NSW Forest Products Association, Timber Communities Australia Ltd, and the Victorian Association of Forest Industries.

54 Rivers and Red Gum Environment Alliance, 2008, *Conservation and Community: A Community plan for the multiple use management of public lands in VEAC's River Red Gum Forests Investigation area*. Submission to the Victorian Environmental Assessment Council's River Red Gum Forests Investigation.

55 Department of Agriculture and Water Resources website: Regional Forest Agreements – Protecting Our Forest Environment – Nationally agreed criteria for the establishment of a CAR reserve system for forests in Australia, accessed March 2016.

56 P. Attiwill et al, 2001, op. cit., p. 148.

57 Figures contained in papers by Bunnell & Kremsater (British Columbia), Angelstam (Sweden), and Niemela (Finland) In *Towards Forest Sustainability* Eds. D. Lindenmayer, J.Franklin, CSIRO Publishing, 2003.

58 P. Attiwill et al, op. cit., p. 154.

59 J.E. Hickey and M.J. Brown, 2003, *Towards ecological forestry in Tasmania*. In: *Towards Forest Sustainability*. Eds. D. Lindenmayer and J. Franklin, CSIRO Publishing.

60 Australian Government, Department of Agriculture/ABARES, op.cit., Criterion 1, Table 1.6, p. 41.

61 David Yencken and Debra Wilkinson, 2000, *Resetting the Compass – Australia's Journey Towards Sustainability*. CSIRO Publishing (2001 Updated Edition), p. 206.

62 David Thomson, 2006, *Native forestry nonsense reaches new heights*. In *The Forester* newsletter, Institute of Foresters of Australia, Vol. 49 No. 1 (March 2006).

63 Nick Cameron, 2015, *River red gum forests in the Murray Valley – a look at past and current management*. In *The Forester*, newsletter, Institute of Foresters of Australia, February 2015.

64 Victorian Tourism Industry Council 2013, *Tourism industry angered as Opposition goes cold on significant investment opportunity*. Media Release, 27 June 2013.

65 Lucy Shannon, 2015, *Tourism projects for Tasmanian WHA and national parks move to next approvals phase*. ABC News Online, 14 February 2015.

66 Commonwealth of Australia Senate Environment, Communications, Information Technology and the Arts Committee Inquiry 2007, *Conserving Australia: Australia's National Parks, conservation reserves, and marine protected areas*, Chapter 12 – Funding.

67 A. Campbell, 2012, "Thinking corporately: Getting national parks on national balance sheets", *The Conversation*, 13 July 2012.

68 J. Gordon, 2016, "Victoria's national parks in jeopardy following deep funding cuts" *The Age*, 4 January 2016.

69 A. Livingston, 2017, "Parks Vic staff spend up on credit cards", *The Australian*, 9 January 2017.

70 US National Parks Conservation Association 2015, *Background: The Economics of National Parks*, 19 October 2015.

71 Bloomberg View 2016, *A Little Commercialisation can help National Parks*, Editorial, 7 June 2016.

72 NSW Parliament, 2013, Legislative Council General Purpose Standing Committee No. 5 Inquiry into the Management of Public Land in NSW, Chapter 15, pp. 289-314.

73 Professor J. Vanclay, 2013, submission to the Legislative Council General Purpose Standing Committee No. 5 Inquiry into the Management of Public Land in NSW.

74 NSW Parliament, 2013, op. cit.

75 B.T.W. Coetzee, K.J. Gaston and S.L. Chown, 2014, "Local scale comparisons of biodiversity as a test for global protected area ecological performance: A meta-analysis", *PloS ONE* 9(8): e105824 doi:10.1371/journal.pone.0105824

76 Ibid.

77 P. Kareiva, M. Marvier and R. Lalasz, 2012, "Conservation in the Antropocene: Beyond Solitude and Fragility", *The Breakthrough Journal*, Winter 2012.

78 B. Pressey, E. Richie and P. Visconti, 2014, "We have more parks than ever, so why is wildlife still vanishing?" *The Conversation*, 12 November 2014.

79 A. Flanagan (undated), *Australia's import trends for forest products 1994-2010*. Unpublished report prepared as part of Masters in Forestry being undertaken at the Australian National University.

80 United Nations FAO 2011 *State of the World's Forests 2011*. Annex: Table 2, *Forest Area and Area Change*.

81 Australian Bureau of Statistics *2002 Australia's Environment: Issues and Trends, 2001*, 4613.0, pp. 18-19.

2

Environmental activism: Building the 'conservation culture'

"Culture is the sum of all the shared, taken-for-granted assumptions that a group has learned throughout its history."
Edgar Schein[1]

In the World Wildlife Fund's (WWF) 1984 documentary, *Amazonia: A Celebration of Life*, viewers were assailed with the claim that 400,000 acres (160,000 hectares) of rainforest would be cleared in the 22 minutes taken to watch it.[2] If true, this would have equated to a staggering 10.5 million hectares cleared each day. At such a rate, the whole of Brazil's Amazonia rainforest would have been gone in just 46 days!

Obviously this bears no resemblance to the truth, and today, more than 30 years after the documentary was made, around 82% of Brazil's Amazonia rainforest remains intact.[3] Despite the enormity of rainforest clearing which has occurred during the intervening period, this should still be welcome (and surprising) news to those who have spent years living under a misapprehension of its imminent disappearance.

In the 1990s, scientists concerned with the anecdotal nature of wild claims about tropical deforestation used Landsat imagery from 1978-88 to show that the rate of Amazon deforestation was more like 3.7 million acres (or 1.5 million hectares) per year at the time the WWF documentary was making its impossible claim.[4] Another analysis of satellite imagery by the Brazilian Government suggested that the rate of Amazon deforestation at that time was around 2.1 million hectares per year.[5] These more credible estimates – although still of grave concern – were respectively 2500 and 1800 times smaller than the WWF's claims.

Although their enormity may have diminished, substantial disparities between many environmental activists' claims and the reality have continued over the past 30-years. For example, in 2015, the WWF was still claiming that 36 football fields of trees were being lost every minute across the globe. Equating to a deforestation rate of around 19 million hectares per year (conservatively assuming an average football field is one hectare), this is over two and a half times greater than the United Nations' Food and Agriculture Organisation (the FAO) estimate of 7.3 million hectares per year.[6]

That reality is often far better than the catastrophic claims regularly made by the more extreme elements of the environmental movement, is not meant to downplay the significance of impacts associated with issues such as deforestation. However, it does demonstrate the long-standing propensity for environmental activists to inflate the truth, often into sensational but unrecognisable disaster scenarios, to draw greater public attention to the issues it decides to highlight.

Decades of such misrepresentations – which have rarely been exposed – have helped to create the populist 'conservation culture' which has now infected all levels of society with exaggerated notions of on-going forest loss and degradation. While such alarmism may have relevance in some developing countries, it is more often grossly at odds with the reality in developed countries, including Australia.

When confronted with evidence of gross disparities associated with their alarmist claims, the usual response of environmental activists and their associates and supporters has been to dismiss statistically accurate facts as being either irrelevant or unimportant. The following comments from several Tasmanian anti-logging campaign supporters exemplify this attitude:

> Facts are one thing, but facts are used to prop up a distorted picture of the world, to cement it, until it congeals into something imagined to be true.
>
> ... I have seen enough changes over the years to convince me that it isn't sustainable whatever figures you come up with, especially percentages.
>
> I'm not sure of the percentages of each type of conversion and I don't really care ... Your arguments are spurious.[7]

While dismissing the veracity of facts and statistics may fit an underlying philosophy that any environmental impact is too much, real knowledge is highly relevant and integral to striking a manageable balance between sensible human use with its attendant environmental impacts, and the needs of ecological conservation.

That too many within the environmental movement seem to lack sufficient compunction to campaign honestly by acknowledging the reality, is indicative of an 'end-justifies-the-means' dogma that excuses behaviour which society normally considers unethical. On the other hand, it can be argued that such an approach may be excusable in relation to developing countries that have traditionally lacked the political will and/or the cultural strength to effectively manage environmental threats. Without the environmental movement's exaggerations of rainforest loss, it may have been much harder to attract the global attention that has arguably helped to underpin recent improvements such as, for example, an 80% reduction in the rate of Amazon deforestation since 2003 largely due to a stronger regulatory and enforcement regime instituted by the Brazilian Government.[8]

However in developed countries, the 'end-justifies-the-means' approach to environmental campaigning has been problematic. This includes Australia which already has strong political and bureaucratic institutions governing forests, including highly developed land use policy and planning mechanisms that have minimised or contained most environmental threats. While there has been an issue with woodland clearing/re-clearing for agriculture in parts of inland NSW and Queensland for many years, the greater focus of Australia's environmental movement has always been timber harvesting in southern Australia's publicly-owned native forests.

Unlike the permanent forest loss associated with land clearing, timber harvesting is a renewable activity whereby forests are regenerated and regrown after use. Especially over the past 30-years, it has evolved to be arguably the most highly regulated land use in Australia and is now limited to only a small minority of the nation's forests and woodlands. Under these circumstances, environmental campaigns which grossly inflate minor environmental impacts into dire threats have often manufactured needless conflict on spurious grounds.

An example is the furure that periodically erupts over timber harvesting in Melbourne's extensively forested domestic water supply catchments. These catchments contain 157,600 hectares of forest of which just 200-300 hectares (i.e. < 0.2%) have been annually harvested for decades. Just 12% of the total catchment area is being managed on an approximately 80-year cycle of harvest and regrowth, with the other 88% being left untouched in perpetuity.[9,10] Despite such relatively minor usage, environmental campaigns regularly accuse timber harvesting of dire consequences such as supposedly drastic reductions in water quality and water yield, and even causing bushfires – claims that are grossly disproportionate to the actual impact.

While today's forest-based environmental campaigns are intent on turning a well regulated minor activity into a major issue, it should be acknowledged that Australian forestry activities were more extensive and environmentally significant in the late 1960s and early 1970s. Consequently, concerns about its impacts at that time were more understandable. Ultimately, this early opposition to forestry became a major driver of the birth and evolution of Australia's environmental movement.[11]

Opposition to forestry grew slowly from the mid-1960s soon after a major national program of softwood plantation expansion had been launched by the Australian Government. These plantations were intended to future-proof the nation's wood supplies in the face of expected population growth, and reflected a realisation that slower-growing native forests would be incapable of fully meeting the nation's projected future wood needs. Not long after, fledgling export woodchip industries were established in NSW, Tasmania and WA to utilise waste-wood generated from native forest sawlog harvesting, sawmill processing and clearing for plantation expansion and agricultural development.

The resultant sudden acceleration in the extent of forest use, disturbance, and replacement with plantations coincided with the maturing of the post-war 'baby boomers' generation which had already been politicised by its opposition to the Vietnam War and was less fearful of social change than its predecessors. Commenting on the simultaneous surge in environmental awareness in 1960s America, environmental philosopher Alston Chase described these 'new environmentalists' as a generation "that found their identity in opposition".[12]

In the US, the 1962 release of Rachel Carson's book, *Silent Spring*, was a key factor in initiating a culture of activism based around preserving the environment. While her book's revelations about the troubling impacts of pesticide use were undoubtedly also influential here, the greater driver of concerted forest-based environmental activism in this country was the 1974 publication, *The Fight for the Forests*, by Australian National University (ANU) social scientists Richard and Val Routley.[13]

Prominent 'environmental thinker', Peter Hay, has contended that the beliefs of those who identify with the 'green' ideals of environmental preservation are triggered by "deep felt consternation at the scale of destruction wrought in the name of … human progression, on the embattled life forms with which we share the planet".[14] The stark initial appearance of the earliest, very large coupes which were woodchipped, cleared and burnt for subsequent plantation establishment could not help but fuel such consternation in spite of its envisaged longer-term benefits. When combined with the Routleys' aggressive questioning of forestry professionals as appropriate managers of a public resource and their unflattering critique of 'exploitative' wood production industries, the images of more intensive timber harvesting and plantation development created a powerful case for 'preserving' Australia's native forests.

Arguably, the Routleys' harsh characterisation of our then forestry authorities reflected an under-appreciation of the pressures created by the unrelenting political and material demand for wood during WWII and the subsequent post-war rebuilding period. This was recognised as a time of unsustainable production that was to have a profound influence on the later management of native forests, which were required to be Australia's major source of wood and timber right up until the mid-1990s.

Indeed, it was foresters' concerns about meeting future wood requirements which underpinned the Australian Forestry Council's 1964 recommendation to the Federal Government on the need for an ambitious national target of establishing 1.2 million hectares of new softwood plantations by 2000.[15] That the urgency to kick-start this plantation expansion program resulted in extensive areas of public native forest being cleared for conversion to plantation must be considered against the alternative of substantial delays, heavy financial costs, and likely social conflict if – as the Routley's and others were advocating – the program had been based from the start on acquiring already cleared marginal farmland for the purpose.

That the siting of these new plantations on public lands would eventually generate such intense societal angst due to concern over native forest loss was clearly an oversight. This was gradually addressed by progressively greater efforts to establish the plantations on acquired freehold lands (also, as expected, with considerable social conflict), until by the mid-1980s all states, except Tasmania, had ceased to clear public native forests for this purpose. Despite its problems, if government forestry authorities had bowed to the bitter opposition that the plantation expansion program faced, Australia would not now be in the fortunate position where around 80% of its sawn timber requirements can be met from outside the nation's native forests.

While the conflict associated with the plantation expansion program has now been largely forgotten, export woodchipping has continued to galvanise opposition to forestry authorities and timber industries based on widely promoted misconceptions that it amounts to selling-off precious forests for very little return, and does little more than sate government and industry greed.

These claims have been somewhat enhanced by the reality that the woodchip market for low-grade wood has enabled the sawn timber industry to access previously uneconomic stands of low sawlog productivity. That such areas naturally yield greater volumes of woodchip-grade material gave rise to the still frequently-voiced claim that timber production is 'woodchip-driven' contrary to the assurances of forestry authorities that the primary aim is to produce higher value sawlogs.

An under-appreciated aspect of early export woodchipping was that it provided a means to fund the productive (and often ecological) improvement of large areas of forest degraded by generations of ill-conceived selective sawlogging conducted without sufficient regard for satisfactory regeneration. It was envisaged that these reinvigorated forests would be needed to produce future sawn hardwood. Their improvement was achieved by either clearfelling and regenerating poor quality stands primarily comprised of sub-sawlog grade wood, or by commercially thinning advanced regrowth to speed its path to maturity. That this also enabled access to low volumes of previously unavailable sawlogs made sense at a time of high demand, as did the financial return generated from wood that had otherwise been traditionally burnt either in the forest as unmillable logging residue, or at the sawmill as unusable off-cuts.

As sawlogs were being produced from poorer quality stands, it was unsurprising that much larger areas had to be harvested to meet sawmill quotas. Accordingly, export woodchipping became synonymous with increased rates of harvesting that further heightened the opposition to it. Unfortunately efforts to explain this have invariably been complex and too easily dismissed as defensive or self-serving.

Accordingly, the perception that stopping woodchipping would 'save' native forests became gradually embedded in the community psyche. This perception has endured as a key ingredient of forest-based environmental campaigns despite the reality that woodchipping for domestic paper production has been occurring in parts of Tasmania and Victoria since the 1930s without any loss of forest area. Indeed woodchipping was widely accepted as socially beneficial prior to the 1970s, and still is in regional Victoria where Australia's last domestic paper manufacturer continues to operate.

Aside from forestry issues, wider public support for environmental causes also gained momentum in the early 1970s through the ultimately failed campaign to save Tasmania's Lake Pedder from drowning under a hydro-electric dam. Though this was a painful set-back at the time, leading environmentalists now regard it as the point where their fledgling movement learnt the powerful lesson that industrial development could only be stopped by committed fighting. This evolved into the uncompromising 'whatever it takes' approach that eventually led to landmark campaign successes against wood production at Terania Creek in northern NSW in the late 1970s, and later on, in preventing the planned building of a hydro-electric dam on Tasmania's Franklin River in the early 1980s.

At Terania Creek, the NSW Forestry Commission could have been excused for thinking that their planned timber harvesting would be uncontroversial given that the operation's nature and scale differed markedly from the contentious woodchipping and plantation conversion which was underway elsewhere in the State. The headwaters of the creek were contained in a 700 hectare basin that included some rainforest. Around its margins, a relatively small number of eucalypt and brush box trees were earmarked for harvesting. This necessitated the construction of a new road partially through the rainforest. This inflamed local residents (largely relatively new arrivals who had settled in the area as part of a 'back to nature' counter-culture)[16] who had for five years campaigned for the basin to be declared as a nature reserve. With support from hundreds of other protesters drawn to the area, they confronted and physically blocked the path of road building machinery to create an impasse that would last for several years.[17]

Whereas earlier anti-logging protests were often passionate but disjointed, the far superior organisation and administration which evolved at Terania Creek – from the food, accommodation, and child care, to the distribution of press releases and stage-managed clashes with authority – would become a template for the anti-logging campaigns of the future. This, plus the later Franklin River victory and the subsequent election of the 1983 Hawke Labor government, also demonstrated to leading environmentalists the great potential for direct political influence through mobilising tens of thousands of protestors from around the country.[18]

Most significantly, these successes showcased the effectiveness of combining theatrical protest techniques with adept lobbying of government.[19] They demonstrated the critical importance of engendering public sympathy via the impressive symbolism of 'red-neck loggers' or police manhandling passive protesters from barricades or bulldozers – always performed for the media before a backdrop of prominent campaign banners.

These campaigns also highlighted for the first time the phenomenon of environmental activism being primarily driven by newcomers to the regions or States where the conflict occurs. Key figures in the Terania Creek campaign fitted this observation, including Hugh and Nan Nicholson who had earlier moved from Melbourne to live on the last property adjacent to the proposed forest harvesting.[20] In the Franklin River campaign, the most prominent figure was Dr Bob Brown who had moved to Tasmania from mainland Australia in the early 1970s and become Director of the Tasmanian Wilderness Society in 1978.[21]

Others to have become leaders in the environmental movement after shifting to forested regions include Geoff Law, until recently a long-time Tasmanian campaign manager for the Wilderness Society after shifting from Victoria to Hobart in the early 1980s;[22] Jill Redwood founder of Environment East Gippsland; Dr. Geoff Courser of Tasmania's Doctors for Forests;[23] Dr. Beth Schultz of the WA Forest Alliance after arriving in Perth via Queensland, France and the USA;[24] Tasmanian anti-pesticides and plantations campaigner Dr Alison Bleaney;[25] and veteran Tasmanian environmental campaigner and political advisor Bob Burton, who was drawn to Hobart by the Franklin Dam campaign.[26]

In considering the propensity for new arrivals or visitors to be more concerned with 'saving' the environment compared to long-time local residents, it is important to note the disparity between most forestry issues and other environmental issues. The earliest significant environmental campaigns in Tasmania against the drowning of Lake Pedder and the proposed damming of the Franklin River for hydro-electric power generation were railing against a truly permanent loss of irreplaceable environmental values. However, the majority of campaigning against Australian forestry opposes timber harvesting on public lands which does not involve the loss of irreplaceable conservation values because: 1) forests regrow and eventually recoup their values; and 2) the majority of forest is not harvested and so continues to harbour those values.

The primary reason why newcomers to regional areas are drawn to anti-forestry campaigns may be that their long exposure to environmental campaigning has infused them with a misguided notion of the reality. It's understandable that those with little prior knowledge of an issue are more likely to be outraged by emotive campaigns that create an errant perception of eco-catastrophe by, for example, claiming that 'irreplaceable forests' are being 'lost/destroyed/trashed by logging'. On the other hand, long-time locals generally have a more balanced perspective because they've long seen that harvested forests are regenerated and regrown, and are also aware that the majority of forests are either already reserved (formally, informally, or effectively due to various circumstances) and will not be harvested.

Further to this, visitors or recent arrivals with few personal or material ties to a region have little or nothing to lose by campaigning to 'save' the environment. On the other hand, long-time locals are more likely to have grown-up with and come to accept resource uses and their associated industries as integral to the socio-economic fabric of their lives and communities. Not only do these locals have far more to lose, but they are typically far more aware of nature's resilience after having seen first-hand how disturbed, burnt or harvested forests vigorously regrow. Conversely, it seems that many/most environmental activists – especially those drawn from large capital cities – tend to arrive with a pre-conceived misconception of the bush as being exceedingly fragile and little or no understanding of fire history.

Implacable ideals and the era of 'big environment'

After its Franklin River victory, the recognition that success ultimately depends on campaigns being incorporated into the political arena led the environmental movement to initiate close ties with the Labor Party.[27] However, while Labor in the early 1980s was regarded as the best political vehicle for achieving conservation aims, this would become somewhat clouded by the founding of the Australian Greens and its state branches in 1992.

Inspired and gaining confidence, the environmental movement grew increasingly more professional after 1983. Between then and 1990, the larger environmental groups – principally The Wilderness Society, the Australian Conservation Foundation, Greenpeace, and the World Wildlife Fund – developed corporatised organisational structures not unlike the biggest resource use companies that they were opposing. This was facilitated through sufficient membership and philanthropical support to employ researchers, lobbyists, and public advocates to develop and promote their agendas.[28]

In keeping with this increasingly corporatised sophistication, formal policies were developed that are now instructive in exposing underlying conservation agendas. In relation to forests, the Australian Conservation Foundation's 1995 Forests Policy stated that the:

> ACF believes that ... [*there should be*] ... immediate cessation of logging in all forests of high conservation value; ... on environmentally sensitive sites; .... of 'old growth' forests; ... in areas regenerating after clearfelling and silvicultural treatment (that fall within a range of criteria); [*and*] cessation of logging within three years in natural regrowth areas – sooner where there is a regional availability of alternative wood supplies.[29]

In a similar vein, The Wilderness Society's Forests and Woodlands Policy (revised in 2005) stated that the group:

> ... does not support the use of native forests to supply woodchips for pulp, wood for power generation, charcoal production, commercial firewood or timber commodities. [*It*] does not believe that there is a native forest logging system in use in Australia that has been proven to be ecologically sustainable ... [*and it*] ... believes that all of Australia's pulpwood, commercial firewood and timber commodity production should come from extant plantations of softwood and hardwood...[30]

These policies read literally as 'we do not agree with the use of any forests to produce wood'. It is notable that detailed policies such as these are no longer to be found on these group's websites. This suggests that either these groups have mellowed or, more probably, that they are now loathe to publicly state a formal position that can be used by critics to expose their pretensions to objectivity when participating in public forestry inquiries or being quoted in media coverage.

As far back as the 1980s, some within the environmental movement were starting to struggle with such implacable attitudes. While they might have been appropriate for black-and-white issues such as dam building or nuclear bomb tests, their application to renewable resource uses such as forestry raised questions about the flow-on consequences of how then to deal with human demands for essential needs, such as building materials and paper.

Patrick Moore, a co-founder of one of the world's biggest environmental groups, Greenpeace, was one who was troubled by the culture of uncompromising opposition thereby shutting-out the possibility of meaningful resolutions taking account of human needs. He would ultimately abandon the environmental movement in the mid-1980s to become a strident critic of its protest ideology and an advocate for balanced solutions taking account of environmental as well as social and economic needs.[31]

Another high profile environmental activist to take a similar path was Tricia Caswell, a former President of the Australian Conservation Foundation, whose subsequent transformation into an advocate for balanced and sustainable solutions saw her appointed as CEO of the Victorian Association of Forest Industries in 2004. This brave attempt by both the industry and Ms Caswell to bridge the polarised divide between environmental ideology and commercial forest use was pilloried by the environmental movement, with Ms Caswell unmercifully vilified by her former environmentalist colleagues as a traitor who'd 'sold her soul to the devil'. Understandably then, the three-year experiment to use a respected environmentalist to heal the wounds of incessant conflict and find an agreed common ground for the polarised protagonists, was doomed to failure virtually from the start.

In recent years, the Wilderness Society (TWS) endured the savage ire of much of its supporter base for its role in Tasmania's 'forest peace deal' (negotiated from 2010-13) which ultimately agreed to a compromise whereby a smaller native forest timber industry would be accepted alongside a proposed plantations-fed pulp mill, albeit in return for a huge increase in forested national parks. Although the purported conservation benefits far outweighed the concessions allowing a continuing low rate of native forest harvesting, the relationship between the TWS and Tasmania's wider 'environment community' was fractured to such an extent that celebrity activist, Richard Flanagan, predicted that it may take a decade for divisions to heal. Ultimately, the environment movement was soon reunited in 2014 when the incoming Hodgman Liberal Government effectively dissolved 'the deal' and created a target that all environmentalists could agree to attack.

The superficiality of the environmental movement's implacable ideology is exemplified by its strident opposition to timber harvesting on the grounds that eco-tourism is the natural socio-economic successor to closed timber industries. Almost without exception, when eco-tourism developments are subsequently proposed for forests that have been supposedly 'saved' by the closure of timber industries, environmental groups hypocritically refuse to countenance them. This has occurred recently in both Victoria and Tasmania when tourism operators were invited to submit development proposals for national park tourism.[32, 33] It seems that to most environmental groups and their supporters, only basic walking tracks are acceptable despite the reality that, except in rare cases, such ventures are only seasonal attractions that generate little income — and are therefore far from the financial replacement for former timber industries that environmental activists had claimed them to be.

Given its past policies and continuing actions, it is logical to conclude that most of the environmental movement and its supporters and associates (such as Greens politicians) view the 'protection' of forests as being incompatible with any commercial human use. This is the essence of their message which has ultimately evolved into Australia's flawed 'conservation culture' of environmental salvation through national park declarations.

Despite the popular perception of the environmental movement as a collection of small voluntary local groups it has, since the mid-1980s evolved into a mix of large wealthy corporations complemented by smaller grass-roots groups. In 2009, it was revealed that Australia's four largest environment groups (listed earlier), collectively spent more than $70 million during the previous financial year, with 60% going to lobbying, fundraising, membership drives and office expenses.[34] A more recent examination of their balance sheets, found that these four groups earned a combined $72 million during the 2014-15 financial year, mostly (75-90%) from philanthrophic donations. To attract these donations, over $23 million was spent on fund-raising. A further $36.4 million was spent on 'campaigns' and with administrative and other costs, very little was used for actual on-ground conservation works.[35] Instead, on-ground conservation work is almost exclusively the preserve of small voluntary site/area-focussed groups.

The reality that the largest and most prominent environmental groups engage almost exclusively in lobbying and campaigning designed to shape public opinion and influence political outcomes has raised concerns that they are in breach of the requirements of the Register of Environmental Organisations. Inclusion on the Register affords gift recipient tax deductibility status in return for a commitment to desist from political advocacy. In response to these concerns, the Abbott Government launched a parliamentary inquiry into the Register of Environmental Groups during 2015, amidst predictable protests from some of the largest registered groups and their supporters that this amounted to an attack on free speech.[36] Despite finding evidence that the most prominent environmental groups routinely fail to meet the Register's requirement to desist from political activism, no action had been taken against those groups at the time of writing.

The rise of 'big environment' – as the large corporatised environmental groups are now often collectively referred to – has to a large degree transformed environmental activism from the province of directly affected locals striving to protect their own patch, to that of besuited career activists working the boardrooms of business corporations, media organisations, and political parties. This is particularly the case in regard to forests.

Where disputes may once have involved direct negotiation with local

Forests, fire and a flawed conservation culture

forestry authorities and/or the timber industry, the days when environmental activists regarded compromise as a satisfactory outcome seem largely gone and unlikely to return given the recent Tasmanian experience. Especially in relation to the use of forests for timber production, environmental activism has achieved so much over the past two decades by shaping public opinion and driving political outcomes, in large-part through media-based campaigns of misinformation, that there is clearly little reason for them to accept compromise.

Small grass-roots battles over local forests still occur, but they now rarely stand alone. Instead, they are part of a much bigger ideological struggle for the complete cessation of forestry activities at a national, State, and regional scale. Seeming exceptions, such as the recent campaign by local residents to stop the harvesting of a 49 hectare regrowth coupe dubbed the 'Lapoinya Forest', in north-western Tasmania, are in fact covertly supported by 'big environment'. This is exemplified by reports that the Tasmanian Wilderness Society lent its expertise to the campaign by training 40 local protesters in the 'dark arts' of non-violent civil disobedience, while Greens luminary Bob Brown conducted his own 'Support Lapoinya Rally' on the lawns of Hobart's Parliament House.[37] Soon after the harvesting commenced in late-January 2016, Brown visited the coupe and was duly arrested in a coordinated photo opportunity to raise wider awareness of the campaign.[38]

Even though environmental campaigns are now more centrally coordinated and configured to meet a wider agenda, they still need periodical injections of the theatre, colour, and movement of in-field protest to maintain the interest and support of the media. For example, during the campaign conducted against Tasmania's native forest timber industry from the mid-2000s, a rough bush camp of forest activists stood for many years in the Florentine Valley in Tasmania's Southern Forests; and even during the later so-called 'forest peace deal' negotiations, a young woman sat on a platform high up in an old growth tree for more than a year garnering media interest while broadcasting her personal anti-logging invective to the world via the internet.[39]

In the late 2000s, the writer, Anna Krien, embedded herself amongst

Tasmania's forest activists while researching a book about the island's forests' conflict. She found that many of these activists were visitors from interstate or overseas who typically knew precious little about what they were opposing.[40] This somewhat fits with pre-existing perceptions of the front-line protesters as mostly young and/or naive followers primarily attracted to causes by social and lifestyle priorities.

Over the years, the so-called 'non-violent protest' tactics used by these activists have been widely publicised. Simple blockading of timber harvesting coupes by locking-on to logging machinery has been value-added by introducing practices such as cementing old car bodies into roads, and erecting cables that attach people to blockade structures and distant trees making their removal a slow, laborious and dangerous process to both the protesters, timber industry contractors, and the police.

Training in the use of these methods has been freely accessible for years via the internet, including how-to-manuals such as the North East Forest Alliance's *Intercontinental Deluxe Guide to Blockading* which purported to provide "everything you wanted to know about stopping the Earth-raping juggernaut but had nobody to ask".[41] In the early 2000s, the south coast NSW group, ChipStop, was promoting the manual *Ecodefense – A Field Guide to Monkeywrenching*, as "an essential classic".[42] Monkeywrenching was described in the manual as the 'art' of destroying logging or mining equipment to financially cripple contractors.

Another WA-based group, Forest Rescue, for around 15-20 years acted effectively as a 'hired gun' contracted by local activists to build elaborate blockade structures.[43] In 2004, its website described the group as "a national movement coordinating non-violent direct actions to protect threatened high conservation value forest" and effectively identified it as the operational protest arm of The Wilderness Society, "courageously standing between the forest and the chainsaws and bulldozers of the woodchip companies."[44] In 2002, it was suspected of using monkey-wrenching techniques taken from the *Ecodefense* manual to damage logging equipment in the Badja State Forest.[45]

What is not so widely appreciated is that these supposedly 'non-violent protests' have often involved irresponsible and potentially life-threatening tactics arguably designed to anger workers and incite violence. For example in southern NSW, protesters campaigning against woodchipping on three occasions (in 2001, 2003, and 2007) strung steel cables (twice) and a heavy duty nylon rope (once) across the Edrom Road leading to the Harris-Diashowa woodchip mill, just south of Eden. This was designed to impede log trucks and mill workers, but without any traffic control or warning it was a miracle that no-one was killed. As the road was also used by tourists and local residents, there was potential to injure or even kill innocent bystanders unconnected to the conflict.[46]

Indeed, anti-logging protesters were strongly implicated in the death of timber industry contractor, John Creighton, who was struck by a falling tree branch at a timber harvesting operation at Whian Whian, in northern NSW, in October 2013. A report by NSW Coroner, J.A. Linden, in February 2015 noted that:

> The necessity of the deceased being on the site at all is related directly and solely to the presence of unlawful protesters. The logging contractor was acting lawfully pursuant to licences to fell timber on the property. The protesters were conducting guerrilla tactics part of which involved an activity described as a "black wallaby". This action placed the protesters in the vicinity of trees being felled thereby endangering their lives.
>
> The deceased was brought on the site as a lookout for such activity to ensure the safety of all persons.
>
> Following the tragic accident that saw the death of Mr Creighton some unlawful protesters were heard by police and workers to comment that it was "Karma". This is of course a cynical and disgusting excuse and/or justification for their unlawful activities which led to his death.[47]

Overly provocative protest tactics used by environmental activists have also included smearing faeces under the door handles of contractors' machinery or inside their cabins, urinating in their safety helmets and/or water containers, polluting oil and fuel, putting sand or stones in hydraulic fuel lines, and verbally abusing workers.[48] The intention to incite forest workers to violence that can be filmed by the activists (and in more recent times, posted to YouTube), clearly contravenes the concept of non-violent protest.

The cost of dealing with such behaviour has been excessive for both the public and private sector. In particular, criminal vandalism of logging equipment has run into millions of dollars over the years and that's without including the value of lost work time and productivity. In arguably the worst instance, an estimated $4 million worth of damage was done to logging machinery, buildings, and stored timber in Tasmania's Southern Forests over a four-month period starting in late 2001.[49] In its wake, there was speculation of increased insurance premiums and the need for 24-hour security which would have substantially increased the operational costs of wood production – undoubtedly a key aim of the perpetrators.[50]

Even legal and genuinely non-violent protests incur substantial taxpayer expenditure. In the early 2000s, the Victorian Association of Forest Industries estimated that the costs of dealing with anti-logging blockades in East Gippsland ranged between $7000 and $11,000 per day, including the wages of police and Departmental staff, the cost of crane hire to remove tree-sitting protestors, and the costs of shifting contractors' machinery if necessary. In some years the actual costs have clearly been higher. For example in the 2001-02 financial year, it was reported that the Victorian Government had spent $2.5 million (not including police costs) on dealing with anti-logging activists on 141 protest days.[51] Nowadays it could be expected that these daily costs would be considerably higher.

Other protest strategies, the internet, and the rise of 'mouse-click activism'

The frequency of in-coupe anti-logging protests has declined in recent years as environmental activist groups have realised that media interest has waned and that other strategies can more effectively generate support for their agenda. Environmental campaigns have always endeavored to shape the community psyche though simplistic messages that cut through the public's typical aversion to complexity. However, they've increasingly based these messages around self-commissioned research, loaded public opinion poll findings, or their own (often skewed) analyses of scientific studies and/or factual data.[52]

To magnify their message, environmental groups have over the years forged alliances with favoured academics to increase their credibility, enlisted celebrity campaigners in an appeal to populism, conducted paid advertising campaigns, fostered relationships with favoured journalists to encourage media support, and mobilised their members through mass letter/submission writing campaigns. Whilst most of these strategies have relied on mainstream media exposure, the advent of the internet has added another dimension which has undoubtedly fostered the growth of the 'conservation culture' based on catchy slogans and simplistic one-paragraph protest claims endlessly repeated on a vast array of linked websites.

By providing easy access to detailed information about virtually any topic, the internet has become the medium that can add flesh to the bare bones of concern. With respect to political issues, it provides a facility that can in a few seconds transform anyone from being an 'arms length' concerned citizen to an active protest campaigner. Amplifying the power of this capability is the expectation that the internet will increasingly supplant newspapers and television as the most popular source of news and current affairs.

By 2010-11, 83% of Australian households had at least one computer and 79% had home internet access.[53] The steadily increased use of the internet over the past 20-years has facilitated a dramatic increase in the reach of environmental campaigns, particularly amongst sections of society not normally reached by conventional mainstream media. This has heightened awareness of forestry issues from amongst those already passionately involved, through to school children and other net 'surfers' who just happen upon the huge volume of material promoting particular viewpoints.

However, despite its effectiveness as an information superhighway, the internet's inherent lack of filtering allows irresponsible sources to easily spread erroneous misinformation. This brings to mind a quote from the early 1900s often attributed to Mark Twain: "a lie can travel halfway around the world while the truth is still putting its boots on". If that was true over one hundred years ago, it is an infinitely greater truth today.

The ease with which misleading or untruthful claims can be spread via the internet is painfully obvious when perusing the websites of environmental groups pursuing an anti-logging agenda. As presenting a fair and balanced portrayal of the issue runs counter to achieving their agenda, these sites are typically littered with misconceptions, exaggerations, or deliberate misrepresentations unashamedly presented as though they are indisputable facts. Unfortunately these are rarely challenged as only the relative few working in or associated with the forestry and timber sectors are fully aware of the facts, and there is an absence of available, appropriate, and effective avenues for correcting such misconceptions.

In 2003, Paul Sheehan was one of the first commentators to articulate concerns surrounding the use of the internet as a protest tool:

> The Internet is harnessing the enormous intellectual power of tens of thousands of minds working on the same problem at the same time. Add these rapid evolutions to the billions of personal emails, the millions of home pages, the rising ocean of spam, the infinite web of Internet links, then filter it all through the evolving moral universe of cyperspace with its anonymity and subversive tendencies, and you have a huge new social machine driving at high speed.[54]

As Sheehan pointed out, the ease with which the internet can allow anyone to become a committed activist from the comfort of their home office is a questionable development. By simply pushing a few computer keys anyone can support a campaign, donate money for a cause, or propel a standard template protest letter to the linked e-mail address of a targeted politician or media outlet. This requires no commitment to research a range of views – just an uncritical and potentially dangerous acceptance of what is laid-out online. On activist websites, such as *GetUp!* and change.org, this is often little more than a one line slogan which provides nothing on which to base an informed judgement.

E-activism that fosters superficial support for simplistic 'feel good' ideas is an anathema to practical, workable resolutions based on weighing-up all the factors needed to ensure the best outcome. That function is supposed to be performed by governments, but it has become increasingly obvious that political and bureaucratic decision-making on forests (and other issues) can be inordinately influenced by the numbers of people registering support for idealistic slogans, such as GetUp's simplistic "Save Tassie's Forests" campaign in 2011.

The internet also facilitates easy linkages between protest groups. As well as amplifying their respective messages and thereby improving perceptions of their credibility, this has fostered cooperation that can be called upon to assist with logistical functions in organising field protests or rallies, and the production of promotional material for the media.

These enhanced linkages were evident in the *Eco-Kit for Law Firms* produced by Victorian group, Lawyers for Forests, as far back as 2003. This document acknowledged the input of seven contributors from other groups, including Doctors for Native Forests, The Wilderness Society, the Australian Conservation Foundation, Concerned Residents of East Gippsland, and the Friends of the Earth.[55] Few groups now think or act independently or locally – instead the internet has enabled them to be participants in a wider struggle occurring at state, national, and even international levels.

When accessed during mid-2006, the Wilderness Society had one of the more sophisticated protest web sites. It not only provided interested browsers with information on a range of environmental topics (including forest campaigns), but also direct access to a wealth of information on a further 170 linked State, national, and international websites. These included:

- activist resources sites providing how-to guides for environmental campaigners, including hints on letter writing that occasionally result in anti-forestry letters from overseas correspondents being printed in Australian newspapers;
- media resource sites that could aid campaign promotion;
- national and international environment group sites that could be accessed for assistance and to broaden the message – one was a

Japanese group with information designed to inform the Japanese public about the so-called destruction of Australian forests;
- political sites, including the e-mail addresses of prominent politicians;
- ethical investment sites, including a site that enabled investors to make donations to environment groups "by making small payments through your workplace payroll";
- Australian and state government land management agency sites;
- outdoor and eco-tourism industry sites, including one site that was an alliance of companies that "supports The Wilderness Society through grants for our projects – please support them";
- 'green' book and art sites; and
- sites offering environmental goods and services.

In addition to the websites of the major environmental groups, there have also been anonymous privately-constructed internet sites dedicated to an agenda of stopping native forest use. Some early examples were the *Forest Letter Watch* site[56] which recorded newspaper letters and articles protesting about logging mostly taken from Melbourne's *The Age* newspaper; the *Forest Fact File*[57] which presented misleading information about native forest logging; and the *Boycott Woodchipping* site[58] which provided a how-to guide to anti-woodchip letter writing. At one time some enterprising Tasmanian activists even created an anti-logging internet site that for a time mimicked the official Tourism Tasmania website, thereby deceiving potential tourists into entering a site that proclaimed "Discover Tasmania – Australia's pretender state … see the forest pictorial of blowing our future."[59] Whilst it is rarely obvious who creates and manages these websites, they are presumably off-shoots from the major environmental groups.

A more recent development has been current affairs weblogs that often provide another online platform for environmental activism. It has been estimated that the worldwide number of web blogs grew from less than 50 in 1999 to 50 million by 2006, of which the majority were little more than personal diaries. However, a significant number of politically-orientated blogs are said to have "reshaped the media environment, rejuvenated political discussion, and created a generation of amateur journalists and pundits",[60] even though there is no clear definition of what constitutes a political blog.[61]

While there are weblogs that favour or promote conservative views, such as *Quadrant Online*, the majority of the best-known sites are repositories of 'progressive' left-leaning thinking that typically includes unquestioning acceptance of the messages spread by environmental activism and strong support for their simplistic agenda of environmental salvation through national park declarations. Such sites include *Crikey*, *New Matilda*, *The Drum*, *Huffington Post*, *Guardian Australia*, *Independent Australia* and the more parochial *Tasmanian Times*.

Weblogs generally work by posting articles or opinions either purpose-written for the site or re-published from the mainstream media, which can then be discussed in an attached online forum. Along with interactive social media tools now routinely used by environmental groups, such as Facebook and Twitter, they effectively act as gathering points for a like-minded community where pre-existing group-think can be massaged and strengthened, largely free from any factually-based challenge.

The internet's effectiveness as a protest tool is potentially matched by the opportunities it provides to government forestry agencies and/or industry bodies to present factual information that could counter environmentalist rhetoric. Most state land management authorities have websites that provide access to statistics, management plans, and reports that can provide a factual overall picture of the extent and nature of forest management activities such as timber production, fuel reduction burning, and bushfire control. It is somewhat of a paradox that although information about forest management is now more prevalent and accessible than ever before, there is little to suggest that environmental groups and their members and supporters take any notice of it ... and so their campaigns continue.

The internet has also undoubtedly played a major role in changing perceptions about environmental activism. By developing highly professional web-sites, environmental groups have gained access to society's most educated and articulate citizens and enabled them to become advocates for their causes without having to dirty their boots. This has helped to change the formerly popular stereotype of activists as zealous, barely rational, unemployed 'ferals' to being more normal, thoughtful, sophisticated and responsible citizens. This image recasting has largely lifted the support base for environmental causes out of the narrow province of those perceived to be radicals, enabling it to become conventional, fashionable and politically correct. Consequently, those still working in natural resource use management or industries which have an inherent environmental impact have largely assumed a mantle of 'political incorrectness' in the eyes of polite society.

From placards and blockades … to boardrooms

While forcing change through political influence has been the major aim of environmental activism in Australia since the early 1980s, a persistent feature of 'green' politics is dissatisfaction amongst activists and their political associates that they either aren't being listened to, or that the pace of change is too slow. In observing the same frustration in relation to tropical rainforest conservation in the Amazon Basin, American political scientist Margaret Keck noted that it fuelled a determination to force change through by-passing political organs.[62]

This is also being observed in relation to Australian forests through a range of often inter-related strategies such as corporate bullying, social licence, certification, litigation, and direct negotiation with industry. While forestry authorities and associated industry sectors have played along with, or been seduced or forced into engaging with environmental activists through these strategies, they have maintained their enduring concerns about misinformation used to exaggerate or create imaginary environmental impacts.

Corporate bullying describes the practice of environmental activist groups in pressuring companies through their shareholders and/or their financiers, and attacking their customers through marketplace threats and boycotts. With regard to Australia's forests, one of the most high profile victims of environmentally-based corporate bullying was the

Tasmanian company, Gunns Ltd, which announced in late 2010 that it would divest itself of its native hardwood interests to concentrate on being a plantations-only wood producer.

In announcing the company's intention to shift away from native forests during a speech to an industry conference in 2010, Gunns CEO Greg L'Estrange, gave an impression that the company had come to a pragmatic realisation that there was no future except as a plantations producer, while admitting that "Our customers, shareholders, employees, contractors and other stakeholders have given us a clear message". However, the manner in which former Gunns' CEO, John Gay, had been forced out by a campaign orchestrated by businessmen and self-styled eco-crusaders, suggested that this message had been foisted upon the company.

In an interview on ABC's *Lateline* program in June 2010, one eco-businessman was quite forthright in explaining how he and the Wilderness Society had "put pressure on the financiers of the company … we put pressure on the customers … we put pressure on the shareholders, finally". For these efforts in challenging a hated company, they were lauded by left-leaning commentators and, for a time, rewarded with guest appearances on ABC current affairs programs, including *Q and A*. More recently, such eco-warrior status has earned one businessman and one-time novelist the presidency of the Australian Conservation Foundation.

For their part, the Wilderness Society's Annual Review outlined how it had also pressured a company being courted by Gunns as a financial partner in its planned Tasmanian pulp mill:

> In June 2009, our supporters contacted Swedish company Södra stating our concerns about their possible support for the mill. This was a great success and an example of the kind of campaigning at which the Wilderness Society excels…We will continue to make sure that any potential financiers of a Tamar Valley pulp mill understand the devastating effects it will have on our native forests, oceans and our fragile climate.

In reality the pulp mill was to be fed only by plantation-grown wood and would have had no impact on native forests or 'the fragile climate'. But facts are seemingly immaterial when companies are threatened with reputational and/or market damage unless they cave-in to activists' demands.

In the USA, the practice of targeting stakeholders to undermine and discredit companies has a longer history and is termed 'brand-mailing'. While it could be justifiable where Western corporations are complicit in serious environmental degradation or human rights transgressions in developing countries, its applicability to very highly regulated activities in a developed nation is inappropriate.

Perhaps it could be appropriate if there were real problems rather than just objections to legitimate activities because they don't fit the prevailing environmentalist ideology. However, those engaged in the campaign against Gunns typically portrayed Tasmanian forestry as something fantastically different to what it actually was. For example, the Rainforest Action Network's March 2006 Tasmanian forests campaign, conducted at the behest of Australian activists, orchestrated protests at various Australian embassies, including in Japan and the UK. During the campaign, RAN spokesman, David Lee, ranted that "everything about the situation on the ground in Tasmania defies belief for anyone who respects democracy and the rule of law" and that "Gunns is trashing a global treasure and ... turning paradise into a toxic Hell on Earth in the process."[63]

More recently, a group dedicated to driving change through attacking the markets of Australian forest products emerged in 2011. At its launch, 'Markets for Change' urged Australians to boycott locally-made hardwood furniture, flooring, and paper products largely on the basis of the grossly exaggerated premise that "logging is still permitted in 76% of Australia's native forests". In fact, only around 5-8% of the nation's forests and woodlands are being managed for long term wood supply.

The campaign mounted by Markets for Change against the retailer Harvey Norman was found to have at least 10 gross distortions or errors of fact.[64] Clearly corporate bullying of this nature can be unnecessarily tragic for resource use industries and their employees when political decisions that affect their future are forced on the basis of false premises.

Integral to the strategy of corporate bullying is the concept of a 'social licence to operate'. The social licence concept was reportedly first coined within the mining industry in the late 1990s and has since been used extensively across all sections of industry and government to drive community engagement over social and environmental concerns. In 2011, Black and Bice of the Australian Centre for Corporate Social Responsibility (ACCSR) defined 'social licence' as being the level of acceptance or approval continually granted to an organisation's operations or a project by the local community and other stakeholders. They went on to explain that "the social licence is a perception of legitimacy – does the company go about its business in a proper way?" They regarded this as being distinct from a company's reputation which is "the overall favourability of the image of a company or project ... it's more of an emotional like and dislike".[65]

There are several obvious flaws with this concept. Firstly, 'social licence' is a metaphor rather than a real licence, so it's hard to say whether it has been granted or not. It's easy for stakeholder opponents to claim that a company doesn't have a 'social licence', and equally easy for that company to claim that it does. Hence, it is quite difficult for an impartial observer to make a balanced judgement. Such a judgement will not be based on fact, but on the subjective opinion of each party based on their interests and assumptions. Furthermore, even if the majority of local community members or broader society withdraws acceptance of a company's presence, the company is not obliged to cease operations, as it can point to its regulatory licence to operate. Finally, it is important to note that the concept is typically only ever applied one way. Those who are opposed to a resource use company, project or activity are never required to prove whether they have a 'social licence' to damage communities affected by business closures, unemployment, and the myriad of related personal and societal issues which typically flow from the closure of resource use industries.

Forestry projects differ substantially from mining. Unlike the largely quite localised impacts of a mine, a proposed pulp mill for example, includes wood harvesting operations perhaps several hundred kilometres distant in the forests and/or plantations from where the mill's feedstock is drawn. Accordingly, the stakeholders to be considered in any assessment of 'social licence' for a forestry development are often far more extensive than just the local community living in close proximity to it. These include the myriad of environmental groups that are intransigently opposed to forestry and, in the case of larger groups like the Wilderness Society, have a majority of their members and supporters living in cities around Australia, far beyond the bounds of any contentious region.

Forest management authorities and timber companies have for some time now routinely addressed social issues and community concerns in regards to their operations. However, even if this is done in an exemplary manner there is no guarantee that it will translate into a 'social licence' to operate. Largely this is because – contrary to the notion of 'social licence' being something distinct from the reputation of a company, agency or activity – most of the community see these two things as being inextricably linked. Therefore for most people, a proposal put forward by a company with a poor public image is likely to be viewed less favourably compared to the same proposal put forward by a well-liked or hitherto unknown company.

A good example is the Gunns pulp mill proposal in Tasmania's Tamar Valley from the mid-2000s. As Tasmania's largest forestry company, it was unanimously hated by everyone opposed to timber production. In regard to its proposed pulp mill, the company reportedly spent $250 million on project development including meeting the requirements implicit to gaining its regulatory approval in State and Federal Government processes over a 4-year period. These processes added enforceable constraints and conditions to the project which would reportedly have made it the world's most environmentally-friendly pulp mill. Despite these impressive and arguably unprecedented efforts to address community and societal concerns, there is nothing to suggest that this did anything to even moderate, far less overturn, opposition to the pulp mill amongst detractors who, in many cases, were actively campaigning to kill off the company itself.

An important question in relation to the Gunns pulp mill example was why 'social licence' was even necessary for such a project with state-wide implications under a democratic system where voting patterns arguably already gave a solid indication of community support? At the 2010 Tasmanian state election, around 80% of voters supported either Labor or the Liberals despite both parties having policies endorsing the pulp mill. It could be argued that such a result should be sufficient to confer a 'social licence' to any project.

The greatest flaw of the nebulous 'social licence' concept is that it essentially requires a company or agency to convert its detractors into supporters. This relies on its critics being reasonable and open-minded enough to be capable of changing often entrenched attitudes. Given the formal, rigid forest policies of the major environmental groups and the uncompromising nature of the 'conservation culture' engendered by decades of anti-forestry activism, there is nothing to suggest that more than a handful of those opposing Australian forestry activities are reasonable and open-minded.

It is interesting to note that calls reiterating the importance of a company or activity gaining a social licence almost always emanate from its opponents. Clearly from an environmental activists' perspective, a sure-fire strategy to impede, undermine, or kill-off a disliked company, project, or activity, is to repeatedly emphasise the critical need for a 'social licence' while having absolutely no intention of granting one.

The notion of a 'social licence' is also intertwined with the concept of forest certification which was pioneered by the international environmental movement specifically to combat forest exploitation in developing countries by providing a market-driven incentive for timber companies to operate in a socially and environmentally-responsible manner. The concept works by developing a consumer preference for wood products that carry labels identifying them as being sourced from forests which have been assessed and certified as being sustainably managed. In time, any uncertified produce from illegal or unsustainable sources becomes socially unacceptable and is excluded from the market.

In 1993, an alliance between two global environment groups – the Rainforest Alliance and the World Wildlife Fund (WWF) – led to the formation of the Forest Stewardship Council (FSC). The FSC was the first body created specifically to audit and accredit natural forest owners and wood processing companies and their operations. Subsequently, a Rainforest Alliance off-shoot, called Smartwood, became the first company accredited to conduct independent third party audits of forestry practices in order to certify companies on behalf of the FSC.[66]

In 1999, an alternative body representing European forest owners and industrial entities was established to develop a Pan European Forest Certification Scheme. This overarching body – now known as the Programme for the Endorsement of Forest Certification (PEFC) schemes – acts as "a global umbrella organisation for the assessment of and mutual recognition of national forest certification schemes in a multi-stakeholder process."[67] Australia's national forest certification scheme – the Australian Forestry Standard (AFS) – was developed under the auspices of the PEFC and launched as a formal, government-approved Australian Standard in late 2002.

Whereas the FSC certification scheme is a generic, Euro-centric international standard administered from Germany; PEFC national schemes (such as the Australian Forestry Standard) specifically take account of local conditions, including acknowledging the pre-existing government land use planning regime which has already evolved to dictate the balance between forest conservation and use. This is an important distinction between the two schemes. [Note: At the time of writing, the FSC was part-way through a process of developing an FSC Australian standard].[68]

Both FSC and PEFC certification schemes have their supporters. Many FSC-certified forests or plantations are located within PEFC member countries that have developed national certification schemes in accordance with PEFC guidelines. Australia is a good example where, despite developing an endorsed national certification scheme (the AFS), most major plantation companies have also attained FSC certification. Similarly, Canada, Sweden, the UK, and Brazil all have millions of hectares certified by the FSC, despite also having national certification schemes developed through the PEFC. This may be partly due to the earlier availability of the FSC scheme, but also because the different schemes may suit the market-place agendas of particular companies and growers in specific ways.

FSC and PEFC certification schemes compete with each other both globally and locally. Unfortunately in Australia, this competition reflects the polarised conflict between environmentalists and the timber industry as each has adopted their own preferred certification scheme. The environmental movement supports the FSC scheme that was developed at its instigation, while frequently deriding the alternative AFS scheme because it takes account of input from state governments and forest owners and is, therefore supposedly, a 'business-as-usual' approach that is less environmentally stringent.[69]

Environmental campaigns depicting FSC as 'the only truly sustainable certification' option stem right back to the Australian Forestry Standard's developmental phase from late 1999 to July 2003, and have continued to dominate discussion about the merits of forest certification in the Australian context.

Even though environmental group representatives participated in the development of the AFS, they were asserting even then that it was a government/industry initiative which would fail to go "beyond superficial improvements and integrate the needs of stakeholders other than forest managers and owners". In a paper released after the 18th National Forest Summit attended by 35 environmental non-government organisations in April 2001, the summit's certification spokesman, Tim Cadman, expressed concern that the AFS would allow existing forestry operations to continue.[70]

In March 2002, the two environmental group representatives on the AFS Technical Reference Group – one from the World Wildlife Fund and one representing a consortium of other environmental groups – resigned when "... it became clear that the new Standard would certify bad logging practices and manifestly fail to improve forest management."[71]

In October 2002, soon after the AFS was officially launched as an Australian Standard, the World Wildlife Fund released a statement decrying as "ludicrous" the claim that AFS would make Australian forest products more appealing to consumers – "nothing could be less appealing to discerning consumers than a bogus claim to sustainability."[72]

It seems that criticism of the AFS by environmental activists is based around expectations of forest certification that fit within their ideological agenda to substantially reduce or eliminate all Australian native forest wood production. This agenda was ostensibly articulated at the 1996 National Forest Summit where a joint statement issued by the participating environmental groups proclaimed that the development of certification and labelling with respect to Australia's native forests must be contingent on an end to certain activities (e.g. woodchipping); the establishment of a reserve system to their satisfaction; and that ecological sustainability rather than sustainable forest management be made the basis for forest policy formulation.[73]

Because the AFS recognises existing government land use plans and policies, such as Regional Forest Agreements, environmental activists claim that it reinforces a status quo which they have long rejected as unacceptable.[74] Accordingly, the AFS has been attacked for allowing activities that were sanctioned by the RFA's such as limited conversion of native forests to plantations in Tasmania (since ended), and the continuation of logging in claimed 'high conservation value' forests in all states.[75] This was the central theme of The Wilderness Society's *Certifying the Incredible, the Australian Forestry Standard – Barely Legal and Not Sustainable*, which appeared on its website in late 2005. Amongst a poisonous diatribe, it referred to both the AFS and the PEFC as "… a sham designed to pass off wood and wood products as legal and sustainable in markets where concerns over environmental and social justice have never been higher".[76]

The discrepancy between the AFS and FSC schemes in Australia is reflected in the respective areas certified thus far. The FSC scheme was originally developed for natural forests with high biodiversity albeit in developing countries. Yet in Australia, FSC has only ever certified a small area of privately-owned native forest, and has almost exclusively certified plantations. On the other hand, the nation's major native forest managers and hardwood producers such as Forestry Tasmania (now Sustainable Timbers Tasmania), VicForests, and the WA Forest Products Commission have been certified by the alternative AFS scheme.[77] Two of these major native hardwood producers have made hitherto unsuccessful attempts to attain FSC certification presumably in a bid to secure greater market acceptance.

While the environmental movement has sought to trash the reputation of the AFS national standard at every opportunity, the AFS and its certified members and supporters have, arguably to their detriment, largely maintained a dignified silence. Perhaps even worse than this is that those Australian forestry authorities and timber companies which have tried to embrace FSC may have inadvertently given it a legitimacy that it doesn't deserve in relation to native forests. While Australia's native forest management practices are of a far higher standard than most other countries, the fact that they are not FSC-certified invariably fuels media coverage that wrongly describes Australian forestry practices as being poor, unsustainable, and akin to Third World standards. This skewed publicity gives unwarranted credence to environmental campaigns that have always been substantially based on such misinformation.[78]

Arguably, in large-part due to this reluctance of the AFS to challenge its critics, FSC has become the most publicly recognisable forest certification scheme in Australia. As Rainforest Alliance spokeswoman, Anita Neville, boasted on the ABC's *7:30 Report* in October 2014: "Really, FSC certification is almost becoming an essential in the forestry industry in order to do business."

The end result of this may well be the creation of a market demand for FSC certified wood products that Australia's major native hardwood producers are unlikely to be able to access. This is because the way the FSC operates allows environmental groups which are stridently opposed to native forest wood production, to effectively act as gate-keepers with the power to veto who can or can't be FSC certified.

The close link between environmental activism and FSC certification was exemplified by the so-called Tasmanian 'forest peace deal' process of 2010-13. The deal was predicated on approximately 500,000 hectares of new forested national parks being created in return for a range of commitments including that environmental groups desist from protesting for the several years it would take for these reserves to be created; and that all signatories (including three environmental groups) support the FSC certification of Forestry Tasmania. Indeed, FSC-certification was regarded as the key to maintaining the durability of the deal by allaying the potential for any future anti-logging protesting after the planned reservations had consigned timber production to being an ongoing but minor forest use.

What can be deduced from this is that environmental group support is needed in order to get native forest wood production FSC-certified; and that FSC-certification will supposedly satisfy environmental groups that future forestry operations are sufficiently constrained because they see it as their own standard. There are obvious problems with this when such groups are implacably opposed to any commercial use of native forests. Indeed it seems that FSC certification in Australia is more about placating an environmental ideology than it is about lifting the quality of forest harvesting practices and management protocols long acknowledged as akin to world's best practice.

Forest certification may be a laudable concept for use, as it was designed, in developing countries which have fewer regulatory controls over their forest use. However, as FSC certification has demonstrated, it can be problematic when applied to a highly developed nation. FSC's refusal to acknowledge pre-existing government forestry controls is problematic where such controls are strong, such as in Australia where the publicly-owned State forest wood production zones are essentially what remains after decades of land-use policies, detailed investigations and planning undertaken by democratically-elected governments. Refusing to recognise this during the FSC certification process for a particular wood production forest typically confers an unwarranted additional layer of forest reservation without any recognition or consideration of the extent to which conservation has already been catered for in the surrounding landscape.

Whereas a developing country may have real and obvious problems with uncontrolled and unsustainable forest use which certification can help to address, such problems are largely absent in developed countries. Accordingly, the environmental movement's endeavours to create a market preference for their preferred FSC-certification scheme in relation to Australia's native forests has been largely based on arguments used in developing countries which, when applied to the Australian context, become over-the-top exaggerations and distortions of forestry impacts that are often hugely disproportionate to the reality.

To the extent that this has forced Australian native forest timber producers to seek FSC certification in order to maintain access to markets, it exposes them to the likelihood of either being severely economically handicapped by having to accept over-the-top or unwarranted conservation demands, or to being wedged out of their markets for either being unable or refusing to accept such demands. This arguably carries a perception of extortion that effectively enables an unelected and unaccountable international body and its environmental activist associates to override a country's sovereign, democratically-sanctioned resource use.

The problem with applying FSC certification in a developed country has been recognised in Canada in-part due to the troubled Canadian Boreal Forest Agreement (CBFA). The CBFA was signed in 2010 between a cabal of environmental groups – most notably, Greenpeace – and 21 members of the Forest Products Association of Canada. It set aside around 40% of the Canadian boreal forest for conservation in return for a commitment by environmental groups to desist from market boycott campaigns against the timber industry.[79]

Once signed, there was varying commitment to the CBFA. Eventually in late 2012, Greenpeace backed away from it by accusing a large CBFA industry partner, Resolute Forest Products, of illegal logging. When Resolute established that these accusations were false they threatened legal action which was initially met with an unprecedented apology by Greenpeace. However, as Greenpeace then resumed its market boycott campaigns against Resolute, the company responded by launching a $7 million lawsuit for "defamation, and intentional interference with economic relations".[80]

Prior to the CBFA, Resolute, like most other Canadian forestry companies, had been severely wounded by market boycotts orchestrated by environmental groups largely on spurious grounds. After eventually realising that it couldn't win a public relations war even if the facts were on its side, Resolute had embraced third party FSC certification.[81] When the company launched its lawsuit against Greenpeace, the World Wildlife Fund (WWF) joined in the attack against it. As both Greenpeace and the WWF were founders of FSC, it seems hardly coincidental that two of Resolute's FSC certifications were then withdrawn on what again seemed like spurious grounds. In the wake of this, one commentator noted that:

> FSC is in many ways inseparable from WWF and Greenpeace. Indeed, the executive director of FSC International, Kim Carstensen, who recently wrote a letter criticising Resolute, used to be head of WWF-Denmark. Its previous executive director, Andre de Freitas, now heads ENGO the Rainforest Alliance, which just happens to be the auditor Resolute uses for its FSC certifications. ENGOs [*environmental groups*] claim to be independent organisations but are basically one hydra-headed group with revolving doors between their executive suites.[82]

While this conflict played out, the FSC had been concurrently moving to introduce what has been described as more draconian control over forest 'development', including a process of development veto for indigenous tribes. This has met with resistance from both communities whose welfare and jobs were under threat, and also Canadian provincial governments.

At FSC headquarters in Bonn, Germany in December 2015, Quebec's Minister of Forests, Wildlife and Parks, Laurent Lessard, reportedly told the certification body that they were going too far and that the government needed to intervene to avert what was looking likely to be 'absolutely devastating' economic implications being visited upon the province's citizens by an unelected, self-appointed group. He concluded by asserting that: "FSC should not be a substitute for State's legislation".[83]

In the wake of this dispute, Canadian timber companies have reportedly begun to divest themselves of FSC certificates once they expire, in favour of the PEFC-accredited national scheme, the Sustainable Forestry Initiative. This is prompting the sort of attacks on non-FSC certification schemes from environmental activists that have been commonplace in Australia.[84]

As previously acknowledged, there is little doubt that the certification concept has aided forest conservation in Third World countries with serious environmental problems. However, in developed countries, although arguably well-intentioned, it has in-part been misappropriated by environmental ideologues as a vehicle for displacing timber industries despite forests being far better managed with only comparatively minor or imagined environmental problems. In particular, the FSC certification scheme operates outside the political sphere thereby enabling unelected and unaccountable groups to deny a democratic voice to those who stand to be disenfranchised by an agenda to end or substantially reduce timber production. That it is largely being forced on the community by campaigns of misinformation is arguably another reflection of the market-place fear of controversy, irrespective of its veracity.

From boardrooms to court rooms ...

Litigation is another emerging strategy being used to force conservation agendas. Ironically, in relation to forests, the increasing propensity for environmental groups to launch legal challenges to support their campaigns may have been sparked by the 'Gunns-20' case in 2004, when Tasmania's largest timber company attempted to prosecute twenty people for allegedly waging reputation-eroding campaigns of misinformation against it. While the case was eventually thrown out of court before it could proceed, the episode undoubtedly heightened the contempt of environmental activists for the timber industry and presumably alerted them to the potential for using the law to achieve their own agendas.

One of the most significant legal challenges against timber production was launched the following year by Australian Greens' Senator, Bob Brown. The case centred on the protection of endangered species in regard to timber harvesting planned within the Weilangta State Forest, east of Hobart. Initially the Tasmanian government dismissed the Federal Court action as a stunt, but later decided to support its responsible agency, Forestry Tasmania. The Commonwealth was also granted limited input to the case because of its potential implications for the Tasmanian Regional Forest Agreement to which both governments are signatories.

The case extended over 38 sitting days in the Hobart Federal Court from December 2005 to the end of August 2006. Its major focus was the supposed exemption of Tasmanian forestry operations from the Commonwealth's *Environment Protection and Biodiversity Conservation (EPBC) Act 1999* in accordance with the Tasmanian Regional Forest Agreement. This 'exemption' is in fact a recognition that, unlike most other resource use proposals (such as a mine), forestry activities are already governed by a series of landscape-scale policies, management plans and operational regulations that meet the requirements of the *EPBC Act*. The case effectively rested on Brown's refusal to accept that Tasmania's extensive conservation reserve network, and Forestry Tasmania's Code of Forest Practices and associated environmental management prescriptions, were sufficient to 'guarantee' the protection of the endangered Weilangta stag beetle and swift parrot.

Although the case was ostensibly about the fate of two threatened species within one defined area of forest, environmental activists anticipated that a favorable judgement would be a substantial step towards achieving their aim of stopping native forest wood production throughout Australia. According to a website established by Brown's supporters to publicise the case, it will "… expose Forestry Tasmania and its practices to public scrutiny as never before", as its alleged exemption from such scrutiny had supposedly "… bred a secretive closed institution" that is part of the "… culture of intimidation, deception and lack of transparency in the forest industry and its regulatory bodies" According to the website, the court case "… has the potential to revolutionise the protection of endangered species" and "… has relevance for all Australian forests covered by Regional Forest Agreements".[85]

In early 2006, the Australian Greens website was referring to the case as 'Big Business versus The Australian Bush' despite the fact that no large corporate entity was involved and the Weilangta forest is just a tiny part of Australia's public forest estate.[86] To further increase the profile of the case, Senator Brown conducted a public auction of personal memorabilia at a Melbourne art gallery in early February 2006 which raised $75,000 to fund his legal costs.[87] Funds were also being raised online through a purpose-built website with facilities for accepting online donations.[88]

The efforts of the Greens and their associates and supporters in publicising this case suggests that, for them, legal actions are also largely about providing a platform for increasing the profile of forestry issues and generating more public sympathy for the cause. This seems to be predicated on the notion that, even if a case is ultimately lost, the public relations benefits may ultimately lead to a bigger win somewhere down the track.

Subsequently, on 19 December 2006, Justice Shane Marshall found that Forestry Tasmania's so-called exemption from the *EPBC Act* did not apply at Weilangta, meaning that further timber harvesting would be illegal. Whilst environmentalists celebrated this judgement and called for its wider application to all Australian forests, Federal Forestry and Conservation Minister, Senator Eric Abetz, observed that the judgement created a definition of 'protect' that extended far beyond that envisaged by Commonwealth and State governments. Further to this, Forestry Tasmania's managing director, Bob Gordon, warned that the decision "… would have far-reaching implications for all land use nationally".[89]

In February 2007, Forestry Tasmania appealed against the judgement citing legal advice that no land use would be unaffected if it was allowed to stand. For farmers, meeting its requirements would be impractical due to the need to apply for exemption from the *EPBC Act* every time they wanted to plant a crop or plough a paddock. Furthermore, it was impossible to 'guarantee' that such activities would protect animals and enhance their survival. As the appeal was expected to take until the end of the year to conclude, the Australian government amended the Regional Forest Agreements to allow timber production to continue during the interim.⁹⁰ In November 2007, the Federal Court unanimously upheld Forestry Tasmania's appeal thereby overturning the initial judgement that had banned timber harvesting from the forest.⁹¹

Although, the Weilangta forest case represents one of the most significant attempts by Australian environmentalists to use legal intervention to achieve a wider anti-logging agenda, civil lawsuits brought by the environmental movement against government land and resource management agencies have long been commonplace in the USA. Most have involved suing agencies for failing to meet statutory deadlines to give species 'threatened' or 'endangered' status, or to secure 'critical habitat' for threatened species. One lawsuit that led to the listing of the northern spotted owl as a nationally threatened species dramatically reduced timber production in the Pacific Northwest. Whilst other suits have led to 5.4 million acres being designated as 'critical habitat' for the threatened California red legged frog, and 39,000 square miles of Alaska declared to be critical for a duck species, the spectacled eider.

There has been concern for decades in the USA about the extent of environmental litigation brought under the nation's Endangered Species Act (the ESA). The ESA's statute deadlines and other provisions reportedly incentify lawsuits by allowing environmental groups to access taxpayer funding for the purpose of prosecuting government agencies. This concern peaked in 2011 when it was revealed that the US Federal government had defended more than 570 cases bought under the Act in the previous four years, and that this had cost US taxpayers more than $15 million in attorney's fees.⁹²

This led to public hearings held by the US Congress to examine concerns that the ESA had become litigation-driven and that the costs of defending often frivolous lawsuits was diverting precious monies away from recovery programs and policies that could better manage endangered species needs and provide a balance with economic activity. It was noted that the bulk of these lawsuits, or petitions threatening lawsuits, were being filed "… by special interest groups to force government agencies into agenda-driven decisions not based on verifiable evidence or sound science or priority, but to block economic and job-creating activity". One group alone – the Centre for Biological Diversity – had filed 117 of the 570 cases launched during the preceding four years.[93]

By US-standards, environmental litigation in Australia is a relatively rare occurrence. However, it may well play a more significant role in the environmental movement's future strategies. In 2003, the litigation-focussed environmental group, Lawyers for Forests, signalled the potential for litigation to play a greater role in Australian forest use conflicts due to "serious concerns … that the laws applicable to the native forest sector are deficient and inadequate". The group was committed to an "analysis of the laws and regulatory framework and advocating law reform to ensure the conservation and better management of our native forests."[94]

To date most litigation with respect to Australian forest management has revolved around 1) prosecutions of protesters in relation to logging coupe blockades; or 2) specific claims bought by protesters or environmental groups against the timber industry or responsible government agencies in relation to specific operations. More recently, a focus for litigation brought by environmental groups has been on lapsed administrative requirements, such as failing to undertake a scheduled periodic review of a Regional Forest Agreement, in hope of invalidating the legality of timber production. This is the subject of a Victorian case (Friends of Leadbeater's Possum Inc. versus VicForests) which is underway at the time of writing.

In Victoria, there has been a notable recent increase in litigation brought by environmental groups against the State government's commercial forestry agency, Vicforests. That these cases are mostly being launched by small grass-roots groups with few obvious assets raises questions about behind-the-scenes backing, including pro-bono legal representation that confers advantages not available to government's required to use taxpayers' money to defend these actions; as well as concerns about the capability of these groups to pay if they lose and are instructed to pay their opponent's legal costs.

Whatever these backroom arrangements are they've been good enough to encourage one group, Environment East Gippsland (EEG) – which was founded and is apparently still largely overseen by a woman based in a remote hamlet an hour's drive from the nearest shop – to launch no less than six separate legal cases against VicForests and/or the Victorian Government since 2009.[95] The cost of defending these cases has been significant at times. In 2011-12, Vicforests reportedly spent $2 million of taxpayers' money to defend itself in two court cases.

There may also be additional taxpayer costs incurred if environmental groups access government-funded legal aid to launch the cases. Further to this, the lawyer group, Environmental Justice Australia (formerly the Environment Defenders Office (VIC)) – which in late 2017 was prosecuting two cases against Victorian timber harvesting – is on the Register of Environmental Organisations and therefore can receive tax deductible donations which could be construed as a taxpayer subsidy in the form of foregone tax revenue. A more perverse example from Victoria has been where the State Government has effectively sued itself when its environmental department (DELWP), which oversees the regulatory compliance of commercial forestry, has taken legal action against another government agency, VicForests, for allegedly breaching timber harvesting prescriptions.

The cases brought by EEG have largely focussed on the adequacy of government agencies in meeting their own management prescriptions or legal obligations. One example was their 2013 writ against the Victorian Government for failing to adhere to the legislated requirement to prepare an Action Statement under the *Flora and Fauna Guarantee Act 1988* "as soon as possible" after a species is listed as 'threatened'. In this case, EEG put forward four species which still did not have an Action Statement despite being listed as threatened species between 10 and 18 years ago.[96]

While such an inadequacy may be a legitimate concern, writing a plan doesn't necessarily mean that anything much will change on the ground, particularly where current land management and forestry practices already take adequate account of environmental values, or where activities are so proportionally small and scattered across the landscape that they realistically have no significant impact. Arguably this is the case in East Gippsland where around 85% of the forests were already formally or effectively reserved at the time the writ was launched, and where it is likely that timber harvesting – which was essentially the target of the litigation – was having no discernible impact on the survival of the frog, small skink, potoroo, and cockatoo species named in the writ.

During the discussion surrounding this case (which was settled out of court), it was revealed that there is a backlog of 370 Action Statements needing to be prepared for threatened species and that, although work is continuing on them, the responsible government department simply didn't have the resources to resolve this any time soon. Under such circumstances, forcing a government to expend potentially millions of dollars in defending one of its agencies in court, including effectively funding legal aid that is then used to underwrite legal action against itself, must reduce the amount of public money available either for preparing plans or undertaking any meaningful on-ground conservation works. This is somewhat counter-productive to improving real conservation outcomes.

In early 2015, former Greens leader Bob Brown issued a High Court challenge to the Tasmanian Government's *Workplace (Protection from Protesters) Act 2014* which was introduced to ensure the safety of both forestry workers and protesters, and the rights of workers to continue to earn a living when protests take place. This challenge was precipitated by the arrest of Brown and five protesters plus the fining of more than 30 other people at an in-coupe anti-logging rally at Lapoinya in north western Tasmania.[97] Again, it is highly questionable as to whether the time, energy, and money expended in this case will be in the public's best interests.

Direct negotiation sans Government

In recent years, there have been several attempts to resolve forestry disputes by taking governments out of the equation and setting up direct negotiation processes between environmental and timber industry interests. The Canadian Boreal Forest Agreement which was signed by industry and environmental group representatives in 2010, free from any formal government input, arguably created a buzz that this might be the best way to resolve the protracted 'war' between forest use and conservation.

Soon after, the so-called Tasmanian 'forest peace deal' negotiations were initiated. At various stages of that process, environmental groups were calling for similar processes to be instigated in other Australian states, and even advocating it as a means of resolving some contentious mining issues. Immediately after its election in 2014, Victoria's new Labor Government announced that it would establish a taskforce of industry and environmental group representatives to "… reach a consensus on proposals … about the future of the timber industry."[98]

In reality, these processes are never free of political input because public lands are managed by government agencies that are ultimately responsible for any proposed changes to land use or management practices that arise from them and may have consequences for the wider community. Nevertheless, such processes are undoubtedly politically attractive because they largely remove the onus from a government to make difficult decisions, and shield it from the controversy inevitably generated by major policy changes. It's far more difficult for forestry workers to complain about job losses arising from the declaration of a new national park if their own industry has agreed to it, and so the industry (rather than the government) cops the wrath that inevitably arises from disaffected workers and their supporters.

There are similarities between the Canadian and Tasmanian processes but also important differences and cautionary tales. Both were precipitated by years of campaigns designed by environmental activists to undermine or destroy timber industry markets and public support. However, the Tasmanian situation was additionally compounded by other significant market factors, such as the sudden and unexpected withdrawal of the biggest local native hardwood processor from the market in order to concentrate on plantation production and the building of a proposed pulp mill.

The Canadian boreal forest process has arguably been far more even-handed given that timber harvesting is still able to occur within a massive resource despite the industry agreeing to give-up its production rights on a huge slice of country – the best information suggests there is still a roughly 50:50 split between wood production and conservation in the affected forests. In contrast to this, in Tasmania, the negotiations were conducted against a back-drop of desperation amongst industry negotiators, and the results were firmly biased towards conservation interests. In terms of publicly-owned forests, only around one-quarter were being managed for long-term wood supply even before the 'peace deal' negotiations began, and only about one-eighth (around 12%) after the peace deal was signed.

Both agreements have also struck trouble. The Canadian Boreal Forest Agreement appears to be still holding together despite two of its environmental group signatories – Greenpeace and Canopy – having walked away from it in 2012, ostensibly due to impatience at the slow progress of change. Subsequently Greenpeace accused the largest forest industry signatory of illegal logging (later found to be false) causing that company to also walk away.[99] While the publicity generated by this has apparently had a negative impact on public perceptions, it seems that the Agreement is still holding together, although one commentator recently referred to it as 'rancid'.[100]

The Tasmanian 'forest peace deal' may not have even been signed if it wasn't for then Federal Environment Minister, Tony Burke, offering tens of millions of dollars in additional compensation for the timber industry just as their negotiators were set to abandon it. It is also notable that although the state's three largest environmental groups were signatories to the deal, four other smaller groups had signalled an intention to continue protesting even if it was signed. Furthermore, opposition to the deal had also been backed by the then current and former leaders of the Australian Greens. It is therefore arguable as to whether the 'peace deal' would ever have had lasting durability – in any case, it didn't survive the politics for very long.

The Tasmanian 'forest peace deal' was legislated in the *Tasmanian Forests Agreement Act 2013*. However, being an exceedingly contentious agreement, it ultimately only passed into legislation after the Tasmanian Upper House had added amendments that phased-in the reservation of 500,000 hectares of new forested national parks in separate tranches over several years. Less than a year later in March 2014, the state's Labor-Green Government was replaced by a Liberal Government that was openly opposed to the agreement and soon repealed the Act. This immediately stopped any more reservations and has left approximately 400,000 hectares of the formerly proposed reserves in limbo, pending a future return to their former status as productive working forests.

Whereas the Tasmanian Forest Agreement failed, the Canadian Boreal Forest Agreement may yet endure as a workable compromise despite its early hiccups. If it does survive, it will arguably be because timber is a proportionally very significant industry for the country in economic terms, and this somewhat tempers the acceptance of conservation demands by the government-of-the-day. Conversely, direct negotiation between industry and environmental groups is less likely to create an enduringly peaceful compromise where timber industries are proportionally small economically and/or are vulnerable at that point in time due to prevailing market or other forces. Under those circumstances, environmental groups and ideologically supportive governments are more likely to view such negotiations as an opportunity to kill-off timber industries or restrict them to tiny cottage remnants by effectively 'blackmailing' them into agreeing to huge conservation demands with lucrative exit compensation.

Insights into the 'deep green' mindset

No discussion about the role of environmental activism in instigating, promoting, and building a populist 'conservation culture' would be complete without examining the underlying mindset that drives 'green' campaigns aimed at 'saving' the environment. In this respect, the *Tasmanian Times* (TT) weblog and online forum makes an excellent case study. As it has become a mouthpiece for the state's 'green-left' demographic, examining its articles and the associated forum comments posted mostly in support of them, provides a barometer of extremist thinking particularly in relation to forestry which has, until recently, been the site's major focus.

TT was established in 2004 by a veteran journalist who described it as

"... a forum of discussion and dissent – a cheeky, irreverent challenge to the mass media's obsession with popularity, superficiality and celebrity". However, with a contributing editor who has been a significant figure in Tasmania's environmental community for over 30 years, including a period as a Greens political advisor; as well as a consulting editor who is described as "a writer and advocate on a number of environmental and social issues in Tasmania and elsewhere", it isn't hard to understand why the site evolved as it did.[101]

To log onto TT is to open a window to a culture of complaint. Any open-minded person viewing the array of articles posted on the blog at any given time, could be forgiven for believing there is little that is good about Tasmanian life. According to the TT group narrative, the state is blighted by a dysfunctional society mired in a myriad of problems, including corruption at every level of governance as it supposedly acquiesces to corporate greed and largesse in a manner akin to a Third World country.

In 2010, at the height of Tasmania's conflict over native forest use, an informal survey of TT found that 43% of the nearly 1,000 articles posted on the site during the first half of that year were about forestry. In 2012, a more detailed survey of 50 articles about forestry posted during a randomly selected period, found that although 200 different people posted a comment in response to those articles, a small core of 40 regulars were responsible for around 75% of the forum comments which those articles attracted.[102]

Amongst those 'regulars' at that time were a select group of seven who posted comments to 20 or more of the 50 assessed forestry articles. They included a Hobart veterinarian (who also was also leading a local environmental group); an operator of a 'wildlife shelter' who regularly has letters about Tasmanian forestry published in Australian newspapers; a former bus driver and would-be organic farmer living on a small rural block; a retired 'independent security and investigations professional' living in a small West Coast mining town; a cartoonist, graphic designer and former local government councillor living in the Tamar Valley; and a retired social science teacher who had become a wood craftsman.

Around 50% of the forestry articles surveyed in 2012 were authored from amongst the site's 40 regular forum participants, with a further 16% being regurgitated media releases from either Greens politicians or environmental groups. An additional 24% of the articles had been reproduced from local or interstate newspapers which, in regard to forestry issues, are usually written from a forest activists' perspective.

The 50% of articles authored by the TT regulars were all strongly opposed to Tasmania's forest management regime. They typically included ideological advocacy of new forest management policies and/or practices from an emotional, but often ill-informed standpoint, supported by home-spun philosophies or unwarranted conspiracy theories. Some examples were:

- *Clean water needs revised forestry operations*, jointly authored by a veterinarian; a self-professed 'red-green' socialist who had been referred to as "the state's public face of communism for 30-years"; and a writer and retired academic.
- *Forestry Tasmania: A compelling case for reform*, by a former national forests campaign director for The Wilderness Society.
- *Forestry Tasmania, the time is nigh*, by an accountant who for 5-years worked as an economist.
- *A case for restoration forestry*, by a retired social science teacher.
- *Defining waste in old growth forests*, by a writer with a focus on the environment and social issues.
- *Plantation Isle forever*, by a former teacher and historian who is an olive grower in the West Tamar Valley.

This phenomenon of narrowly-focussed laypersons striving to effectively 're-invent the wheel' is not only confined to forestry. For example, also in 2012, an academic in the field of oil and gas sub-surface engineering was leading TT's campaign against the state's commercial fishing industry by purporting to unpick the science behind the determination of annual seafood catch quotas.

Amongst the blog's 40 'regulars', 32 were derisive of Tasmania's native forest and plantation-based timber industries and were responsible for 75% of all comments emanating from that group. Much of this was abusive and/or accusatory of the current management regime including routine allegations of corruption and mismanagement. Some was more thoughtful, considered and factual although (somewhat irrationally) still usually advocating a complete overhaul of current forest management systems, including the abolition of the state's commercial forest management agency and closure of the timber industry. While there was often some reference to factual data selectively appropriated from annual reports or other published reports, it was typically misused to draw unwarranted or wrong conclusions that support an anti-forestry agenda.

As much as the articles, it is the nature of forum participation that distinguishes TT as an exemplar of the 'deep green' mindset. Although its founder may have envisioned something different, TT has evolved into a vehicle for airing the views of an 'online community' with a common ideological stance on most issues and little apparent interest in genuine debate. Although posts from others with alternate views are not uncommon, they are usually greeted with barely muted tolerance which at times strays into abusive put-downs. Despite this, it appears that these challenging views are required to galvanise, strengthen, and confirm the dominant positions of the blog's 'online community'. In their absence, forum discussions quickly peter-out after an initial round of mutual backslapping.

This concurs with 2007 research which examined several of the USA's most popular political blogs and found them to be spaces where participants with similar ideological viewpoints conversed. The lack of dissenting opinion on these blogs was attributed to the tone of discussion whereby those expressing a minority view were often targeted as though they were 'trolls' with their views caricatured as being stupid or irrelevant. Accordingly, it was found that opinion quality was low on blog sites where people are less concerned about explaining or accounting for the other side's viewpoint.[103]

This resonates with observations of TT where forestry discussions are routinely punctuated by references to negative caricatures such as Tasmania's 'failed forestry model', or the supposedly 'corrupt' or 'morally bankrupt' behaviour of the responsible government agency and its staff, and timber industry companies, as though such concepts are irrefutable and beyond dispute. For example, a recent comment to a TT forestry article described the government agency, Forestry Tasmania, as "... the worst GBE (Government Business Enterprise) in Australia's history (which deserves) a Gold Medal for unsustainability, intransigence, obstinacy, failure to pursue corporate governance and due diligence, environmental vandalism, and just plain commonsense".

The tendency for the overwhelming majority of TT's 'online community' to dismiss (rather than argue against) opposing views fits with the research of Haite who found that there is a natural human tendency for people to bind themselves into groups that share common narratives, and that once an individual accepts this narrative, they become blind to alternative views.[104] Kahneman has also attributed the tendency for a like-minded group to avoid constructively arguing against alternative views to the dominance of conclusions over arguments which is most pronounced when emotions are involved.[105] Slovic describes this as people letting their likes and dislikes determine their beliefs about the world.[106] Accordingly it seems, if someone who doesn't like the idea of trees being cut down, for example, is presented with information that justifies why it occurs and explains its benefits, they are naturally predisposed to reaffirm their pre-existing attitude that it as wrong, rather than to seriously reconsider it on its merits.

Walker found that political blogs typically involve a considerable amount of information-sharing with forum participants often referring to each other by name, quoting each others posts, and asking questions or giving feedback to each other.[107] This is certainly evident on TT, where it is apparent that the site's 'online community' learns far more about forestry from each other than by researching relevant information which is freely available from government sources. This was exemplified by a comment to a forestry article by a woman who had moved to northern Tasmania from Melbourne only a few years earlier. While she initially admitted that she was "... a reader endeavouring to make sense of the conflict, armed with little more than their own research, and lacking the long-term personal experience of native Tasmanians, ...", she nevertheless went on to demonise Tasmania's forest management:

> ... For many decades, and at least since the days of the Forestry Commission, the state's forest administration has been characterised by greed, mismanagement, and a cavalier approach to the state's natural environment and its land-owning citizens.
>
> Very few of those citizens complain publicly of Forestry Tasmania's unpalatable activities. For reasons which I am yet to fully appreciate, FT and its contractors hold many Tasmanians in their thrall, whether by intimidation, or through a generations-old subservience to the power of the wealthy.

It is safe to conclude that such an attack on a government agency by a person who has only recently arrived in the State can only have been inspired by reading and believing environmental campaign material, as well as presumably, TT articles and forum comments, which overwhelmingly articulate similar views.

This example also fits with a reality that environmental activism in Tasmania has traditionally been substantially driven and/or instigated by immigrants to the state, mostly from mainland Australia. On TT it is apparent that many of the regular forum participants are relatively new arrivals many of whom appear to be financially-secure 'sea changers' or retirees who have imported mainland inner urban values to a state with an economy which has traditionally relied on rural primary industries based on activities that they may abhor, such as cutting down trees, digging up minerals, or using pesticides. People of this ilk are generally articulate and 'time-rich', but they hardly deserve to be annointed as credible arbiters on the evolution and conduct of Tasmanian industry.

An analysis of the commentary on TT during a randomly selected period in 2012 suggests that extreme anti-forestry views are founded on:

- A total distrust of responsible government agencies and industry bodies, and a strident refusal to countenance that anything they do can be good.
- An unsubstantiated insistence that Tasmanian/Australian forestry practices are as bad or worse than those in developing countries.

- A refusal to acknowledge the century-long evolution of forest science and practice which underpins contemporary forest management.
- A determination to "re-invent the wheel" particularly in relation to timber production and plantation establishment.
- An inordinate focus on any instances of sub-optimal forest management performance and a determination to portray it as the 'rule' rather than the 'exception'.
- A refusal to acknowledge evidence of scale and proportion that would otherwise significantly weaken over-the-top claims about supposed environmental impacts.
- A lack of understanding of the fundamental basis of multiple-use forestry as a landscape-scale concept which considers individual coupes and localised impacts as only a small part of a much broader perspective.
- An inordinate focus on clearfelling during or immediately after its conduct, and a refusal to acknowledge what the same site may look like after regenerating 10, 20, or 30+ years later.
- A determination to label all timber harvesting as clearfelling by ignoring the reality that less intensive harvesting methods are more widely used.
- A reluctance to accept that different forest types require different silvicultural approaches.
- A determination to judge the economic viability of forestry purely on the profitability of the state government agency which sells logs and a disregard or strategic lack of acknowlegement of the far greater socio-economic activity that ensues once those logs are taken from the forest and processed.
- A selective reliance on science based on taking notice of studies which support pre-conceived anti-forestry beliefs, while ignoring other studies which may challenge those beliefs.
- Advocacy of eco-tourism as a replacement for timber industries.[108]

That such an array of problematic judgements/beliefs is so strongly held by mostly arms-length observers with such limited personal exposure to the target of their indignation, can arguably be attributed to the evolution and spread of a 'conservation culture' derived from decades of environmental campaigns and supportive media coverage repetitively espousing the same themes.

Despite such blinkered thinking, TT's regulars clearly believe that the site is highly influential in political circles, as was articulated in a recent forum post:

> This article and the one above ... exemplify the service that Tasmanian Times and its readership provide to the people of Tasmania. ... This forum provides a platform currently denied to the informed by a gutter press controlled by their servicing of the lowest common denominator, reliant on the power of the cash flow derived from conforming advertisers.
>
> On TT, the thoughtful may expound their views to peer criticism for they may write from an informed position, armed with the voice of reason all bereft of vested interest.

Such self-aggrandisement concurs with research findings showing that those who establish blogs or are regular participants in blog forums tend to overstate their importance and influence. In reality, their self-righteous notion of providing a community service is often at odds with the reality of a blog 'group-think' which is typically far from open-minded, thoughtful or well-informed.[109]

As TT's treatment of forestry issues is typified by aggressive cheerleading by like-minded laypersons promoting concepts from articles largely authored by each other, it would be disappointing if the site had any significant political influence. By comparison, other national political and current affairs blogs such as *Online Opinion* or the ABC's *The Drum* may be more influential because their articles are purpose-written and are more likely to be authored by those with a higher level of expertise in their chosen topic, including academics, practitioners, or politicians.

What passes for debate on TT is also somewhat problematic. Kelsey points out that intellectually dishonest debating tactics are rife on political blogs and identified six commonly used ploys which diminish the quality of discussion - all are commonly used by those opposed to Tasmanian forestry:

- Name calling to diminish or draw attention away from an opponent's argument.
- Use of irrelevant logic or facts (also referred to as 'straw manning') in which a debater bases statements on 'facts' or concepts that are unproven, especially by re-casting the views or position of his/her opponent.
- Changing the subject when a debater doesn't have a response or declines to respond.
- Employing hearsay in place of factual evidence to support a position.
- Questioning an opponent's motives to discredit their logic or factual evidence.
- Misrepresenting unqualified 'expert opinion' as factual.[110]

That the quality of debate about forestry on TT is still blighted by such tactics after years in which a small number of more informed commentators (mostly local foresters) have attempted to provide an insiders perspective to the blog's coverage of this issue, exemplifies the apparent futility of trying to change thinking in relation to environmental issues at the extreme end of the 'green-left' spectrum.

This raises the vexed question of whether there is any point in mounting evidence-based challenges to the environmental movement's 'group think'. It has generally been assumed by forest scientists and practitioners that putting factual information into the public arena is a pre-requisite for turning around adverse opinions of forestry, and that the general failure to do so (or do enough) has been a primary factor in the success of environmental campaigns in fostering a populist, but flawed 'conservation culture' amongst the wider community.

However, a study undertaken in relation to climate change and forests in Oregon suggests that this isn't necessarily the case. Contrary to expectations, this research found that people with the highest degrees of scientific literacy and technical reasoning were often less concerned about climate change than other groups. Instead, they were the group that exhibited the greatest level of cultural polarisation, thereby suggesting that access to and understanding of the science may have little to do with how personal beliefs are formed and maintained.[111] Although this is only one study, it concurs with a more recent acceptance that different segments of the public fit their interpretations of scientific evidence to their own cultural philosophy rather than absorbing it to perhaps rethink their position.[112]

According to noted psychologist, Edgar Schein, culture is "... the sum of all the shared, taken-for-granted assumptions that a group has learned throughout its history." Accordingly, a group with a common purpose develops a strong culture as it achieves success, with its power stemming from the unconscious minds of the group acting in concert. New members are effectively inducted in the manner of learning a new skill, such as driving a car, which can eventually be done automatically without any procedural thinking. Eventually the group develops a shared mental template that its members take for granted.[113]

Viewed in this way, a strong culture is obviously stable and resistant to change because it is a way of feeling, thinking, and seeing the world derived from accumulated shared learning. It is largely immune to any challenge emanating from the provision of new information that contests the group's core assumptions and beliefs, particularly if it comes from outside and from a group that the culture perceives as its enemy.[114]

Accordingly, if we accept that environmental group activists, members, and supporters strongly share a particular 'conservation culture' – i.e. that 'saving' forested environments hinges on evicting human resource uses and declaring national parks – than it is virtually inconceivable that this could be turned around by the provision of factual information by the forestry profession.

Equally, this accusation has been levelled at the forestry profession by critics alleging it to have a strong and implacable culture stubbornly resistant to changing its practices in response to suggestions by environmentalists and the broader community. Superficially this argument is supported by the commonality of learning and training undertaken by forestry students, and reinforced by their superiors once working in the field. However, this is surely no different to the work-place culture engendered within any scientific or professional discipline given that, for example, lawyers, doctors, or teachers are all collectively shaped by similar training and workplace experiences.

The 'culture of forestry' has been informed by science and over a century of practice often involving trial and error, and – far from being implacable – has constantly changed and evolved in accordance with improved knowledge, as well as in response to valid community concerns and political directives. On the other hand, the 'conservation culture' cultivated by environmental activism and embraced to varying degrees by the wider community has – as has been described earlier – been largely founded on exaggeration and embellished with emotion. Asking the former to bow to the latter is a question that could be posed in many other analogous situations, but typically isn't because of the obvious absurdity of, for example, allowing criminals to run the prison system, or students to run schools. Somehow, forestry is regarded differently, presumably because it is not so readily apparent that removing those with training, knowledge and practical experience from forest management would be fraught with disaster. Ironically this further exemplifies the existence of a pervasive 'conservation culture' that has fostered a general unwillingness amongst the media to examine the complexity of forest management.

Another important question in regard to whether forestry should surrender to the 'conservation culture', is to what extent do environmental activists and their supporters actually represent the wider community as they like to presume? They themselves are often referred to as 'dark greens' or 'deep greens' and are estimated to comprise only a minor portion of the population. In 2001, a survey by the National Association of Forest Industries found that 14% of the population were deeply committed to saving forests; with a further 51% estimated to comprise the 'light greens' who are inclined to agree with the 'dark greens' but are not bound nearly so tightly to a shared group ideology. The remaining one-third of the population are either not interested in forestry issues, or are opposed to 'deep green' views of forestry, arguably including the majority of rural Australians.

More recently the Australian Bureau of Statistics found in 2007-08 that, amongst Australian adults older than 18-years, 14% had donated money to protect the environment; 10% had expressed concern for the environment through sending a letter, an email or talking to responsible authorities; and 2% had participated in a protest rally or demonstration in regard to an environmental issue.[115]

If we accept that in many (or perhaps most) cases, a citizen with deep concerns for the environment would partake in several or all of these activities, the 14% figure arguably quantifies the upper limit for the proportion of the population that can be classified as 'dark green' – as it did in 2001. This suggests that the 'dark greens' equate approximately to the body of voters who typically support the Greens, and accords with Steffen's definition of a 'dark green' as someone who believes that environmental problems are inherent to industrialised civilisation and therefore supports radical political change.[116]

While the 'dark greens' would appear to be a lost cause in terms of rational engagement about forestry, Kahneman believes that loosely held beliefs can be changed (at least a little) if people are made aware that the environmental risks associated with a disliked activity are lower than previously thought.[117] While this potentially represents an opportunity to steer the 'light greens' to a more considered understanding of forest management, the greater likelihood is that the 'dark greens' – through ongoing environmental campaigning and progressive infiltration of educational institutions and media outlets – will counteract this by continuing to bombard the community with messages that reinforce their manufactured 'conservation culture'.

In 2010, Sheil and Meijaard noted that the propensity for committed environmentalists to display a judgemental bias was concerning because it encouraged them to overlook or act against major conservation opportunities. They contended that a better appreciation of the tricks of the human mind would make environmentalists more open to appreciating different viewpoints that could ultimately lead to better environmental outcomes.[118] Unfortunately, the strength of the prevailing, overly simplistic 'conservation culture' which has been engendered by the environmental movement, is likely to prevent optimal forest conservation outcomes. As Sheil and Meijaard observed in a later interview:

> ... Conservation needs to change. We need to recognize that pragmatic conservation solutions aren't about black and white, good and evil, or nature versus non-nature. Long-term conservation solutions have to involve compromises, otherwise we will just be wasting our time ...[119]

Chapter 2 Endnotes

1 E.H. Schein, 1999, *The Corporate Culture Survival Guide: Sense and Nonsense about Culture Change*, San Franciso: Jossey-Bass.

2 *The Amazon rainforest is not in danger of being destroyed – Opposing Viewpoints in Context*, Gale Databases, Massachusetts Board of Library Commissioners and Massachusetts Library System; From: Morano, Marc, *Rain Forests*. Edited: HaiSong Harvey, San Diego: Greenhaven Press, 2002.

3 Brazil: national website.

4 CFACT 1997, *The Rainforest Issue: Myths and Facts*, Committee for a Constructive Tomorrow Briefing Paper #102.

5 Brazil's National Institute for Space Research publishes annual Amazon deforestation rates based on an analysis of satellite imagery.

6 Livescience 2015, *Deforestation: Facts, causes and effects*, by Alina Bradford, 3 March 2015.

7 M. Poynter, 2012, *Forestry and 'Green-Left' Thinking: The Tasmanian Times case study* (unpublished manuscript submitted to the journal, *Australian Forestry*).

8 Chris Arsenault, 2015, "Brazil on right track for reducing deforestation rates: study", *Reuters*, 30 November 2015.

9 M. Poynter and G. Featherston, 2008, *Review of timber production in Melbourne's water catchments*, unpublished report prepared for the Victorian Association of Forest Industries; and MBAC Consulting 2006, *Feasibility of plantations substituting for timber currently harvested from Melbourne's water catchments*, unpublished report prepared for the Victorian Department of Sustainability and Environment.

10 *Native Timber Harvesting in Melbourne's Water Catchments*, VicForests Fact Sheet, VicForests website, accessed March 2016.

11 D. Hutton and L. Connors, 1999, *A History of the Australian Environmental Movement*, Cambridge University Press, p. 14.

12 A. Chase, 1986, *Playing God in Yellowstone – The Destruction of America's First Park*, Harcourt, Bruce & Company.

13 R. Routley and V. Routley, 1974, *The Fight for the Forests – the takeover of Australian forests for pines, woodchips and intensive forestry*, Research School of Social Sciences, Australian National University, Falcon Press.

14 P. Hay, 2002, *Main Currents in Western Environmental Thought*, UNSW Press, p. 3.

15 F.R. Moulds, 1991, *The Dynamic Forest – a History of Forestry and Forest Industries in Victoria*, Lynedoch Publications, p. 118.

16 V. Bible, 2010, *Aquarius Rising: Terania Creek and the Australian Forest Protest Movement*, Thesis submitted for a Bachelor of Arts Degree, University of New England, Armidale.

17 J. Dargavel, 1994, *Fashioning Australia's Forests*, Oxford University Press, p.184

18 D. Hutton and L. Connors, 1999, op. cit., pp. 155, 165.

19 Ibid, pp. 118-124, 149.

20 G. Borschmann, 1999, *The People's Forest – A Living History of the Australian Bush*, The Peoples Forest Press, pp. 111-114.

21 Wikipedia 2016, Bob Brown.

22 A. Hay, 2002, *Gum – The Story of Eucalypts and their Champions*, Duffey & Snellgrove, p. 179.

23 G. Courser, 2003, *Sawbones speak up for forests – and health*, Australian Medicine, Vol. 15 No.12, p. 13.

24 Ecological Society of Australia, 2016.

25 "Something in the Water", *Australian Story*, ABC TV, Part 1-15 February 2010 and Part 2 – 22 February 2010.

26 Hutton, D. and Connors, L. 1999, *A History of the Australian Environmental Movement*, Cambridge University Press, p. 160.

27 Ibid, p. 165.

28 Ibid, p. 167.

29 From the Australian Conservation Foundation website (accessed October 2005).

30 From The Wilderness Society website (accessed 2007).

31 P. Moore, 2010, *Confessions of a Greenpeace drop-out: The making of a sensible environmentalist*, Beatty Street, pp. 137-138.

32 D. Benuik, 2015, "Milne plea to UN heritage committee over push for tourism projects in state's wilderness areas", *Hobart Mercury*, 25 April 2015.

33 J. Gordon, 2016, "Alarm at alpine trail plan", *The Age*, 27 December 2016.

34 R. Beeby, 2009, "Green groups' spending hots up to $70 million" *Farm Weekly*, 23 December 2009.

35 J. Kelly and D. Shanahan, 2016, "$685m windfall for greenies", *The Australian*, 1 November 2016.

36 House of Representatives Standing Committee on the Environment inquiry into the Register of Environmental Groups, initiated on 26 March 2015 after being referred by the Minister for the Environment, the Hon. Greg Hunt MP.

37 M. Bulky, 2016, "Lapoinya: The bulldozers move in ...", *Tasmanian Times*, 20 January 2016.

38 "Bob Brown arrested at anti-logging protest in Tasmania", *The Age*, 26 January 2016.

39 Miranda Gibson sat in the so-called 'Observer Tree' for 15 months starting in December 2011.

40 A. Krien, 2010, *Into the Woods – The Battle for Tasmania's Forests*, Black Inc, pp. 34-44.

41 The *Incontinental Deluxe Guide to Blockading* was produced by the northern NSW environmental group, the North East Forest Alliance and was formerly available on their website.

42 Timber Communities Australia 2002, *Green group promotes sabotage*, media release, 9 April 2002.

43 Badja State Forest action was documented on the Forest Rescue website; and the Marysville blockade accusation was contained in: Hodge, A. 2002, "Union lashes Greens' forest fortress", *The Australian*, 21 February 2002.

44 Forest Rescue website, accessed October 2004.

45 Timber Communities Australia 2002, op. cit.

46 South East Timber Association 2016, *Activists endanger road users*, posted to the SETA website in February 2016.

47 NSW Coroners Report 2013/350429 into the death of John Creighton, 17 February 2015.

48 M. Duffy, 2005, "Under the Gunns", *Sydney Morning Herald*, 25 March 2005.

49 *Logging contractors hit hard by vandalism*, transcript from the ABC's *7.30 Report*, 19 March 2002.

50 *Chipman foresees manned gates*, *Hobart Mercury*, 26 February 2002.

51 *Logging protests cost state $2.5 m.*, by Melissa Fyfe, *The Age*, 9 August 2002.

52 During the lead-up to Victoria's November 2017 Northcote by-election, the Greens commissioned an opinion poll which they claimed showed that almost 90% of voters supported their policy of declaring the Great Forest National Park. In December 2017, NSW Greens spokeswoman, Dawn Walker, cited a ReachTEL poll conducted on the NSW North Coast which showed overwhelming public support for new national parks to protect declining koala populations.

53 Australian Bureau of Statistics 2011, 8146.0 Househaold Use of Information Technology, Australia, 2010-11.

54 P. Sheehan, 2003, *The Electronic Whorehouse*, Pan Macmillan, p. 12.

55 *Becoming Forest Friendly – An Eco-Kit for Law Firms*, October 2003, Lawyers for Forests website.
56 *Forest Letter Watch* website, accessed mid-2006.
57 *Forest Fact File* website, accessed mid-2006.
58 *Boycott Woodchipping* website, accessed mid-2006.
59 *Discover Tasmania* website, accessed late 2004.
60 G. Lewis, 2007, *Blogging Democracy: The contribution of political blogs to democracy*, by Gareth Lewis, Winner of the Dalton Camp Award, Friends of Canadian Broadcasting.
61 A. Veenstra, 2010, *The effects of reading political blogs*. Accessed via the *Snurblog* website.
62 M.E. Keck, 2002, *Amazonia in Environmental Politics*, In: *Environment and Security in the Amazon Basin*, edited by J.S. Tulchin and H.A. Golding, Woodrow Wilson Centre Reports on the Americas #4.
63 *Global outcry over falling forests and failing democracy in Australia's island state of Tasmania*, Rainforest Action Network media release, 6 March 2006.
64 Submission by Mark Poynter to the House of Representatives Standing Committee on the Environment's Inquiry into the Register of Environmental Organisations, May 2015.
65 L. Black and S. Bice, 2011, *Defining the essential and elusive social licence to operate*, News, Australian Centre for Corporate Social Responsibility (ACCSR), August 2011.
66 *Smartwood*, Rainforest Alliance website, accessed April 2005.
67 *About PEFC*, PEFC website, accessed April 2005.
68 FSC Australia website.
69 *Other certification schemes*, Forests and the European Union Resource Network (FERN) website, accessed April 2005.
70 T. Cadman, 2001, *The Australian Forestry Standard (AFS): another PEFC in disguise?* by Tim Cadman, Native Forest Network spokesman on certification for the National Forest Summit, May 2001.
71 WWF 2002, *Australian standard lacks credibility in the marketplace* by Michael Rae, World Wildlife Fund Media Release, 17 October 2002.
72 Ibid.
73 T. Cadman, 2000, *The Development of an Australian Forestry Standard: An Environmental NGO Perspective*. Preliminary submission to the Ministerial Council on Forestry, Fisheries, and Aquaculture (sic), prepared on behalf of the Conservation Council of Western Australia, Friends of the Earth, Native Forest Network, Rainforest Information Centre, and West Australian Forest Alliance.
74 *Forest Product Certification*, Native Forest Network Australia website, accessed April 2005.
75 *Australian NGO's reject AFS certification*, EU Forest Watch, Issue No. 68 (November 2002), reproduced on the Friends of the Earth website.

76 Wilderness Society 2005, *Certifying the Incredible, the Australian Forestry Standard, Barely Legal and Not Sustainable,* accessed from the Wilderness Society website, May 2007.

77 H. Crawford, 2006, *A review of forest certification in Australia,* prepared for the Forest and Wood Products Research and Development Corporation, Project PN05.1025 (September 2006).

78 Stephen Cauchi, 2014, "Audit criticises VicForests application for FSC rating", *Sydney Morning Herald,* 5 July 2014.

79 Canadian Boreal Forest Agreement website.

80 P. Foster, 2016, "Resolute Forest's day in court promises to expose global anti-development agenda", *Financial Post,* 5 January 2016.

81 "Why Resolute can't win in an ever-changing bush war", *Globe and Mail,* 4 January 2016.

82 P. Foster, 2016, op. cit.

83 Ibid.

84 Joshua Axelrod, 2016, *The world needs sustainable forestry: Efforts to greenwash harmful logging needs to be stopped,* Natural Resources Defense Council blog, 16 April 2016.

85 *On trial – Australia's endangered species law,* On-trial website, accessed February 2006.

86 Australian Greens website, accessed February 2006.

87 "Brown raises cash", *Sunday Age,* 5 February 2005.

88 *On trial – Australia's endangered species law,* op. cit.

89 M. Denholm, 2007 "Court ruling could stop Tassie logging", *The Australian,* 10 February 2007.

90 Ibid.

91 "Court allows Weilangta logging", *SBS News,* 30 November 2007.

92 US Congress House Committee on Natural Resources, Oversight Hearing: *The Endangered Species Act: How litigation is costing jobs and impeding true recovery efforts,* 6 December 2011.

93 US Congress House Committee on Natural Resources, Oversight Hearing: *Taxpayer-funded litigation: Benefitting lawyers and harming species, jobs and schools,* 19 June 2012.

94 *Law and Policy* section, *Lawyers for Forests* website, accessed October 2004.

95 Environment East Gippsland website, accessed March 2016.

96 *EEG successfully takes legal action - #3,* Environment East Gippsland website, accessed March 2016.

97 *High Court challenge to Tasmanian Government's protest laws,* The Bob Brown Foundation, media release, 10 March 2016.

98 *Forest industry taskforce terms of reference released,* media release from the Hon Daniel Andrews, Premier of Victoria, 20 November 2015.

99 Fraser Los, 2014, "Boreal truce", *Canadian Geographic,* January-February 2014.

100 P. Foster, 2016, op. cit.

101 Tasmanian Times website – About Us.

102 M.W. Poynter, 2012, op. cit.

103 D.A. Walker, 2007, "Blog commenting: A new political information space", by Dana M. Walker, University of Michigan, School of Information, In: *Proceedings of the American Society for Information Science and Technology*, 43:1.

104 J. Haidt, 2013, *The Righteous Mind – Why Good People are Divided by Politics and Religion*, Vintage Books, New York.

105 D. Kahneman, 2011, *Thinking Fast and Slow*, Farrer Strauss and Giroux, New York, p. 103.

106 P. Slovic, 2000, *The Perception of Risk*, in Risk, Society and Policy Series, London and Sterling Va., Earthscan Publications Ltd.

107 D.A. Walker, 2007, op. cit.

108 This analysis was undertaken by the author's examination of 50 Tasmanian Times forestry articles and their associated forum discussions taken from a randomly selected period during 2012.

109 D. Brown, 2006, *Book Review: "Blog" by David Kline and Dan Burstein*, by Dan Brown, Online Editor, Free Press (Canada), 11 May 2006: On the Future-of-Journalism blog.

110 Kelsey 2012, *Guest Post: Intellectually dishonest debate tactics*, Political Friends Blog, 15 July 2012.

111 T. Satterfield, 2002, *Anatomy of a conflict: Identity, knowledge, and emotion in old growth forests*, Vancouver BC: UBC Press.

112 T. Beath, 2016, "Science and controversy", *The Forester*, February 2016

113 E.H. Schein, 1999, op. cit.

114 T. Beath, 2016, op. cit.

115 Australian Bureau of Statistics collected data about the level of interest in environmental issues in their publication: 4102.0 Australian Social Trends, June 2010 – Environmental awareness and action.

116 Alex Steffen, prominent social scientist and futurist cited in "Bright green environmentalism", *Wikipedia* (accessed, March 2016).

117 D. Kahneman, 2011, op. cit.

118 D. Sheil and E. Meijaard, 2010, "Purity and Prejudice: Deluding Ourselves About Biodiversity Conservation", *Biotropica*, Vol 42:5, 566-568, Sepember 2010.

119 Erik Meijaard quoted in: Hance, J. 2010, "Environmentalists must recognise biases and delusions to succeed", *Mongabay*, 18 October 2010.

3

Media: Entrenching the culture

"A job well done is about more than just handing someone a microphone. It is to probe, and to question, and to dig deeper, and to demand more"

US President Barack Obama

When incumbent US President Obama berated the media for substantially creating the 'Donald Trump phenomenon' during the 2016 US presidential election campaign, he was unwittingly speaking for many others who could lay claim to being similarly frustrated by the media's preference for whipping-up sensationalist hysteria over fully reporting the often boring facts. Chief amongst them would be resource use industries which have long despaired over context-free media portrayals based primarily on an unquestioning acceptance of environmentalist rhetoric. Arguably this has played the greatest role in manufacturing a popular opposition to their very existence, amongst some elements of society.

Typically, the media bristles at any suggestion that it contributes to adverse societal outcomes. As *The Australian's* Janet Albrechtsen retorted, Obama's portrayal of Trump as a media construct was indicative of his 'cluelessness'. Far from being the beneficiary of "billions of dollars of free media, minus any serious accountability", Trump was, in her view, a product of disaffected voters reacting to "a failed, insular political class."[1]

While there is undoubtedly some truth to her latter assertion, the shortcoming of such a defence is its blindness to the media's primary role in initiating and then enhancing public dissatisfaction. As most voters lack the time and/or inclination to follow issues closely enough to make their own informed judgements, their political leanings are invariably shaped partly or wholly by media reporting and analysis. As Obama noted, "real people depend on you [*the media*] to uncover the truth". When the media doesn't report fully or objectively and/or pushes its own agenda, it is inevitable that public perceptions are shaped accordingly.

This is irrefutable when it comes to environmental issues and is being regularly demonstrated by media coverage of recent episodes such as the 2016 Tasmanian bushfires and widespread coral bleaching on the Great Barrier Reef. In both cases irresponsible reporting misrepresenting the reality sparked serious concerns of tourism downturns.[2] In Tasmania, Premier Will Hodgman took the unprecedented step of inviting journalists to take a helicopter flight with him over the fire area to show that, contrary to reports, only a small proportion of the state's World Heritage landscape had been fire-affected.[3]

The predilection for many journalists to report complex environmental issues as alarming eco-catastrophes is no better illustrated than with regard to forestry activities – especially timber harvesting and fuel reduction burning. It is easy to dislike an activity that involves cutting-down trees or deliberately firing the bush. But it shouldn't be too difficult to engender a high level of community acceptance if the public are fully informed about its rationale, its scale and proportional extent, its inherently high level of planning and regulation, and its benefits.

Unfortunately in relation to forestry and many other environmental issues, the city-based media which reaches the great majority of the population, has shown little interest in reporting such critically important context and has typically paid little more than lip-service to its ethical requirement to report on issues with appropriate balance.

An example is a 2015 article in *The Weekend Australian* which addressed the allegedly dire threat posed by timber harvesting to a threatened species of giant freshwater crayfish in Tasmania's north-west. Almost all of the article's 790 words dealt with the supposed threat to the species from the perspective of one conservationist fighting to protect it; while just 33 words in the final two paragraphs quoted an alternative view from forestry authorities that "current practices provide sufficient and rigorous protections for the species".[4] The article displayed an appalling lack of balance by virtually ignoring the fact that timber harvesting is a limited and highly regulated activity founded on practices which already mitigate those fears based on hundreds of thousands of research hours and observations by scientists and forestry practitioners across several decades and countries.

Having monitored media coverage of forestry issues for several decades, it is clear that such grossly unbalanced articles are to be expected as 'the norm'. It seems that most journalists – particularly those working for the largest city-based media outlets – have no compunction in crossing the line from presenting news to conveying an opinion when reporting on forestry (and other environmental) issues. Despite usually having no personal experience or hard knowledge of these topics, they misuse their position to support their personally favoured side of a complex argument by giving greater weight to preferred views and/or either ignoring or minimising alternative information or views that would challenge their pre-determined 'angle'.

Presumably due to having grown-up with the pervasive influence of an exaggerated and distorted pro-environment narrative, today's journalists seemingly start from a premise that environmental campaigners are the 'good guys', whereas resource use industries are inherently bad. At its worst, the media openly describes or implies that such industries are corrupt, greedy, out-of-control profiteers from a war against nature. It seems that these pre-conceived convictions can be hard-wired within affected journalists to such an extent that no amount of discussion or demonstration of the limited extent and/or highly regulated nature of resource use can ever hope to shift them.

In the public discourse over forests, there have been many examples of journalists shelving any pretence to professional objectivity to essentially become a mouthpiece for environmental campaigns. Their reporting reflects the environmental activists' rhetoric whereby context and complexity is ignored in favor of simplistic championing of new reservations, primarily as a vehicle for banning human resource use.

Reporting of forestry issues (especially in 'opinion' pieces) is typified by a range of stereotypical 'angles' which reflect the major campaign themes of environmental activism, such as:

- a refusal to acknowledge the low proportional impact of timber production at a landscape scale;
- questioning the need for native hardwood timber and claiming that a direct transition to plantations is immediately possible;
- ignoring, downplaying, or undermining the views of government forestry agencies and the timber industry on the basis of their economic self-interest;
- implying that a government/industry conspiracy is masking endemic corruption;
- blithely accepting the claims of supposed industry insiders such as 'former loggers' or other disaffected whistleblowers based on unquestioning belief in their honesty and righteousness;
- referring to public opinion based on poll results with no explanation of the questions asked or the survey population;
- dismissing or downplaying the notion that timber industries are associated with significant socio-economic benefits; and
- glibly promoting the supposed socio-economic advantages of tourism as an alternative industry.

Occasionally there are surprises in how forestry issues are reported. An example which both affirmed and challenged some aspects of these stereotypical angles was a major feature article by John van Tiggelen published in the July 2014 edition of *The Monthly*. It investigated the plight of the Triabunna woodchip mill on Tasmania's east coast, which is somewhat emblematic of the disaster that has befallen Tasmania's forestry and timber sectors since 2010.[5]

The Triabunna mill had operated since the early 1970s, but in 2011 was sold by cash-strapped forestry company, Gunns Ltd, to two wealthy 'green' entrepreneurs – Graeme Wood and Jan Cameron. The sale came with a proviso that the mill may be reopened in the future in accordance with the Tasmanian Forests Agreement then being brokered by environmental activists and timber industry representatives.[6] However, as *The Monthly* reported, the mill's new 'manager' – former Wilderness Society boss, Alec Marr, who'd been rather cynically appointed to that role by Wood and Cameron – had taken advantage of the distraction of AFL Grand Final day in late September 2013, to systematically vandalise the facility. As intended, this killed-off the possibility of it ever re-opening.

It is hard to decipher whether van Tiggelen's unnecessarily detailed expose of Marr's destructive rampage was meant to invoke hero worship amongst a readership that largely supports environmental causes at any price, or was more indicative of a determination to report the full story without fear or favour. Whatever the reason, he ventured into largely unchartered territory by providing instructional insights into the literal and metaphorical thuggery of radical environmentalism. This was a stark contrast to traditional media portrayals characterising such ugly behaviour as being the sole province of angry 'red-neck loggers' perpetrated against 'harmless, gentle greenies'.

Despite its unexpected revelations, van Tiggelen's article still dismayed many within Tasmania's forestry and timber sectors by repeating in my opinion to be the stereotypical errors, misconceptions, and unsubstantiated allegations that have been used to demonise them for at least the past 20-years. Chief amongst these were frequent references to clearfelling and the "old growth industry" as though this accurately described a native hardwood sector that was mostly neither clearfelling nor working in old growth forests. In 2008-09 – a year before the Tasmanian timber industry was plunged into so-called 'peace' negotiations with its environmentalist opponents – only 6% of the state's annual native forest harvesting involved the clearfelling of old growth trees.[7]

But such inconvenient truths have long been missing from media coverage of Australian forestry issues. According to long-time timber industry advocate, Alan Ashbarry, who has spent many a day with journalists attempting to inject factual balance into the media's coverage of

Tasmania's forestry conflict, most are uninterested in documented forestry facts and figures.[8] It is then hardly a surprise when their published articles leave out key contextual information such as the pre-existing level of forest reservation or how State forests are being managed. Typically in their place are the usual sensational myths peddled by the environmental activist groups who had presumably incited the media to investigate the story.

While this has long typified the city-based media's reporting of Australian forestry issues, it may have been even more pronounced in this case as van Tiggelen's article was written for an inner-urban 'green-left' audience which reportedly comprises the vast bulk of *The Monthly's* 120,000 readers.[9]

A previous anti-forest industry article in *The Monthly* by celebrated Tasmanian novelist, Richard Flanagan, had also been spectacularly sensational. Published in May 2007 during the lead-up to a Federal election, Flanagan's, *Gunns Out of Control,* contained passages from his earlier article – *The Rape of Tasmania* – published by *The Bulletin* in December 2003.[10]

Subsequently described as "igniting a nationwide debate about woodchipping in the island state", Flanagan's original article, *The Rape of Tasmania*, had colorfully embellished all the stereotypical anti-forest industry themes. That it was written without any consideration of opposing views was obvious, but was nonetheless confirmed by a prominent figure from the Tasmanian timber industry who had approached *The Bulletin* three times prior to publication in an optimistic bid to have forestry context and industry perspectives included.[11]

In both articles, the major theme explored by Flanagan was alleged corruption and greed associated with Tasmanian native forest wood production. His later article in *The Monthly* had been broadened and updated to include the proposal by Gunns Ltd. to build a pulp mill on northern Tasmania's lower Tamar River. In supporting his claims of supposed secrecy, lies, conspiracies, cover-ups and fear, Flanagan denigrated the integrity of an array of prominent Tasmanian government and industry figures. This would later prompt Tasmanian Senator, Eric Abetz, to remark that the article had cemented Flanagan's standing as an outstanding author of fiction.[12]

While Flanagan's articles represent one 'media opinion', their impact may be somewhat muted because the identity of the author generally makes readers aware of either or both a blatant agenda or an obvious lack of professional expertise in the subject matter. Far more dangerous are articles written by zealous journalists with similarly strong personal convictions and agendas and no reservations in displaying them. Because it seems that a large segment of the community still naively regards all journalists as impartial and objective, their distortions or exaggerations may be more widely accepted as gospel truth.

Currently, few city-based journalists working on resource use issues for the mainstream media are truly objective and free of strong pro-environment convictions. This includes some journalists working for News Ltd media outlets that are usually perceived to have conservative or right-wing biases that would typically result in more balanced portrayals of resource use and rural industries. On the other hand, the ABC and Fairfax media outlets have long displayed a default position of opposing aspects of natural resource use on the presumption that it is environmentally destructive.

In relation to forests, those inclined to dismiss the influence of the media on informed thinking and public policy often argue that it is, in an overall sense, dominated by right-leaning conservative outlets which tend to support industry over environmental care. While it is probably true that the majority of the population does favour commercial television and radio, as well as tabloid-style newspapers and magazines which are collectively categorised as right-leaning/conservative; it is important to note that these outlets generally don't cover forestry issues to any significant degree. Most media interest in forestry issues emanates from left-leaning media organisations, such as the ABC and Fairfax, which typically report from a pro-environment, anti-industry perspective. The major exception to this is *The Australian,* which is widely regarded as a right-leaning conservative media outlet but has displayed considerable interest in forestry issues which it typically covers from a pro-environment, anti-industry perspective.

Further to this, the supposedly dominant right-leaning media outlets are typically more focussed on sport and entertainment with largely only a superficial current affairs coverage. Accordingly, the more detailed analysis of current affairs and politics which characterises the left-leaning media outlets is presumably preferred by a more thoughtful audience that is arguably more likely to contain our current and future political leaders and decision-makers. Unfortunately, while they may believe they are being fully informed they are, at least with regard to forestry issues, mostly being fed a jaundiced version of the reality containing little more than a nod to objectivity and balance.

This may well be largely due to close linkages which have evolved between environmental activism and the media. This was exemplified by an ABC TV *Australian Story* episode about Richard Flanagan which was screened in late 2008. As expected, it lauded Flanagan's celebrity-brand of environmental activism as a positive force. However, of particular interest, was the program's revelation of Flanagan's close associations with prominent media figures, Morrie Schwartz, publisher of *The Monthly*; Charles Wooley, well-known Channel Nine reporter; Matthew Denholm, the Tasmanian correspondent for *The Australian;* and (Richard's brother) Martin Flanagan who writes for *The Age*.[13]

Flanagan's bitter antipathy towards Tasmania's forest industry is, in my opinion, reflected in journalism which creates grossly unfair perceptions of the industry's environmental impact thereby hardening community attitudes against it. This was exemplified by *The Australian*'s coverage of the Tasmanian forests 'peace deal' process from 2010-14 which routinely omitted the critically important point that the negotiations were only over a 20% portion of the state's forests, thereby fostering a widely-held misconception that the future of Tasmania's forests was dependent on a deal to greatly restrict or end timber production. The ease with which errant perceptions of forestry can be manufactured is also illustrated by the regular use of throw-away lines, such as the opening sentence of an article in April 2014, in which *The Australian's* Matthew

Denholm wrote: "The debate around whether Tasmania should log or protect its remaining wild forests used to be so simple ..." – thereby implying such falsehoods as that 1) few wild forests remain; 2) that they aren't currently protected; and 3) that they could all potentially be lost to timber production.[14]

Veteran broadcast journalist, Charles Wooley, has also exhibited a disturbing philosophical approach to stories about environmental issues. In 2004, whilst defending a controversial ABC's *Four Corners* episode about Tasmanian timber company, Gunns Ltd, he arguably encapsulated the cynical attitude of many of his peers when relating the story of an editorial mentor who, early in his career, had advised him to always listen to what interviewees said when researching investigative stories, but to then take note of who pays them.[15] While such a 'guilty-until-proven-innocent' approach may in some instances expose activities that are contrary to the public interest, its presumption of government or industry wrong-doing is more likely to unfairly malign legitimate and sensible activities. More disturbing is that it effectively dismisses or downplays the informed perspectives of those who should know the most through working daily within and around resource use issues, in lieu of the opinions of critics such as environmental activists who, while certainly independent in a pecuniary sense, are usually merely arms-length observers less able to offer credible insights.

The mainstream media's predilection to ignore industry or Government expertise while uncritically accepting the views of their critics has arguably been most prevalent on current affairs television. That it can grossly distort the public record was exemplified by the ABC's *Four Corners* episode, '*Lords of the Forest*' which was broadcast in 2004.

In response to angry complaints about this program, an investigation by the media's Independent Complaints Review Panel concluded that it contained "instances of serious bias, lack of balance and unfair treatment" and that "*Four Corners* broke from its constraining guidelines... its enthusiasm to canvass the logging-in-Tasmania issue ... compromised the program that resulted". Special mention was also made of journalist, Ticky Fullerton's use of emotive language such as "... voracious appetite for timber ... overwhelming devastation ... absolute assault on the landscape and senses ... corruption and cronyism ... aggressive forest policy ... and mushroom cloud" which were found to leave "the reasonable viewer with the impression that the program is anti-logging, i.e. seriously lacking in balance and fairness".[16]

'*Lords of the Forests*' was not the first *Four Corners* episode to seriously misrepresent Australian forest management due to an unquestioning reliance on the blinkered views of environmental activists. In June 1990, its supposed expose of Western Australian forestry entitled, '*The Wood for the Trees*', drew an outraged response from the state's forest management agency, professional foresters, and the timber industry. The WA Department of Conservation and Land Management vigorously pursued the ABC for retribution, eventually forcing *Four Corners* to broadcast a substantial apology in which it admitted to 44 false assertions, bias and incorrect data.[17]

It was apparent in both instances that *Four Corners* had been 'invited' to undertake these programs by local environmental activists who saw it as a means of discrediting forestry authorities and the timber industry in front of a national audience. In the WA example, the *Four Corners* team apparently worked with these activists for several weeks before even contacting the Government department responsible for forest management, and it was revealed later that one of WA's most high profile activists had mapped out the program's interview schedule including suggested interviewees, lines of questioning, and the locations of field stops for filming.[18]

When the *Four Corners* team eventually met with the Department of Conservation and Land Management (CALM), it was clear that the mind of the program's lead journalist/presenter, Mark Colvin, had already been made up after such prolonged exposure to the typical untruths, half-truths, and conspiracy theories espoused by the local forest activists. According to former senior CALM officer, Roger Underwood, who was present, the subsequent four-hour interview of CALM Executive Director was aggressive and unrelenting. Far from an exercise in gathering information or seeking understanding, it was a hunt for 'gotcha' moments that could be used to demonise him and his organisation. After editing, only a few minutes of carefully selected snippets from this expansive interview actually made it into the final program.[19]

The ABC's *Four Corners* website, describes the program as "… Australia's premier television current affairs program. Since 1961, it has been exposing scandals, triggering enquiries, and firing debate, …" Its "current team of … maintains a proud tradition of investigative journalism and rigorous analysis" and "… its consistently high standards of journalism and film making have earned international recognition and a swag of Walkleys, Logies and other national awards".[20]

On the evidence of it's *'Lords of the Forests'* and *'Wood for the Trees'* episodes, these are somewhat hollow claims. In fact, these *Four Corners* episodes exemplified the extent to which environmental ideology had already penetrated ABC current affairs as far back as twenty-eight years ago. That this has persisted is evident from more recent coverage of forestry issues, such as the reporting of water quality concerns in north-eastern Tasmania by the network's flagship human interest program, *Australian Story*, during 2010.

These water quality concerns had been first raised by a local rural GP, Dr Alison Bleaney, who was convinced that plantation forestry was contaminating the St Helens' town water supply. In July 2004, she and a Sydney-based marine scientist, Dr Marcus Scammell, released a joint report claiming that pesticide-spraying of young plantation trees in the George River catchment was responsible for both damaging the local aquaculture industry in George Bay and contributing to human health problems throughout the region.[21]

A subsequent independent academic review of Bleaney and Scammell's report commissioned by the Tasmanian Government concluded that it was not scientifically sound and, that by mixing science and policy, it "… creates a set of illusory relations based on improper conclusions". It further described their report as "… an opinionated manifesto" that was primarily a vehicle for promoting an anti-forestry agenda.[22] Despite this expert opinion, Channel Nine's *Sunday* program promoted Bleaney and Scammell's claims in a September 2004 episode, just weeks before a Federal election in which Tasmanian forestry was a very prominent issue.

By 2009, Bleaney and Scammell reported that oyster formers and other concerned parties were speculating that the toxins released from the leaves of the plantation tree species, *Eucalyptus nitens*, were responsible for shellfish deaths in George Bay, local human health problems, and possibly a facial tumour disease that was decimating the Tasmanian Devil. As there are widespread eucalypt plantations, this new claim was highly concerning to the large proportion of Tasmanians who rely on drinking water from catchments which may be partially forested with these trees.

In February 2010, a double-episode of *Australian Story* entitled, '*Something in the Water*', presented this new claim to a national television audience just weeks before Tasmania's State election in which forestry was, as usual, a major issue.[23] The program sparked hysterical reactions across Tasmania and also raised concerns in Victoria where natural stands of *Eucalyptus nitens* (as opposed to plantations), have always been present in Melbourne's domestic water supply catchments.

However, it was soon apparent that '*Something in the Water*' had ignored, downplayed, dealt with improperly, or failed to fully present key evidence that would otherwise have put the supposed threat of plantation forestry into its proper perspective.[24] Not least amongst this ignored evidence was that just 3-4% of the George River catchment was actually under plantation, and that ongoing regular water monitoring by the responsible agency was finding nothing to indicate that it posed any threat to human or environmental health.

In addition, although the program had reportedly been filmed some months earlier, the ABC's screening of it just weeks before the Tasmanian state election put the incumbent State government into an invidious position. Mud sticks, and despite the government's previous efforts to investigate and invalidate claims made about the St Helens water supply since 2004, this new bout of whipped-up hysteria demanded a further response to mitigate potential electoral damage.

The Tasmanian government immediately appointed an independent panel of expert scientists – the George River Water Quality Panel. However, as it would be incapable of delivering any meaningful findings prior to the election, the government also immediately installed a carbon filter at the St Helens water treatment plant. Whether or not this was actually necessary, it would at least reassure panicky locals that their government was prepared to do whatever it took to safe-guard human health.

While the government later estimated the combined cost of these measures at a conservative $400,000, other incalculable costs arose from the fears unleashed by *Australian Story*, including a substantial reduction in tourism and a devastating loss of revenue for the region's aquaculture industry due to reputational damage. But arguably the program's most critical impact was the political fall-out caused by a rising Greens' vote at the state election presumably due, at least in-part, to an enhanced anti-forestry fervour fuelled by the prospect of 'toxic plantations'. This would ultimately force Labor into an uncomfortable minority government with the Greens which in time would prove to be disastrous, not only for Tasmania's forestry sector.

Just how much of this political fall-out was attributable to the *'Something in the Water'* program is impossible to quantify. However, it is notable that senior Labor MP and incumbent Minister for Health, David Llewellyn – a local member in the electorate encompassing the St Helens region – lost his seat in the ensuing State election just weeks after being poorly portrayed in the program.

In late June 2010, three months after the state election, the George River Water Quality Panel released its final report. This confirmed that *Eucalyptus nitens* plantations were not polluting the water supply and were not responsible for damaging aquaculture or causing human health problems in and around St Helens.[25] This prompted then Tasmanian Premier, David Bartlett, to write to the ABC demanding an apology on behalf of the people of St Helens. He would have been equally justified in demanding an apology to all Tasmanians for the unnecessary hysteria created by the program, as well as the political upheaval and substantial economic costs.

However, by then, the ABC had already conducted its own internal investigation in response to 13 formal complaints about the program, including one from the Tasmanian Premier. It concluded that "… the program in some respects fell short of ABC editorial standards relating to contextual accuracy and balance", but, perhaps unsurprisingly for an internal investigation, it "... found no breach in relation to the proximity of the program to the 2010 Tasmanian election". The ABC's Managing Director ordered that a summary of the investigation's finding be placed on the ABC website and that *Australian Story* broadcast a short 45-word statement outlining the investigation's findings.[26] This did not include an apology.

Beyond giving a brief sense of satisfaction, neither an admission of unprofessional conduct nor an on-air apology can undo the serious damage caused by a poorly researched and/or biased media broadcast. As the environmental activists who encourage these programs must be all-too-aware, what matters is the initial impression, and this won't be undone by a correction or apology that is highly likely to be screened or published months later. In this instance, the perception of eucalyptus plantations being toxic to human and environmental health was adopted by environmental activism as soon as the program was broadcast and continued to feature in scare campaigns against the nation's forestry sector.

Furthermore, the sincerity of any admitted mistakes or apologies offered by media organisations must be highly questionable given the acclaim that has at times been afforded to sensational environmental stories irrespective of their accuracy. Despite – or perhaps because of – its 13 complaints, including from leading political and industry figures, the ABC's '*Something in the Water*' program was nominated as a finalist in the environmental journalism category of the 2010 Australian Museum Eureka Awards.

In 2004, the ABC's highly controversial '*Lords of the Forests*' episode of *Four Corners* had gone one better and been honoured by the Australian Museum with an Excellence in Environmental Journalism award. Surely this was highly inappropriate for a program derided by so many complainants whose concerns about serious bias, unfairness and lack of balance were largely upheld by three separate investigations.[27,28] While such an award says much about the poor understanding (and perhaps the agenda) of the wider scientific community with respect to forestry issues, it also encourages poor journalistic standards that place greater importance on creating and maintaining controversy than accurately informing the public.

Unfortunately, these two programs are far from an aberration on ABC TV. Similar instances of bias and lack of balance have been commonplace on the *7:30 Report* and *Stateline* where the views of government or forest industry spokespersons are usually sought but often presented as little more than after-thoughts in narratives based primarily on the spurious, but largely uncritically accepted views of environmental activists.

Celebrity activist, Richard Flanagan, has in the past written that Tasmanians opposed to the state's forestry practices were forced to live in constant fear of retribution from disaffected forestry workers or their supporters. If this was ever really the case, it arguably reflects the festering anger and frustration amongst those with 'skin in the game' whose livelihoods were constantly under threat due to media coverage dominated by the skewed opinions and conspiracy theories of agenda-driven activists with nothing tangible at stake.[29]

Such frustration has undoubtedly been exacerbated by the failure of the media to question the veracity of even the most outrageous anti-forestry claims and the seeming impossibility of ever bringing such unfair journalism to account. It is notable, for example, that there is nothing to suggest that ABC journalists such as Tikky Fullerton and Mark Colvin (deceased) were ever disciplined despite presiding over such blatantly biased *Four Corners* programs that arguably played a significant role in creating unwarranted social and economic upheaval for many forestry workers and their families. Indeed, both went on to attain more senior roles within the ABC.

Indeed, it seems that a different set of journalistic standards applies to environmental issues. There can be no other explanation given the sanctions recently applied to the ABC's *Catalyst* program and its reporter, Maryann Dimassi, for her part in presenting two controversial episodes about human health which raised the ire of the medical profession. These episodes were soon pulled from the *Catalyst* website, and the program was slated for a restructure, with Dimassi initially suspended and likely to lose her job.³⁰ In contrast to this, similarly flawed episodes of *Four Corners* have been lauded and their presenters promoted.

Further magnifying the hurt within the forestry sector has been the tendency for the ABC, in particular, to make heroes of ideologue eco-warriors such as Flanagan and Geoffrey Cousins by parading them as 'forestry experts' on programs such as *Q and A*, or *Lateline*, while those with real inside knowledge have been mostly ignored.

The failure of the media to routinely exercise objectivity, rigour and balance in the reporting of Tasmanian forestry issues was undoubtedly a major contributing factor in why Tasmanian timber company, Gunns Ltd. felt compelled to launch a lawsuit in late 2004 against 20 environmental activists, including then Greens leader, Dr. Bob Brown. This motivation was effectively confirmed by the then Gunns Ltd. chairman, John Gay, in a November 2005 article in Melbourne's *The Sunday Age*, in which he acknowledged that his company had been adversely affected by "… bad publicity generated by environmentalists and hostile sections of the media".[31]

Dubbed by environmental activists as the 'Gunns20 Case', it was ultimately dismissed before being heard due to a reportedly sub-standard prosecution brief. Irrespective of this, it had been widely condemned by the media as an attempt to restrict the fundamental rights of protest and freedom of speech. Alternatively, it can be viewed as an attempt to meaningfully enforce some standards of honesty and integrity to the conduct of protest and free speech which, at least in relation to environmental issues, has become mired in untruths, half-truths, distortions and exaggerations promulgated by a 'whatever-it-takes' style of activism.

The concept of using the legal system to bring to account those (perhaps including journalists) who deceptively campaign to force unwarranted environmental and socio-economic outcomes would undoubtedly be supported by rural Australians who are ultimately at the butt-end of these campaigns. Arguably, Gunns' attempt to go down this path also highlights the abject failure of the journalism profession to meaningfully enforce its own ethical standards.

The Journalism Code of Ethics formulated by the Media, Entertainment and Arts Alliance, requires journalists to:

> 1. Report and interpret honestly, striving for accuracy, fairness and disclosure of all essential facts. Do not suppress relevant available facts, or give distorting emphasis.
>
> 4. Do not allow personal interest, or any belief, commitment, ... to undermine your accuracy, fairness or independence.
>
> Guidance Clause: ... Only substantial advancement of the public interest or risk of substantial harm to people allows any standard to be overridden.[32]

With respect to conflicts over Australia's forests, the environmental movement's many stage-managed protests and press releases have ensured the issue has remained newsworthy and the media would be negligent if it were to be ignored. In addition, as a Tasmanian forest industry identity and former journalist once noted, the time-poor modern media's predilection for presenting simple answers to complex questions is more suited to reporting emotional protest claims than the scientific facts and figures that can counter them.[33]

Whilst this may partly explain the poor standard of journalism which dogs media coverage of forestry issues, there have been far too many examples where journalists have clearly contravened their professional ethics. Accordingly, it can only be presumed that either some journalists are personally unethical, and have concluded that, with respect to forestry issues, ignoring professional ethics is in the public's best interest.

This notion of serving the public interest was articulated by Fairfax columnist, Caitlin Fitzsimmons, in October 2016, when she wrote an opinion article for *The Age* and *Sydney Morning Herald* berating native forest timber production[34] which she also posted to her Facebook page, with the following declaration:

> Here is my column for the op-ed page, ... I talk about my childhood in the bush and then get into the economics of forestry. I feel so privileged to be able to write about stuff that I really care about and the only interest I need to serve is the public interest. This is why I'm in journalism.

Despite the self-aggrandising, a jaundiced personal opinion arising from distant and simplistic childhood memories, embellished with the one-sided rhetoric of agenda-driven environmental ideologues, is hardly serving the public interest. Sadly Ms Fitzsimmons is not alone amongst journalists in failing to appreciate that the public interest is actually best served by factual, objective, and balanced reporting – which should be a major concern for a mainstream media that is already facing a future of irrelevance as online blogs and social media threaten to usurp it.

Arguably this skewed notion of public interest is unsurprising as the media and its journalists are naturally attracted to popular causes, particularly where minority groups are pitted against the supposed might of the government or corporate sector. However, its existence also supports the notion of a wider 'conservation culture' engendered by decades of environmental activism which has pervaded all elements of society, but particularly universities and colleges where journalists and other white collar professionals are trained.

John Henningham, Australia's first professor of journalism and founder of the independent journalism college, Jschool, has lamented that today's journalism and media educators largely share left-leaning values and are teaching an activist, opinionated form of journalism, largely at the expense of the past emphasis on impartial and balanced reporting.[35]

Not withstanding that the media is probably bombarded with requests from environmental activists clamouring for exposure, its willingness to accommodate them generally with minimal critical examination of their message exhibits a tacit support for their causes. Certainly, the mining industry believes that most city-based journalists take a 'green-left' approach to conservation matters thereby making it difficult for opposing views to be heard.[36] The similar trend noted with respect to forestry issues may also reflect the deployment of young, impressionable reporters to environmental rounds, possibly mentored by older journalists unashamedly wedded to 'green' ideals.[37]

Even properly researched articles about forestry typically give greater emphasis to the protest message than to alternate views. Arguably this reflects the time-poor modern media's over-reliance on press releases with considerably less effort spent on investigating issues for themselves. Media reporting that simply reiterates statements issued by protest groups without substantiating their validity results in distorted and false information being disseminated to the public. With regard to the full suite of environmental issues, this is happening on an almost daily basis.

As is typical of environmental activism more generally, the loudest calls denouncing resource use industries ostensibly to protect the environment emanate from the city-based sector of the media that is also most remote from the issue and least knowledgeable of what they are demonising. Generally there is a noticeable distinction between how environmental issues are covered by regionally-based media outlets where resource use industries and their workforces are a far more significant presence within the community. With regards to forestry issues, this is generally typified by a more balanced, thoughtful and questioning approach to environmental activism. Unfortunately this more enlightened journalism is far less influential because regional media generally only reaches the minor proportion of the community which resides outside state capital cities.

At times, regional media coverage of environmental issues has drawn flak from the city-based media for not being sufficiently 'anti-forestry'. In July 2006, the *Launceston Examiner* was lambasted in a five-minute segment on ABC television's *Media Watch* for allegedly failing to sufficiently emphasise reports that were critical of Tasmanian timber giant, Gunns Ltd. However, the ABC's claim that the *Examiner* was giving one side of the debate over a proposed pulp mill a 'head start' was refuted by the newspaper's editor who countered that "our reporters looked beyond press releases put out by groups driving a particular agenda".[38]

There are exceptions to the more balanced coverage typical of the regional media. For example, regional radio station, ABC Gippsland, has over many years given countless opportunities and undeserved credibility to one local environmental activist to such an extent that listeners could be excused for presuming that she, rather than the responsible Government agencies, is the local expert on forestry matters. This disparity was illustrated by a recent search of her name on the ABC Gippsland website which brought up 213 results under 'News' for the previous year, whereas a search of 'Vicforests' – the government agency which actually manages Victoria's commercial timber production – returned just 5 results over the same period.[39]

While allegations of ABC bias routinely sparks strident denials from 'green-left' supporters and commentators, a former ABC personality recently acknowledged the left-leaning bias of the ABC's capital city radio presenters and admitted there is an entrenched culture founded around their personal view that most of the ABC audience thinks as they do.[40] Clearly, the frequent hand-wringing and conspiracy theorising about right-wing enemies trying to bring down a public institution suggests that, after decades of exposure to largely unbalanced media coverage pandering to their core beliefs and values, blinkered ABC supporters lack the objectivity required to see what is obvious to others.[41]

Bias is as much about what isn't being said or shown as what is, and so it can be far more subtle than blatant. With regard to science-based environmental issues, detecting bias requires some knowledge of the topic at hand to be able to appreciate who isn't being interviewed; the known ideology, beliefs or values of who is being interviewed; the nature of the questions being asked and what is not being asked; and the context of an issue that may be being studiously ignored. As most media viewers, listeners or readers have no such detailed knowledge they can be easily duped by a biased media report masquerading under the pretence of balance.

For example, the afore-mentioned ABC *Four Corner's* episode, 'Lords of the Forests', devoted roughly equal screen time to interviews with environmental campaigners and forestry/timber industry spokespersons, thereby giving an appearance of balance. However, comparing the program's on-screen narrative with the full transcripts of pre-program research interviews with key government and industry figures reveals the extent to which selective editing was used to misrepresent Tasmanian wood production as an ecological disaster supposedly driven by systemic corruption and mismanagement.[42]

By storing in their website archives the full research interview transcripts that enables such an analysis to be undertaken, the ABC presumably believes that it is fulfilling its ethical and chartered requirements for balanced journalism despite broadcasting a seriously biased program. Undoubtedly, the overwhelming bulk of viewers do no more than watch an episode and take it at face value, oblivious to the reality that examining all the information obtained by the program's researchers can give a quite different picture.

Just one of the many examples from the program of how excluding key information can skew the portrayal of an issue was Ticky Fullerton's on-screen narrative of what happens after a forest is harvested:

> Taking the place of the forest is a tree farm – plantations of pine or gums which one scientist says has all the biodiversity of a car park. Other areas are regenerated with native trees, but with a short life, as they too are now bred to clear-fell, when the cycle begins all over again.

Evan Rolley, the then Executive Manager of Forestry Tasmania, quite ably addressed this in his pre-program research interview, but it was excluded from the program that was broadcast:

> Well 95% of what we do of course is replacing the same eucalyptus seeds on the sites with the same forests that if you go back and look at them in 70, 80, 90 years which is when they are planned for harvesting, they will be that rich diversity of native forest … About 3% of our public forests are plantations and even if we planted at the maximum rate possible, that would change from 3 to 5%, so 95% of all public forest will be retained as native forest forever.

Bias can also take other forms such as timing media coverage to fit a particular agenda, and controlling the extent of the coverage of specific events to advance or disadvantage that agenda. For example, in early 2012, just days before the start of a court case launched by local environment group, My Environment, against proposed harvesting of three coupes near Toolangi, Melbourne's *The Age* newspaper published two articles totalling over 2,500 words questioning the very legitimacy of Victoria's hardwood timber industry.[43] Six weeks later, when Vicforests won the case and the harvesting was allowed to proceed, *The Age* devoted just

250-words to what it had earlier described as a 'landmark Supreme Court trial'.[44] Clearly, once the court's judgement had gone against *The Age's* hoped-for outcome, the case became almost an irrelevance.

Bias can also be introduced by the media's choice of experts when covering particular events. A forestry example was the ABC Fact Check Unit's report in March 2014 of the attempt by the Abbott Federal Government to revoke the listing of 74,000 hectares of Tasmanian forest added to the Tasmanian Wilderness World Heritage Area (TWWHA) in 2013 at the behest of the previous Labor Government. In dispute was whether the area in question sufficiently met the values and criteria for World Heritage listing after a long history of timber harvesting and associated disturbances, amidst concerns that the listing had been deceitfully nominated to avoid proper scientific evaluation in order to meet domestic political imperatives.

The ABC Fact Check report[45] on this issue was initially released without any input from the most critically important source of historical records, Forestry Tasmania, apparently because they'd failed to meet a deadline to supply information to the fact checkers.[46] This suggests that publishing their report as quickly as possible was more important than having all the facts, presumably because a Senate Inquiry was currently examining the issue. It appears that the exercise was undertaken at the behest of environmental activists and their political allies and was specifically timed to influence that inquiry.

Releasing a report and claiming it to be the final arbiter on a question of public interest without any input from a primary knowledge source, is at best misleading and at worst unethical behaviour from the public broadcaster. Just as damning was the Fact Check Unit's deliverance of a 'verdict' based primarily on the views of three 'experts' (Professor Brendan Mackey, Peter Hitchcock, and Sean Cadman) each with either a history of personal involvement in forest activist campaigns and/or strong linkages to the groups conducting those campaigns; and who had all been intimately involved in the previous Labor Government's nomination to extend the Tasmanian Wilderness World Heritage Area. It almost goes without saying that they would strongly oppose the Coalition Government's plan to delist a 74,000 hectare (or 40%) portion of that 2013 extension – yet their verdict was presented as a supposedly independent expert review.

To its credit, the ABC Fact Checking Unit at least updated its report several days later after receiving new information from Forestry Tasmania. However, by initially publishing its report without this information the interested public had already been misinformed and this was unlikely to be overturned by publishing an updated version several days later. In any event, their updated report didn't even include Forestry Tasmania amongst its 'experts' and it did not alter its initial 'verdict' based on the advice provided by the original three 'experts' pushing their agenda.

These are just a few of the huge number of examples of media bias in relation to forestry issues. It is difficult to specifically quantify the impact of this reporting on politics and public policy, but it is more properly assessed as part of the whole impact of environmental activism as it essentially promotes and reinforces its campaigns. The notion that biased media coverage reinforces and hardens the already skewed conventional wisdom amongst the most environmentally-sensitive 'green-left' demographic is evident from online forums discussing forestry issues where links to newspaper articles or radio/TV interviews are frequently cited as though they provide irrefutable truths supposedly representing the last word in the discussion.

Despite concerns about the unbalanced media coverage of forestry issues, Australia's forestry and related industry sectors have long been divided about how to respond. Notwithstanding the difficulty of convincing the media to publish or broadcast an industry response, one school of thought is that responding to it actually feeds the controversy thereby making it worse than if left alone. This approach may be rooted in the contention that most people retain a healthy cynicism about what they read, hear, or see in the media, and so may not readily accept what is said about forests and forest industries. Supporting this contention are recent US surveys of public perceptions about the honesty and ethical standards of various professions, which rank TV and newspaper reporters about the same as real estate agents and lawyers – which is not much higher than car salespersons and advertisers.[47]

In Australia, a 2013 Essential Research poll found less than half of us had at least 'some trust' in newspaper reporting, while only around 40% felt this way about commercial radio and TV current affairs programs. However, when different media outlets are considered in isolation, around 70% of polled voters had 'some' or 'a lot of' trust in ABC TV and radio current affairs, while Fairfax was found to be publishing the most trusted newspapers.[48] As these are the outlets which most often cover forestry issues, it can be presumed that the segment of the public which is exposed to it affords it a high degree of credibility. On the other hand, the commercial media outlets that are the least trusted typically don't cover forestry issues to any significant extent.

Accordingly, it is likely that city-based media coverage is significantly shaping and reinforcing the conventional wisdom on forests. Even though it appears that only a minority of voters place environmental issues above self-important material concerns, the recent political climate has provided opportunities for minorities to be influential drivers of public policy. Unfortunately, when based on convictions supported by public opinion derived from media coverage which ignores the complexity inherent to forest management, it promotes a very simplistic view of the issues based primarily on the hyperbole and emotional rhetoric of environmental activism. Ultimately, public policy founded on populist views derived in the absence of sufficient consideration of critically important facts can have serious unintended socio-economic and environmental consequences.

Today almost any media coverage of forestry issues – and perhaps other environmental issues as well – needs to be treated with caution. Unfortunately, there have simply been too many examples of editorial trickery, loaded interview questions, uncritical acceptance of absurd nonsense from people representing a favoured ideology, and selective use of inappropriate interviewees to accept such coverage as fair, impartial and balanced. There are exceptions of course, but it generally takes some knowledge of the topic to identify them as such – knowledge that most of the general public doesn't possess.

Although initiated by the environmental movement, it is the prominence accorded to forestry issues through biased or unquestioning media coverage that is largely responsible for millions of Australians who rarely visit forests and who have little idea of their management, being convinced that their survival and integrity is under grave threat. The resultant political imperative to 'save' forests by national park declarations specifically to evict resource uses has been tragically unnecessary in most instances, particularly since the Regional Forest Agreements 15 – 20 years ago, and has chronically weakened the capacity to manage and protect key environmental values – particularly from severe fire – in stark contrast to what was intended.

In recent years, the results of the Brexit referendum in the UK and the US Presidential election of Donald Trump have precipitated a long overdue focus on the behaviour of the media. Both outcomes arguably defied strong campaigns by elements of the media for a different, favoured outcome, i.e., respectively for the UK to remain in the European Union, and the election of Hillary Clinton to the US Presidency. The strongest criticism of the media has emanated from those who have been disaffected by the Brexit and Trump results, and this has given rise to claims that we are now in a so-called 'post-truth era'.

The term 'post-truth' stems back to the early 1990s, but the Brexit and Trump campaigns apparently led to such an unprecedented surge in its usage that it was announced as the Oxford Dictionary Word of the Year 2016. It is dictionary-defined as "relating to or denoting circumstances in which objective facts are less influential in shaping public opinion than appeals to emotion and personal belief".[49] Its usage in relation to politics is defined as "a political culture in which debate is framed largely by appeals to emotion disconnected from the details of policy, and by the repeated assertion of talking points to which factual rebuttals are ignored".[50]

In discussing the 'post-truth' concept, British philosopher, Professor A.C. Grayling, recently cautioned that it harks back to the 1930s and the 'volatile and intolerant' international landscape which existed in the lead-up to World War Two: "These guys have realised that you don't need facts, you just lie."[51] Perhaps that is true in the broad international context, but it is also true that on the Australian domestic front, scientists and their associates in resource-use sectors such as forestry, fisheries, and mining have for several decades been similarly lamenting the appalling lack of honesty in activist campaigns mounted against their activities, and the deplorable lack of media balance in their reporting.

Far from being something new, as has been breathlessly proclaimed by those angered by the success of Brexit and Trump, 'post-truth politics' ably describes the approach of environmental activism since the early 1980s when Greens' founder Bob Brown realised that the notion of 'saving wilderness' could be sold through advertising campaigns.[52] The great irony is that those who are now most loudly denouncing this approach have had no qualms while it progressively achieved their favoured environmental outcomes. Perhaps their outrage reflects a realisation that their own tactics have been turned against them. Indeed, one commentator who refutes the notion of a new 'post-truth era' has described the increased reference to it as "a coping mechanism for commentators reacting to attacks on not just any facts, but on those central to their belief system".[53]

Criticism of the media in relation to the Brexit and Trump results has arguably centred on 1) its failure to detect and reject obvious untruths; 2) its wanton willingness to engage in sensationalist coverage that gave oxygen to lies and misconceptions; 3) its abject failure to predict and help counter rising disenchantment with the prevailing political establishment which was central to these unexpected outcomes; and 4) its role in creating this disenchantment by largely ignoring the concerns of the silent majority in favour of peddling fashionable, but far less relevant causes held dear only by them and society's political and cultural elite.

It remains to be seen whether this will result in an across-the-board media 'shake-up' of how it operates, or whether it will focus its attention only on how it treats so-called 'right-wing' conservative perspectives, given that these have been tagged as the villains that delivered the Brexit and Trump results. If the latter, we can expect the unbalanced media moralising over fashionable social and environmental causes to continue as it always has, with large segments of the community continuing to feel disenchanted and disenfranchised as they were before.

So far, there has been little apparent self-reflection amongst the Australian media over its behaviour. However, one of the more telling observations has, perhaps surprisingly, come from Ita Buttrose – more of an occasional celebrity commentator than a bone fide journalist – who recently noted:

> I think it (the media) has lost its way a bit ... Maybe less opinion and more facts might be a good place to start. ... You earn the right to have an opinion and I think a lot of people in the media have not earned that right yet ... Now whether the media likes how people are thinking is not really the issue here, it is what are people thinking and why.[54]

Chapter 3 Endnotes

1 Janet Albrechtsen, 2016, "Trump is no media monster, the people created him" *The Australian*, 6 April 2016.

2 Jon C. Day, 2016, "Great Barrier Reef bleaching stats are bad enough without media misreporting", *The Conversation*, 27 April 2016.

3 Adam Morton, 2016, "PR war over fires in Tasmania's World Heritage Area takes to the air", *Sydney Morning Herald*, 16 February 2016.

4 Matthew Denholm, 2015, "Rare 'dark knight' under siege from logging", *The Weekend Australian*, 7-8 November 2015.

5 John van Tiggledon, 2014, "The destruction of the Triabunna mill and the fall of Tasmania's woodchip industry", *The Monthly*, July 2014.

6 "Industry warns loss of Triabunna woodchip mill will be forestry death knell", *ABC News*, 4 December 2013.

7 Forestry Tasmania, Stewardship Report 2008-09.

8 Alan Ashbarry, 2014, personal comments.

9 Paul Barry, 2011, Media Maestro No. 10 – Morry Schwartz, *The Power Index*, 5 December 2011.

10 Richard Flanagan, 2003, "The Rape of Tasmania", *The Bulletin*, Volume 121, No. 6403, December 16 2003.

11 Terry Edwards, CEO, Forest Industries Association of Tasmania 2004, In: a letter to *The Bulletin*, 27 January 2004, responding to its article '*The Rape of Tasmania*', December 2003.
12 Senator Eric Abetz, 2007, Letter to *The Monthly* published online, 17 June 2007.
13 "A Letter from Richard Flanagan", *Australian Story*, ABC TV, 3 November 2008.
14 Matthew Denholm, 2014, "A fragile peace of the Tassie forests uprooted", *The Australian*, 12 April 2014.
15 Charles Wooley, 2004, Letter to the editor, *The Hobart Mercury*, February 21 2004.
16 Independent Complaints Review Panel 2004, Report into the compliants made by Timber Communities Australia and Forestry Tasmania against the ABC *Four Corners* program '*Lords of the Forests*' broadcast on 16 February 2004.
17 Roger Underwood, 2007, "Deja vu at the ABC", *Online Opinion*, 24 January 2007. Mr Underwood was a senior officer of the WA Department of Conservation and Land Management (CALM) at the time, and was present during the 4-hour interview of CALM Director, Syd Shea, by ABC *Four Corners* journalist, Mark Colvin, during the filming of the episode. The Department made its own film of the interview and this was ultimately used to prove that the program's on-screen narrative deviated substantially from the facts that had been provided to the journalist.
18 Ibid.
19 Ibid.
20 ABC *Four Corners* website
21 Dr M. Scammel and Dr A. Bleaney, 2004, *Environmental Problems, Georges Bay, Tasmania* (unpublished report), July 2004.
22 Professor Paolo Ricci, 2004, *Review of Drs. A.Bleaney and M.Scammell Report – Environmental Problems Georges Bay Tasmania*, University of Queensland, 27 July 2004.
23 "Something in the Water", *Australian Story*, ABC TV, Part 1 – 15 February 2010 and Part 2 – 22 February 2010.
24 Mark Poynter, 2010, "Somethings in the water at the ABC", *Online Opinion*, 5 March 2010 and "Tasmania fumes over media misconduct", *Online Opinion*, 7 July 2010.

These articles detailed the ABC's non-disclosure that $E.nitens$ plantations occupied just 3 to 4 percent of the Georges River catchment; failure to disclose that a far greater proportion of the catchment is used for agriculture; failure to explain that Government water testing had found safe levels of toxicity; misrepresentation of the state of public health in and around St Helens; failure to acknowledge the Tasmanian Government's past responses to claimed water quality problems; wrongly implying that the plantation trees had been genetically-modified; and failure to acknowledge that an interviewed scientist had debunked the toxic plantations theory.

25 George River Water Quality Panel 2010, *Final Report*, June 29 2010. This report was no longer available online in December 2014, but a very brief summary of its findings is outlined on the Tasmanian Department of Health and Human Services – Public Health Alert: St Helens Drinking Water Quality Update.

Of particular significance was the Panel's finding that the basis of the allegations raised by Drs Bleaney and Scammell was flawed by their use of an inappropriate water sampling technique which had concentrated the level of naturally-occurring organic compounds by up to 1,400-times greater than which they are normally present in aquatic ecosystems. Accordingly, the analysis of their water samples misrepresented the impact of compounds with a naturally negligible level of toxicity in stream water as being highly toxic.

26 ABC 2010a, Managing Director's Summary: Complaints re Something in the Water, 20 June 2010.

27 Christian Kerr, 2006, "Lords of the Forests controversy rolls on", *Crikey*, 3 August 2006.

28 Andrew Bolt, 2004, "Top prize for fiction", *Herald Sun*, 17 December 2004.

29 Richard Flanagan, 2007, "Gunns out of control", *The Monthly*, May 2007:

"It is commonplace to meet people who are too frightened to speak publicly of their concerns about forestry practices, because of the adverse consequences they perceive this might have for their careers and businesses. Due to the forest battle, a subtle (and sometimes not-so-subtle) fear has entered Tasmanian public life; it stifles dissent, avoids truth"

30 Karl Quinn, 2016, "ABC 'saves' Catalyst but staff to go in restructure", *The Sydney Morning Herald*, 3 November 2016.

31 Claire Miller, 2005, "Pulp friction", *The Sunday Age*, 6 November 2005.

32 Media, Entertainment and Arts Alliance website.

33 Simon Bevilacqua, 2004, "Bush Bash", *The Sunday Tasmanian*, 18 January 2004. Reporting on comments made by Bruce Montgomery of the Forests and Forests Industry Council.

34 Caitlin Fitzsimmons, 2016, "Native forests are worth more unlogged, so why are we still cutting them down?" published in *The Age* and the *Sydney Morning Herald*, 5 October 2016.

35 John Henningham, 2014, "Journalism schools need practical focus", *The Weekend Australian*, 18-19 October 2014.

36 Peter Stitt, Secretary of the Mineral Industry Council of Australia 2003, *Mining and the Conservation Movement*, 12 November 2003.

37 Bruce Montgomery, 2004, "War over forests provides Greens with political oxygen", *Australian Forest Grower*, Spring 2004.

38 "The Examiner's Head Start", *Media Watch*, ABC TV, 17 June 2006.

39 ABC Gippsland website, searches of Jill Redwood and VicForests conducted on 17 December 2014.

40 Jonathon Holmes, "ABC radio personalities need to tune out their left-wing bias", *The Age*, 5 April 2016.

41 Martin Flanagan, 2014, "Don't let self-serving News Corp frame the debate on our ABC", *The Age*, 29 November 2014.

42 ABC *Four Corners* archive – *Lords of the Forests*, 16 February 2004.

43 Adam Morton and Melissa Fyfe, 2012, "Industry pushes against the grain" and "Labor eyed shut-down of native logging", *The Age*, 4 February 2012.

44 Tom Arup, 2012, "Possum activists lose logging fight", *The Age*, 15 March 2012.

45 ABC 2014, "Tony Abbott's Tasmanian wilderness claim doesn't check-out", *ABC Fact Check Report*, first released on 26 March 2014. It was later updated to include Forestry Tasmania figures provided by Tasmanian Senator Richard Colbeck without changing the verdict.

46 Mark Poynter, 2014, "ABC Fact Check Unit loses its way in the Tasmanian wilderness", *Online Opinion*, 3 April 2014

47 US Survey of Honesty and Ethical Standards, Gallop Polls.

48 Bernard Keane, 2013, "Trust in media: ABC still leads, Telegraph takes a hit", *Crikey*, 18 December 2013.

49 English Oxford Living Dictionaries – Word of the Year 2016 is ...

50 Wikipedia: *Post-truth politics*.

51 Sean Coughlin, "What does post-truth mean for a philosopher?", *BBC News*, 12 January 2017.

52 Part of a 1983 interview regarding the Franklin River campaign, reported in *The Rest of the World is Watching*, edited by Richard Flanagan and Cassandra Pybus, Pan Books (1990).

53 Alexios Mantzarlis, 2016, *No we're not in a post-fact era*, Poynter Institute, 21 July 2016.

54 From a speech given to the Priceline Sisterhood Foundation Luncheon, which was reported in the *Sunday Herald Sun*, 15 January 2017.

4

Conservation science: advancing the culture

"... much of the scientific literature, perhaps half, may simply be untrue. Afflicted by studies with small sample sizes, tiny effects, invalid exploratory analyses, and flagrant conflicts of interest, together with an obsession for pursuing fashionable trends of dubious importance ..."
Richard Horton, former editor of *The Lancet*[1]

In mid-2015, the Melbourne University Early Learning Centre's '*Voices in the Forest*' project was awarded the Most Overall Effective Interpretation at the Pre K-Kindergarten Level under the International Interdependence Hexagon Project. According to the university's internal E-news bulletin, the Centre's pre-school children visited the forests north of Melbourne under the guidance of an environmental scientist, to learn about "the current logging of old growth mountain ash forests and the Leadbeater's possum's impending extinction" and were shocked to learn that "the 150-year old Mountain ash that cover the hills are logged to produce paper that was used for the children's drawings".[2]

That these simplistic assertions about the forest were wrong should matter a great deal to a respected university.[3] In this instance, the scientist it had engaged to educate pre-school children was an associate of a local environmental group that is campaigning to end timber production in those forests. Although only of minor significance, this episode nevertheless adds to already growing concerns about the relationship between Australia's environmental academe and the nation's forests, whereby:

- Universities can no longer be automatically assumed to be bastions of unbiased truth and objective teaching; and

- Some environmental scientists – despite being privileged with a community-wide credibility for being free from bias and factually correct – appear to have little compunction in pursuing ideological agendas, principally by ignoring inconvenient truths.

Such concerns have been especially prominent since the lead-up to Victoria's 2014 State election when only one non-climate environmental issue seemed to genuinely capture the interest of voters. This was a proposal supported and strongly promoted by the Greens and their environmental activist associates to create a huge new so-called 'Great Forest National Park' in the wet mountain forests, north-east of Melbourne. As the subsequently elected Andrews Labor Government has yet to declare this proposed new park after three years in office, it remains a prominent issue that will feature again in Victoria's November 2018 election.

According to the environmentalists' rhetoric, declaring the 'Great Forest National Park' will bring Victoria's faunal emblem, the Leadbeater's possum, "back from the brink of extinction" while protecting biodiversity, securing water supplies, and transforming the region into "a playground for Melbourne."[4]

The proposal for this new national park had been devised during 2013 and before long had been developed into a major enviro-political campaign focussed on removing the supposedly 'key threat' posed by 'industrial logging'. During 2014, a 'Great Forest National Park' website was constructed by local environmental group, My Environment, and the campaign was soon joined by much larger groups, the Wilderness Society and the Australian Conservation Foundation.

Typical of an environmental campaign, the promotion of the 'Great Forest National Park' proposal has been devoid of contextual acknowledgement of the pre-existing balance between forest areas already reserved for conservation against those designated for renewable use. In reality, the majority of the region's mountain ash forests, preferred by Leadbeater's possum, have long been contained in several national parks, a large area of closed water supply catchments, and a network of State forest reserves. As far back as 1998, 61% of the possum's most preferred mountain ash (*Eucalyptus regnans*) forest was already formally, informally, or effectively reserved by management constraints.[5] This reserved area has since been increased by significant additional reservations especially from 2015-17 so as to provide additional protection for the possum.[6]

It is also noteworthy that timber harvesting occurring within the minor portion of mountain ash forest designated for wood production is already limited to advanced regrowth trees – of mostly 78 to 91 years old stemming from the 1926 and 1939 bushfires – that, according to some scientists, are at least 100 years away from developing nesting hollows suitable for Leadbeater's possum.[7] However, recent survey work conducted since 2015 by Victorian Government scientists using a different methodology, is finding the possum in recovering forest as soon as seven years after fire or logging, thereby suggesting that the possum is far more resilient than earlier research had suggested.[8]

This most recent Victorian Government surveying has overturned previous dire assumptions about the limited habitat requirements and extreme fragility of the possum's population which had been predicated on only around 250 known colonies (of from 3-11 individuals) detected by Australian National University (ANU) researchers in the 16 years from 1998 to 2014. Since the start of 2015, the improved surveying method has led to a further 500 new possum colonies being detected in less than three years within just a 6-10% portion of its potential habitat range.[9]

By strategically neglecting to disclose the wider context of pre-existing conservation reserves and knowledge, the 'Great Forest National Park' campaign has been able to imply that unless the new park is declared, 'industrial logging' will eventually consume the whole forest with the loss of key ecosystem values. This has unsurprisingly generated considerable concern amongst an unknowing public effectively trained by years of one-sided media coverage to be fearful that 'irreplaceable forests' are being lost to logging. Accordingly the campaign has also gained high profile support from international conservationists, such as David Attenborough and Jane Goodall, who may well be embarrassed if they ever become aware of the full story.

In fact the proposed national park would provide only negligible immediate benefit to Leadbeater's possum (despite being loudly proclaimed as its saviour). Further to this, it would not 'secure' water supplies because they are already safe-guarded by extensive closed catchments and stream reserves; nor would it significantly add to the region's pre-existing role as a 'playground' for Melbourne given the extent of already reserved areas and the well-developed tourism sector which has successfully co-existed with the timber industry for generations.

Despite such contrary realities, some ecologists and their research associates, principally from the ANU's Fenner School of Environment and Society, have become vociferous supporters and public advocates for the proposed 'Great Forest National Park'.

Blurring the lines ... academics or political activists?

It is instructive to examine the conduct of these (mostly) ANU forest conservation scientists and how this links with the 'Great Forest National Park' campaign:

- The ANU's Fenner School forest conservation research program has included a focus on biodiversity in Victoria's Central Highlands' region since the 1980s. However since 2008, it has broadened and intensified this focus through the following research themes:
  - The relationship between timber harvesting and biodiversity conservation, particularly in relation to Leadbeater's possum.
  - The relationship between timber harvesting and bushfire threat.
  - The relationship between timber production and climate change mitigation.[10]
- This post-2008 research has almost routinely questioned the existence of the regional timber industry, despite much of it being controversial due to questionable assumptions, factual errors and/or omissions of important context that has often then been repeated in follow-up papers.

- The concept of a huge new national park encompassing the Central Highlands was first raised by prominent ANU ecologist, Professor David Lindenmayer, in an article on *The Conversation* in late 2012.¹¹ It is presumed that he and his research associates then developed this idea into a formal proposal. This was arguably confirmed when Melbourne University academic, Chris Taylor, a regular research associate of Professor Lindenmayer and his ANU colleagues, laid claim to being an 'architect' of the 'Great Forest National Park' proposal during a speech to a Greens Party rally for the proposed new park held a fortnight before the 2016 Federal Election.¹²

- In September 2013, Professor Lindenmayer posts an article on *The Conversation* website explicitly advocating a new 'Giant Forest National Park' to remove the "key process – clear fell logging – that is threatening both Leadbeater's possum and mountain ash forests".¹³ Between then and 2015, Professor Lindenmayer authored or co-authored a further seven articles on *The Conversation* which were routinely critical of Victorian forest management while advocating a cessation or severe curtailing of timber harvesting by reserving more forests in national parks.¹⁴

- During 2013-14, local environmental group, My Environment, develops the national park proposal into an enviro-political campaign, including a dedicated 'Great Forest National Park' website which features specially filmed talks by Professor Lindenmayer and other park advocates.¹⁵

- Professor Lindenmayer (and some of his research associates) speaks publicly in support of the 'Great Forest National Park' proposal at environmental campaign events, including Greens political events. This includes being the Key Note speaker at a 'launch' of the 'Great Forest National Park' in Melbourne's Federation Square just four days before the 2014 Victorian election; and addressing a meeting of ANU alumni just two weeks before the 2016 Federal election in a speech titled "Professor Lindenmayer presents the Great Forest National Park".¹⁶

- From 2013 onwards, Professor Lindenmayer's already high media profile allows him to publicise ANU research and advocate the 'Great Forest National Park' proposal, particularly on the ABC and in Fairfax newspapers.[17] This media coverage has been typified by alarmist headlines and assertions that are not always reflective of more cautious research findings. A good example is the paper, Lindenmayer et al (2009), which was a brief four-and-a-half page literature review citing around 50 references to past research mostly from the wet temperate forests of North America and tropical rainforests of the Asia-Pacific and South America. Only four of these references related to Australian forests and fire management. Yet, despite its relative lack of local context, the paper became a platform for leveraging the powerful message that "Decades of industrial logging in Australia's wet forests have made them more fire prone, raising urgent fire management issues …".[18]

- As the proposed park's most credible advocate, Professor Lindenmayer fronts a Pozible crowd-funding website to raise funds for the 'Great Forest National Park' campaign in the months leading up to the 2014 Victorian election. The website enabled would-be donors to pledge money by clicking a button beneath a photo of the Professor alongside the immortal political slogan, *'Its Time'*. Further to this, he was offering a personally signed copy of one of his published ecology books to anyone donating at least $90.[19] A full page advertisement for the proposed national park which appeared in *The Age* newspaper a day before the 2014 Victorian election was presumed to have been paid for from these donations.

- Professor Lindenmayer is sent a personal email from the then Labor Opposition on the eve of the 2014 Victorian election promising him that, if elected to government, they intend to declare the 'Great Forest National Park'.[20]

- Collectively, the ANU's Fenner School forest conservation scientists have also led processes to secure other outcomes that have put more pressure on politicians to declare the 'Great Forest National Park'. This includes getting Victoria's Central Highlands' mountain ash forests added to the International Union for the Conservation of Nature (IUCN) Red List as a 'critically endangered' ecosystem in 2014; and convincing the Federal Environment Minister to upgrade the threat status of Leadbeater's possum to 'critically endangered' in 2015. The veracity of both processes has since been seriously questioned by forest scientists.

- The IUCN Red listing of the Central Highlands mountain ash forests as a 'critically endangered' ecosystem was achieved by submitting a single scientific paper prepared in accordance with ecosystem assessment criteria purpose-designed by the IUCN.[21] This paper (Burns et al, 2015) was marred by critical errors, misconceptions, or highly questionable assertions. If these had been corrected in a more rigorous peer review, there is considerable doubt as to whether its conclusions could have been justified.[22] Arguably its most significant error was in asserting that 80% of the forest (rather than the official 1998 figure of 39%) was designated for timber production. In view of these flaws, the IUCN listing is at the least, highly questionable, if not invalid, especially when it is appreciated that these forests still occupy 97% of their pre-European range.[23] Other scientists have previously voiced concerns about the IUCN Red List as a vehicle for misusing the precautionary principal for enviro-political advocacy.[24]

- The upgrading of the Leadbeater's possum's threat status to 'critically endangered' by the federal Environment Minister in April 2015, was based on a 53-page Conservation Advice drawn primarily from ANU research provided to the government's Threatened Species Scientific Committee (TSSC). Initial concerns about this upgraded threat listing stemmed from it being drastically out-of-step with other Victorian 'critically endangered' listings, such as the Orange-bellied Parrot (less than 50 individuals) and the Helmeted Honeyeater (130 individuals) which survive in very limited areas of suitable habitat. By comparison, in 2014 the Victorian Government had estimated the Leadbeater's possum population at 4,000-11,000 individuals spread across a 5600 km² range. By 2017, Government scientists using an improved survey method and a more intensified survey effort had found a further 500 new possum colonies in just a 6-10% portion of its potential habitat range. This strongly suggested that even the 2014 population estimate had substantially understated the reality. Accordingly in May 2017, the science within the TSSC's Conservation Advice was reviewed by several scientists on behalf of the Institute of Foresters of Australia with a view to challenging the veracity of the possum's upgraded threat status.[25]

- In late 2015, a book jointly authored by Professor Lindenmayer and three other ANU conservation scientists goes to considerable lengths to denigrate current management of Victoria's mountain ash forests while advocating the 'Great Forest National Park' as a superior alternative.[26] The book has been criticised by forest scientists for an array of reasons, most notably for omitting important context and key facts.[27]

- In 2016, Professor Lindenmayer features (albeit without speaking) in a short film clip advocating the 'Great Forest National Park' produced as an advertisement to be shown in cinemas.[28]

- Also in 2016, it was becoming apparent that a program of nest box installation and artificial tree hollow creation undertaken by Victorian Government scientists was achieving considerable success in housing possum colonies, thereby making it apparent that the possum can readily survive without fully reserving the Central Highlands' forests in a new national park.[29] In response, an article on the ANU's Long Term Ecology Group website dismissed these active conservation strategies: "There is no evidence whatsoever to suggest that nest boxes will be effective for Leadbeater's possum in these forests" and "There is no evidence that creating artificial hollows will be effective for Leadbeater's possum – the technique has not been appropriately trialled in a scientific way."³⁰ Another article by leading Leadbeater's possum researcher, Lachie McBurney, on the same website noted, on the basis of past ANU research into the value of nest boxes, that "Ultimately, it is a distraction from the main game of habitat identification and protection …"³¹

- In September 2017, Professor Lindenmayer featured prominently on ABC television news and a *7:30 Report* feature advocating the declaration of the 'Great Forest National Park' based on the findings of a new report by five scientists from the ANU Fenner School (including himself).³² This media coverage strongly implied that the economic value of the Central Highlands timber industry was just $12 million even though the new report had only valued the in-forest component which supplies logs to the industry, and had therefore excluded the great majority of the industry's economic activity and employment which occurs in log processing and wood products manufacture outside the forest. A previous study of the whole industry by Deloitte Access Economics had valued it as generating over $570 million in economic activity per annum and supporting over 2100 jobs.³³ This renewed advocacy of the proposed national park coincided with campaigning for a by-election in the inner-Melbourne seat of Northcote which the Greens – with their policy to declare the 'Great Forest National Park' – eventually won.

- Also in September 2017, the Federal Environment Minister announced that the 'critically endangered' listing of Leadbeater's possum would be reviewed on the basis of recent Victorian Government surveying showing that it was far more abundant and resilient than was previously supposed.³⁴ Several weeks earlier, the ANU Fenner School had produced its own review in which – despite the huge upsurge in detected colonies – its ecologists had continued to assert that the Leadbeater's possum population had declined by two-thirds over the previous 20 years.³⁵ In a subsequent interview on ABC Radio National to discuss this new ANU review and the upcoming Federal Government review of the possum's 'critically endangered' status, Professor Lindenmayer dismissed the recent substantially increased estimate of the possum's population as having resulted from double and triple counting by Victorian Government scientists.[36] This was followed-up by an article on *The Conversation* several days later in which he reiterated that the possum population was in serious decline and that downgrading its threat listing would be misguided.[37]

To provide some context to this discussion it must be acknowledged that Professor Lindenmayer and his research associates are heavily invested in the welfare of Leadbeater's possum and its key Central Highlands' habitat after decades of researching these forests. According to the website of the ANU's Long Term Ecology Group, "ANU researchers led by David Lindenmayer have published 200 peer reviewed scientific articles and 8 books about the Central Highlands forests and Leadbeater's possum since 1983".[38]

Lindenmayer and his research associates were reportedly dismayed when 43% of Leadbeater's possum's designated 'habitat reserve' (a sub-set of the total possum habitat) was burnt by the 2009 'Black Saturday' bushfires, even though periodic fire is integral to creating future possum habitat as forests regrow. They were then outraged when timber salvage operations (i.e. salvage logging) commenced in the fire-killed ash-type forests, given that they had long expressed concerns about its ecological impacts.[39,40] From the perspective of forest scientists and forestry practitioners, these events seem to have coincided with a greater intensity of ANU research effort in the Central Highlands' forests, especially in relation to the impacts of so-called 'industrial logging'.

As is also the case with the environmental campaigning for the proposed new park, almost all of this research lacks any acknowledgement (or accurate acknowledgement) of the proportional scale and extent of 'industrial logging'.[41] This is a serious oversight given that the landscape significance of an activity such as harvesting and then regenerating forests is directly related to its scale and extent. Accordingly this lack of disclosure implies that 'industrial logging' is unconstrained and will eventually consume all forests despite the reality that it is actually a proportionally small-scale activity limited to a minor portion of the forested landscape.

The issue of salvage logging further exemplifies how a lack of context can foster a gross misunderstanding. While the ecological concerns associated with it may have validity, their landscape-scale significance is directly related to the proportion of the fire-killed forest area that is actually salvage logged. In fact, salvage logging is restricted only to the minority portion of forest already designated for wood production and is further constrained by the relatively short time-frame before dead standing timber degrades, and the limited capacity of the contractor workforce. These regulatory and practical realities are almost always ignored when concerns are expressed about salvage logging, thereby implying that it is far more extensive and ecologically significant than what it actually is. In fact, salvage logging occurred in less than 5% of the 189,000 hectares of ash-type forests which were collectively killed by the major Victorian bushfires of 2003, 2006 and 2009.[42]

The post-2008 research conducted by ANU Fenner School conservation scientists in Victoria's Central Highlands forests – and especially its associated media promotion at times including participation by the scientists – typically alleges that 'industrial logging' in the region's forests is responsible for a range of dire environmental and societal consequences (i.e., wildlife extinctions, damaged water supplies, past and future bushfire disasters, and enhanced climate change). As such claims could not be sustained alongside an admission that a strong majority of the regional forests will never be logged, the propensity for most of this body of research to leave out such important context has naturally created a strong perception that the scientists responsible for it (and often its associated media promotion) are pursuing an ideological 'no-logging' agenda.

We expect environmental activists to pursue their causes via a whatever-it-takes approach which typically involves selectively emphasising some facts while strategically ignoring other inconvenient truths that could weaken their campaign message. However, scientists are expected to produce research findings based on an objective evaluation of the full body of available knowledge and accordingly they enjoy a high level of community credibility as oracles of scientific truth. Unfortunately, if personal convictions are allowed to dictate their work they are more likely to engage in activist-style practices which can be the antithesis of objective science.

The synergy between academia and activism

Arguably, there has always been a natural synergy between academia and environmental activism. Contrary to what might be expected, this hasn't been restricted only to those academics or researchers engaged in environmental sciences. Indeed, it was a pair of social scientists, Richard and Val Routley, who co-authored the first landmark anti-forestry tome – *The Fight for the Forests* in the early 1970s – which in no small part underpinned the beginnings of Australia's environmental movement.[43] Yet, despite this more than 40-year history, it is fair to say that, prior to the campaign for a 'Great Forest National Park', such intense participation of any individual or group of academics in supporting an enviro-political campaign had been unknown or rare.

Formal partnership between conservation science and environmental activism had been evident for some time at the ANU Fenner School of Environment and Society. Since entering into a financial partnership with the Wilderness Society in 2005, the School's forest conservation researchers had, in my opinion, whether by design or coincidence changed the way their findings were being disseminated.

Suddenly, ANU forest conservation research findings were being routinely paired with alarmist and widely-promoted media headlines. These usually made (what have become) predictable calls to curtail native forest timber production and/or broad scale fuel reduction burning while increasing forest protection in new conservation reserves. Supposedly in these new reserves, nature would recover and flourish free from any human interventions beyond the light footprint of eco-tourism. This matches the ideals of environmental activism which for decades has been declaring that Australia's forests are no place for resource use industries or human interference.

The ANU Fenner School appears to have wholeheartedly embraced this ideal given their joint establishment of an ANU Wild Country Research and Policy Hub to be operated in accordance with the Wilderness Society's WildCountry Vision.

The WildCountry Vision is based on the US Wildlands Project[44] which has advocated huge reservations of rural lands for biodiversity conservation – some critics have claimed it to be proposing the reservation of as much as 50% of the land mass of the continental USA.[45] In 2008, the Wilderness Society had described its WildCountry Vision as a "big picture perspective [*which*] underpins all of the Wilderness Society's work". It went on to list five steps describing how they were applying the WildCountry Vision across Australia, including: 1) "Developing a continent-wide planning framework underpinned by cutting edge science ..."; and 2) "Campaigning to protect the last great wild places in Australia from destructive practices – [*including*] our carbon-rich native forests ..."[46]

Around that time, the Wilderness Society's biannual newsletter, *Wilderness News*, described the WildCountry Vision's plans for Victoria in somewhat stronger terms by asserting that "... securing our future starts with protecting our forests, one of the world's biggest carbon stores;" and " ... removing threats like woodchipping".

When accessed in September 2014, the Wilderness Society's website described the ANU WildCountry Research and Policy Hub as being "... designed to build upon existing ANU research activities, promote linkages ... and facilitate the transfer and application of scientific research and development outcomes to the community of conservation stakeholders". From this it can be argued that the Wilderness Society regarded the Hub as a mechanism for increasing the credibility of its forests campaigns which were otherwise problematically rooted in emotional rhetoric.

Higher education and research institutions typically receive funding from a range of sources, including resource use industries and regulatory agencies. Indeed, the ANU Fenner School's facilities have reportedly benefitted over the years from the receipt of hundreds of thousands of dollars of material goods donated by national and international forestry and timber agencies without the scientists who are based there ever being compelled to produce research that only supports these benefactors. Indeed, the ANU's Victorian forest conservation research program has for several decades been in-part undertaken for government forest management agencies which regulate and receive revenue from timber production, clearly without any proviso that the researchers desist from criticising it.

However, the ANU's formal alliance with a major environmental group appears to have taken the typical donor-university relationship a step further. Arguably, forestry agency or timber industry funding has never come with a proviso that the university establish a joint 'research and policy hub' dedicated to an unfettered expansion of timber production. Whereas the ANU's Wildcountry Research and Policy Hub was slated to operate in accordance with a 'deep-green' vision that is essentially a manifesto for reserving virtually all forests so as to prevent human resource use thereby (supposedly) allowing the landscape to recover and revert back to its pre-European wildness.

It can be argued that a science-activist partnership aimed at jointly producing research to drive evidence-based policy change could be an admirable initiative if applied through honest environmental campaigns. However, as honest environmental campaigning is now almost an archaic notion, it is surely problematic for an academic institution charged with conducting research in an impartial and objective manner to make common cause with those pushing ideological and highly politicised agendas. Indeed, it is suggestive of academic research being directed and shaped for political ends albeit that there is little to suggest that those conservation scientists conducting this research are unhappy with such an arrangement, presumably because it accords with their collective conservation philosophy. This is exemplified by the strong support of the ANU's Long Term Ecology Group – to which most of the Fenner School's conservationist scientists appear to belong – for the proposed 'Great Forest National Park'.[47]

The existence of a formal partnership between the ANU Fenner

School and the Wilderness Society first came to wider public notice in late 2007 during the lead-up to the Bali Climate Conference. This was a heady time for Australia's environmentalists. Just a week earlier, the self-styled political messiah, Kevin Rudd, had ridden a populist tide for change to a momentous Federal election win. His was a victory that swept aside a decade of conservative government with the promise of a new era of progressive action on key social and environmental problems.

True to his pre-election hype, the new PM's first official act was to ratify the Kyoto Protocol on 3 December 2007 – the opening day of the Bali Climate Conference. To Rudd's buoyant supporters this commitment to join global action on greenhouse gas emissions signified Australia's refreshing new attitude towards the environment compared to life under outgoing conservative PM, John Howard, who for a decade had steadfastly refused to sign the Protocol on pragmatic economic grounds.

At Bali, a considerable posse of Australian activists, representing an array of environmental groups, lobbied the conference delegates and key decision-makers urging ever greater action to curb greenhouse gas emissions. Greenpeace Australia-Pacific would later laud the effort and influence of "a coalition of local and international groups" for "obtaining the historic Bali roadmap, and a commitment from the Australian Government to the 25 – 40% emissions reduction negotiating range."[48]

However for Australia's Wilderness Society, climate change is not only a problem needing global action, but a gift that can deliver arguably its oldest and most closely-held ambition to 'save' the nation's forests from logging. The Society's *Annual Review for the Year ended 30th June 2008*, effectively articulated this advantageous link:

> Australia's old growth forests are majestic, abundant and an important ally in tackling climate change. ... One of the quickest and cheapest ways to achieve deep cuts in greenhouse gas emissions is to protect the Earth's natural forest.
>
> New science (*Green Carbon*, Mackey et al. 2008) reveals that

> Australia's forests are some of the most carbon-rich on Earth – storing on average three times more carbon than previously estimated. In the case of our south-eastern eucalypt forests, they can store up to 10 times more carbon.
>
> The Wilderness Society has greatly strengthened its engagement with the UN climate processes during 2007 and 2008 ... lead campaigners were able to attend the UN Climate Conference in Bali (December 2007) and key international climate meetings throughout 2008 ... At these international forums, we have been telling the story of Green Carbon and lobbying for practical policies and approaches to protecting and restoring the carbon stored in the world's forests.

This 'new science' was contained in an (at that stage) yet-to-be-published report, entitled – *Green Carbon – the Role of Native Forests in Carbon Storage – Part 1: A green carbon account of Australia's south eastern eucalypt forests, and policy implications*, by ANU scientists Professor Brendan Mackey, Heather Keith, Sandra Berry and David Lindenmayer.

At Bali in late November 2007, just days before the start of the Climate Conference, the Wilderness Society hosted a function to publicly launch the findings of the Mackey et al *Green Carbon* report. This was some nine months before the report would be eventually published in August 2008.

At this launch, lead author, Professor Mackey, was reported as presenting "new scientific research highlighting the critical role of forest protection in addressing climate change". A blog of his presentation by the Zero Emissions Network gushed that this new research showed that "if the [Australian] forestry sector was included in a carbon pricing mechanism ... the native forest industry would collapse overnight".[49]

The *Green Carbon* report's unpublished findings were also appropriated by the Wilderness Society to form the basis of submissions to the Garnaut Climate Change Review during its public consultation phase ending in April 2008.[50] The influence derived from submissions based on the potentially un-peer-reviewed findings of an (at that stage) unpublished report was revealed later when the Garnaut Review's Final Report talkedup the supposedly huge potential for Australia's native forests to help mitigate climate change:

> Mackey et al. [2008] estimate that the eucalypt forests of south-eastern Australian could remove about 136 M t CO2-e per year (on average) for the next 100 years. This estimate is premised on several key assumptions, including cessation of logging and controlled burning over the 14.5 million hectare study area.[51]

The appearance of such an observation in a major Government review undertaken specifically to shape the future direction of Australia's climate change policy, naturally raised concerns amongst forest scientists, forest managers, and wood products industries that stood to be drastically affected if timber harvesting and controlled burning were to suddenly cease. However, their capability to even understand the rationale behind this assertion, let alone respond to it, had been hamstrung by the *Green Carbon* report being unpublished during the Garnaut Review's public consultation phase.

Furthermore, when ANU E-Press did eventually publish the *Green Carbon* report several months later, it did so without including the technical data and associated calculations that had informed its findings, because "a technical paper is in preparation, detailing methods and results, for submission to a scientific journal." Again, this disadvantaged those with 'skin in the game' who stood to be hurt by any resultant policy changes that it may have precipitated.

The absence of this supporting technical information raised considerable conjecture about how such a report could even pass peer review and led some to conclude that it hadn't been properly reviewed. Also deserving of more than passing interest was that the *Green Carbon* report's lead author, Professor Mackey, was at that time the Director of the ANU's Wild Country Research and Policy Hub, and through this partnership, the Wilderness Society had provided funding for the report.[52]

The 'Green Carbon' episode first raised the prospect of forest conservation research being directed in collusion with an environmental group partner, including its findings and eventual publication being timed for release and appropriated to meet its campaign requirements.

Working with the Wilderness Society to drive evidence-based policy change is always likely to be problematic, but especially so in relation to forests because their policies have never been designed to improve contemporary forest management. Instead they aim to largely eliminate it by forcing policy change that leads to ever more national parks and other conservation reserves that would exclude resource uses, constrain sensible fire management, and restrict or end a range of other community uses.

This is apparent from the Wilderness Society's Forests and Woodlands Policy (revised September 2005) which categorically stated that it "does not support the use of native forests to supply woodchips for pulp, wood for power generation, charcoal production, commercial firewood, or timber commodities".[53] In addition, they have been demonstrably less than enthusiastic about broadacre fuel reduction burning, which is a key forest management tool, and are largely only tolerant of small-scale burns adjacent to suburban or township boundaries which would be inadequate to meet broader forest protection aims.[54]

The *Green Carbon* report was just the first in a series of contentious forest conservation papers emanating from the ANU Fenner School since 2008. Collectively, these have reinforced the initial concerns about inappropriate links and relationships with environmental activism and its political agendas, and have increasingly frustrated forest and bushfire science specialists with far better credentials in these areas.

While the scientific methodology used in some recent ANU forest conservation papers has at times been questionable, much concern about these papers has revolved around their refusal to consider a wider context when drawing conclusions that at times defy common sense. For example, it isn't rocket science to appreciate that wildlife extinctions due to timber production are highly unlikely when most forests are already in national parks or other reserves that will never be harvested. Similarly, concluding that carbon emissions will be substantially reduced if native forest timber harvesting ceases is premature in the absence of any consideration of the carbon consequences of then having to import wood products (often derived from SE Asian rainforests) and/or account for a consumer shift to greater use of non-wood substitutes, such as steel or concrete, which embody far greater emissions in their production (i.e. the 'substitution effect').

Indeed, it is widely acknowledged, including by the Intergovernmental Panel on Climate Change, that managed forests and the increased use of wood products is an important part of mitigating carbon emissions.[55] This was reinforced by a recent local study of native forest timber production (Ximenes et al, 2016) which took account of the 'substitution effect' and found that Australia's native forest timber industry is beneficial in mitigating carbon emissions and that this could be further improved if wood waste was also used to produce bioenergy.[56] This is in stark contrast to the findings of ANU forest carbon research which, by typically ignoring the broader context, builds an unwarranted case against continued local timber production.

The findings of several recent bushfire research papers emanating from much the same group of ANU conservation scientists have also been vigorously contested. For example, Lindenmayer et al, 2009 – *Effects of logging on fire regimes in moist forests*[57] – was only a short literature review mostly about tropical forests, but nevertheless became a launch pad for sensational media claims that 'industrial logging' was increasing the fire risk in Australian forests.[58] Such an 'over-the-top' assertion prompted a group of scientists, including two of the nation's foremost bushfire specialists, to prepare a countering paper which was eventually published in the same journal.[59] However, due to a four-year gap since the publication of the original paper, this rebuttal got little publicity and has had a less than hoped-for effect on negating a now widely held misconception about timber harvesting causing more bushfires. This episode somewhat exemplifies the difficulty of challenging research findings once they enter the scientific literature and a dumbed-down media version of their findings has been widely disseminated in the public domain.

A further example of misconstrued ANU forest conservation research which has frustrated forest scientists was the paper by Keith et al, 2014 – *Managing temperate forests for carbon storage: impacts of logging versus forest protection on carbon stocks*, published in *Ecosphere*, an ESA Online Journal.[60] The lead author, Dr Heather Keith, and one of the paper's co-authors had also been co-authors of the earlier *Green Carbon* report.

As with seemingly most ANU forest conservation research, Keith et al 2014 was focussed on Victoria's Central Highlands' montane ash forests. The paper contained several basic conceptual and factual flaws that were then compounded in carbon accounting calculations leading to erroneous results. These included:

- Grossly understating the proportion of montane ash forest that is already reserved, and therefore overstating the future extent of timber harvesting and associated woody residue (slash) burning.
- Presuming a 50-year timber harvesting rotation, rather than the 80-year rotation actually being progressively implemented – thereby presuming a substantially greater frequency of future harvesting and slash burning than will actually occur.
- Misrepresenting the split of forest biomass into wood products removed and harvest residue left in-situ, by inappropriately using data from the harvesting of 50-year old forest which contains significantly smaller trees than the 75-year old trees being harvested at the time.[61]
- Inexplicably presenting the volume of biomass removed in wood products as a proportion of total (above and below-ground) biomass thereby creating a perception of a larger than actual proportion of remnant waste. In other studies (including the cited reference), wood product removals are cited as a proportion of only the above-ground biomass because the below-ground biomass (i.e. roots) is largely unaffected by slash burning.
- Overstating the rate of annual forest harvesting by a factor of approximately three.

- Flowing from the above flaws, the paper substantially overstated the amount of forest biomass left on-site that is subsequently burnt (to secure forest regeneration) thereby releasing carbon. It thereby significantly understated the proportion of harvested wood that ultimately becomes solid timber products which provide long-term carbon storage.

These flaws first came to light when Ms Keith discussed what were then unpublished findings from the research during an interview on ABC radio.[62] At least one of these errors had occurred despite two of the paper's authors having been earlier briefed by the local forest managers who specifically informed them of the true situation.[63] When the paper's problems were formally pointed-out by the Institute of Foresters of Australia in a letter to the Director of the ANU Fenner School about a year later, the paper's authors briefly responded in writing some months later. Without specifically admitting to the errors they nevertheless explained how they had arisen and in-part justified the data they had used. However, they steadfastly maintained that, even if these errors were taken into account, it wouldn't change their paper's conclusion that "greater carbon storage is obtained by conserving native forests".[64]

Arguably, the most significant error from the paper – grossly understating the proportion of above-ground forest biomass that ends up in wood products – would have been avoided if the authors had been willing to constructively engage with the commercial forestry agency, VicForests. This could have enabled them to be supplied with data from the current harvesting of 75-year old trees in place of the older data from harvesting smaller 50-year old trees which the paper relied upon.

Since then the flawed findings arising from Keith et al 2014 have been publicly repeated by co-author David Lindenmayer on ABC Radio, in several newspaper articles, at a Greens-sponsored 'extinction emergency' forum, and in a recently published book.[65] In addition, they have been used in promotional material for a forests and climate change forum organised and sponsored by a number of climate-focussed environmental groups.[66] Significantly, the uncritical acceptance of the Keith et al 2014 findings as a foundation for subsequent research should effectively invalidate later ANU forest carbon research papers.

Similar to the earlier *Green Carbon* report, Keith et al 2014 concluded that "emissions to the atmosphere can be avoided by ceasing logging" and that "changing forest management policy to avoid emissions from logging contributes to the global objective of reducing atmospheric carbon dioxide emissions and to national targets for reducing emissions".[67] Even if the paper had no errors skewing its carbon accounting, it is inconceivable that such a conclusion can be drawn in the absence of any consideration of the subsequent consequential emissions associated with the 'substitution effect' of replacing Australian-grown hardwood products either with hardwood timber imports or greater use of non-wood materials such as concrete and steel.

Peer review and 'publish or perish'

Keith et al (2014) highlights the ease with which errant research can cumulatively misinform the conventional wisdom and potentially drive bad public policy. Further to this, the errors it contains raise grave concerns about the veracity of a peer review process which could miss them. Clearly the paper was peer reviewed by person/s unfamiliar with its Victorian study area and the basic forestry concepts and practices being used within it. If appropriately knowledgeable local peer reviewers had been used, the paper's conclusions could have been corrected prior to publication.

The peer review concept is vaunted as a rolled gold standard by the scientific community, the media, and arguably the bulk of the wider community to whom it largely represents a supposedly unimpeachable guarantee that scientific evidence has been through a process of unbiased testing. However, in a recent article, ANU academics Will Grant and Rod Lamberts, described peer review as merely a test which signifies that "the research is ready to be put out to the community of relevant experts for challenging, testing, and refining".[68]

This description of peer review as not necessarily the last word in any discussion differs from how it is routinely presented in public debate over environmental issues where it is lauded as a marker of irrefutable evidence and a blunt instrument to kill-off further debate. That this misperception is rarely corrected by academia suggests that many scientists and researchers either also believe it, or are happy for it to be regarded this way, particularly in relation to their own work.

Contrary to this view, it is being increasingly recognised across all academic disciplines that the peer review process can be readily biased by who the reviewers are and their human frailties, such as professional, personal, or political agendas and/or the extent to which they are knowledgeable about the topic.[69] The lesson reinforced by Keith et al (2014) is that it is a folly to automatically accept peer review as a guarantee that the factual and conceptual accuracy of a research paper is beyond reproach.

It is also noteworthy that ANU Fenner School forests conservation research papers are at times being published in online open access journals which, for a fee, can offer quick turn-around and peer review which may not be as rigorous as was traditionally the case. Indeed, *ESA Journals*, which published Keith et al (2014), enables submitting authors to suggest their own peer reviewers from an international panel of potential candidates. The use of non-local reviewers, who are less likely to have sufficient background knowledge of the Australian context, is surely problematic.

Recent exposure of the growth of predatory online academic publishers suggests that they have arisen in response to the so-called 'publish or perish' mantra enshrined as a marker of success in modern academia. There is concern that "they allow authors to publish articles that would never survive legitimate peer review" that are then being referenced by political activists to bolster their arguments.[70] This raises the further question of whether or not these journals are being deliberately used by academics requiring quick compilation of a body of research to impress decision-makers, as well as strategic/flexible publication timing to match activist campaign milestones.

One of the more problematic online academic publishers appears to be the OMICS International Group based in India, which has boasted of publishing over 700 open access journals, including the Journal of Biodiversity and Endangered Species.[71] In October 2014, this journal published the paper – *Preventing the extinction of an iconic globally endangered species – Leadbeater's possum (Gymnobelideus leadbeateri)* by Lindenmayer, Blair, McBurney and Banks. Emphasising the advantages and potential pitfalls of this form of academic publishing, just 18-days elapsed between the journal receiving their manuscript and its publication.[72]

Professor Lindenmayer and his ANU research associates collectively boast an impressive publication record with regard to the Central Highlands forests. Their more than 200 published papers and eight books on this topic clearly dazzles most journalists, politicians, and their advisers. However, the propensity for errors, misconceptions and wrong assumptions to cumulatively infect their subsequent papers on the same topics suggests that it is dangerous to elevate any substantial body of work from one research institution above other scientific views. Aside from the ANU's work, there has been a large body of past research into these forests by Victorian scientists such as Ashton, Attiwill, Bassett, Campbell, Loyn, Neumann and Squire which is now being largely ignored.

Nevertheless, the contention that numbers of published papers are indicative of scientific importance is well entrenched as a marker linking academic advancement to research funding and job security. On the other hand, some within the wider scientific community are now questioning whether the 'publish or perish' mantra has effectively incentivised publication numbers over and above research quality. There are concerns that it is leading to sloppy science and declining standards of rigour as evidenced by a 'distressing increase in positive results' that is unlikely to reflect the real world. The need for scientists to maintain research funding may be discouraging negative findings when financial backers are paying for their proposals or hypotheses to be examined. Reportedly this has, over time, generated a substantial body of bad science which in some cases is being used by governments to justify bad policies.[73]

This contention was recently supported by Simon Gandevia, Deputy

Director of Nueroscience Research Australia, in a recent article on *The Conversation*:

> We do not need to look far to expose the fundamental cause for the problematic practices pervading many of the sciences. The 'publish or perish' mantra says it all.
>
> Academic progression is hindered by failure to publish in the journals controlled by peers, while it is enhanced by frequent publication of nearly always positive research findings. ... Starkly put, the rate of publication varies between scientists. Scientists who publish at a higher rate are preferentially selected for positions and promotions. Such scientists have 'children' who establish new laboratories and continue the publication practices of the parent.[74]

Certainly, many Australian forest and fire scientists would agree with the contention that bad science is being used to justify bad policy given the often unwarranted expansion of national parks in southern Australia over the past 20-years. With respect to the campaign for the 'Great Forest National Park' in Victoria, flawed assumptions, misconceptions and critical omissions – cumulatively strengthened by being repeated through a hierarchy of subsequent research papers – have already played their part in creating a politically-correct anti-logging narrative.

Despite its highly questionable basis, this has heightened the political imperative to resolve a supposed 'forestry crisis'. This was exemplified when the Victorian Government established a Forest Industry Taskforce in 2015 to determine the future of the state's native hardwood timber industry – as if renewable timber harvesting restricted to just a 6% portion of the state's public native forests really posed any significant environmental threat.

Publicising research findings ... breaching university standards

While there are questions about the veracity of recent forest conservation research, there is probably greater concern about how research findings are being disseminated through the media because of the greater potential for this to mislead the wider community. Mainstream media coverage of forestry issues has always primarily reflected the perspective of environmental activism, and forest conservation research – especially that emanating from the ANU Fenner School – seems to have neatly tapped into this entrenched bias. The resultant media coverage has been typified by simplistic conclusions that overhype the significance of the research and its implications for the future under sensational headlining.

Arguably, it is this alarmist reportage and the involvement of research scientists in producing and/or promoting it which has provoked most consternation amongst forest scientists and forestry practitioners about the existence of a preconceived academic agenda against forestry. This has been further inflamed by perceptions of inappropriate collusion with environmental activism as some researchers have eagerly launched or promoted their findings at environmental group campaign events. In addition to the *Green Carbon* report, the findings of Taylor et al (2014) were first presented to such an event – the 3rd Australian Forests and Climate Forum – some six months before the paper was published.[75] Similarly, selected findings from Keith et al (2014) were presented to a Greens-sponsored 'Extinction Emergency' forum almost a year before it was published.[76] Potentially, these findings were being publicly presented prior to being peer reviewed.

Public dissemination of research findings is governed by policies and directives set by academic institutions with the intention of maintaining high standards of scientific honesty, objectivity and impartiality. The relevant policies of the Australian National University (ANU) include:

ANU Policy 007403 – Code of Research Conduct[77] which in part requires ANU researchers to:

> 4.7 Explain to any sponsor or media outlet who receives research results prior to peer review that the results are not final.
>
> 4.8 Correct the record as soon as possible, where the researcher becomes aware of misleading or inaccurate statements about their work.
>
> 4.13 Avoid discussion of research findings prior to testing through peer review; or during such discussions clearly advise that the results are interim and are not peer reviewed.

ANU Policy 000359 – Academic Expertise and Public Debate[78] which in-part requires researchers to:

> In public debate, such as opinion pieces or columns in the media, ... it is expected that the position adopted should be defensible and that justification for it should be either available or able to be given at a level which would be of acceptable standard in the field of scholarship.

Arguably, few of the post-2008 forest conservation research papers emanating from the ANU Fenner School would meet all of these requirements, and some may be hard pressed to meet any. In several cases they fall short due to media blitzes which are suggestive of having been orchestrated to meet the agenda-driven requirements of the School's partner, the Wilderness Society, including sensational but errant claims that clearly contravene ANU Policy 000359.

For example, one 2009 paper,[79] which claimed that timber harvesting greatly exacerbated bushfire risk, was promoted by an ANU media release on 1 March 2010, some five months after the paper was published.[80] This had the (presumably hoped-for) effect of immediately spawning sensational media coverage (in the Hobart *Mercury* and on ABC Radio) that neatly coincided with the 2009 Victorian Bushfires Royal Commission's consideration of land management practices and bushfire threat, as well as the Tasmanian state election where forestry was, as usual, a significant issue for voters.[81] It is difficult to believe that such timing could be merely coincidental.

Similarly, the media's promotion of the findings of another paper in 2011[82] coincided with the start of the Wilderness Society's boycott of Officeworks as part of their campaign to end timber production in the ash forests of Victoria's Central Highlands. This included an ANU media release which initiated articles subsequently published in *The Age*, *Sydney Morning Herald* and *Canberra Times*.[83]

In 2014, a then unpublished paper also asserting that bushfire risk is exacerbated by timber harvesting was sensationally publicised in Hobart's *Mercury* newspaper just days after the Tasmanian Government had repealed the so-called 'forest peace deal' legislation to begin the process of restoring a portion of State forest back to being available for future timber production.[84] It would again be hard to believe that the paper's authors and/or the Wilderness Society didn't play a role in this timing.

The findings of unpublished research papers have typically been promoted through the media without any mention of their status, including whether peer review has even been completed, in contravention of ANU Policy 007403 – Clauses 4.7 and 4.13. For example, the findings of Keith et al 2014[85] were being publicised by the lead author on ABC Radio four months prior to the paper's publication and just 10 days after it had been first submitted to a scientific journal.[86] As the paper would undergo two subsequent revisions before its publication, peer review cannot have been completed when the lead author was interviewed, but this was not disclosed to the ABC audience.

Further to this, media promotion of ANU forest conservation research has frequently contravened ANU Policy 000359 by including unsubstantiated and indefensible public assertions made by ecological researchers presumably intent on maximising the impact of their findings even if it means deviating from their special field of expertise. Some examples are:

> ... we have an atrocious forest management policy and as a result of that we will see extinctions within 20 to 30 years. (*Canberra Times*, September 2011)[87]
>
> Clear-felling now loses large amounts of money for the state of Victoria, degrades the forest, erodes water catchment yields, increases fire risks, and is driving Leadbeater's possum – the state's faunal emblem – to extinction. *And* right now, millions of dollars are being lost annually through forest logging in Victoria – there is effectively a government loss-making subsidy to cut down forest. (*The Conversation*, December 2012)[88]
>
> Current logging operations significantly impair the water supply for the city of Melbourne. (ABC Radio National, August 2014)[89]
>
> [*In response to concerns that research findings are being exaggerated*]: The kinds of people that are saying this are staunchly pro-forest industry advocates who are putting the interests of paper and pulp

companies ahead of the health and safety of communities and ahead of the health of the forest itself. (*The Mercury*, September 2014)[90]

We had a sham process in Victoria where the logging industry basically awarded itself $7 million to keep cutting the forest under the guise of protecting Leadbeater's possum. ... All that essentially did was create an even more perverse situation where the logging industry was logging Leadbeater's possum to extinction. (ABC RN Breakfast, April 2015)[91]

Such statements arguably reflect the tone of an agenda-driven activism far more than the measured and moderate prognostications normally expected of an objective scientist.

Science is essentially a contest of ideas. In a bygone era this contest played out and evolved into a firm scientific consensus in the background and, if appropriate, would quietly feed into bureaucratic policy and practice. Nowadays, the publication of particular research papers about trendy causes is often a public spectacle, promoted by its authors and eagerly appropriated by lobby groups to help push their agendas via a generally uncritical media. If such papers have conceptual and factual errors, they are simply overlooked in the rush to publicise and shape their findings into a politically attractive message, and by the time these errors are unearthed, the minds of the interested public and politicians have already been made-up.

Promoting research ... and the university's brand

While forest scientists and forestry practitioners have become frustrated by sensationalised, over-the-top media coverage of often questionable research findings, the ANU have may been inadvertently encouraging it. After the release of the *Green Carbon* report in late 2008, the industry lobby group, Timber Communities Australia, openly queried the university's role in promoting unwarranted conclusions about timber harvesting. In response, the Office of the ANU Vice-Chancellor released a statement which emphasised the university's global academic standing and noted that it "is proud of researchers that challenge current views and develop new ways of understanding our environment."[92]

It could be inferred from this is that the ANU cares as much or more about its scientific researchers being out there promoting their work (and the university's brand), than about the veracity of what they are saying. Certainly, a key ANU goal is to shape public opinion, and this is clearly more achievable when its research generates sensational media headlines.

That Australia's universities are now heavily committed to promoting and enhancing their image is evident in the numbers they employ in communications and public affairs roles. The ANU, which in 2017, was ranked second amongst Australian universities by the Times Higher Education World University Ranking system, employs 18 personnel in these roles.

The ANU's Strategic Communications and Public Affairs Unit puts out a quarterly magazine, as well as three e-newsletters, while managing the ANU Newsroom website, the ANU Experience website and the ANU's You Tube TV Channel. It also organises major events, public lectures, and facilitates other corporate and community engagement, as well as generating media releases promoting ANU research.

The ANU's annual Media and Outreach Awards further emphasise the high priority given to image promotion. They recognise those academics who have: 1) made significant contributions to public discourse; 2) raised the university's public profile; and 3) who achieved wide reach and influence with stories about their work. The 2015 awards included a Vice Chancellor's Award for Advancing the Reputation of the University Through Media.[93] In 2008, there had been 11 different award categories, including for the highest number of media hits on a single media release; the highest international impact judged from coverage in overseas media; and the Vice Chancellor's Award for the best use of media to impact on public debate.[94]

Unfortunately, having an impact on public debate doesn't necessarily equate to adding anything constructive to it. In relation to environmental issues, the greatest impact is usually generated by sensational, narrowly-focussed, and simplistic assertions invariably made in the absence of critically important context or perspective to what are usually complex issues. This is how, for example, timber production can be regularly demonised for supposedly threatening animal extinctions despite it not even occurring in most Australian forests.

Whether the ANU is more concerned with superficial self-promotion than the substance of its research is probably a question being asked of every academic institution at present. In an era of university performance rankings the competition for prestige, research funding, and student numbers has probably never been stronger, and the coercion of researchers to assist by embellishing the university's scientific credentials is probably unprecedented. At the ANU, researchers are advised that:

> Gone are the days where a media release is the central point of your media strategy. Engaging with the SCAPA [*the Media Unit*] early ... gives us the best possible opportunity ... to shape a tailored and targeted strategy that gets you the best possible media coverage.[95]

This rather illuminating advice arguably explains why the release of research findings has often been strategically timed to coincide with significant environmental campaign events. This may indeed engender 'the best possible' media coverage, but it should be no surprise if it also creates a perception of inappropriate academic activism that can ultimately damage, rather than enhance, the university's brand.

The pressure on scientists to publicise and promote their research may have further increased since the emergence of *The Conversation* website. Founded in 2011 by a coalition of Australian academic institutions, it pledges to provide "an independent source of news and views, sourced from the academic and research community and delivered direct to the public".[96] According to its website, a major advantage of *The Conversation* is that it combines academic rigour with journalistic flair to produce "independent, high-quality, authenticated, explanatory journalism ... Our aim is to allow for better understanding of current affairs and complex issues".

The Conversation often provides useful insights into what is being researched and publicises findings which may otherwise struggle for coverage in the mainstream media. However, at times it publishes unduly alarmist articles about environmental topics which raise questions about the wisdom of expecting a combination of academic rigour and journalistic flair to improve our understanding of complex issues. An example was an article jointly-authored by eight of Australia's more prominent conservation scientists which began by asserting that:

> It's make or break time for Australia's national parks. National parks on land and in the ocean are dying a death of a thousand cuts in the form of bullets, hooks, hotels, logging concessions and grazing licences.[97]

Such a level of simplistic exaggeration about very limited proposals to develop and/or use some parts of some national parks wouldn't be out of place in an environmental activists' campaign. But it represents a complete dumbing-down of what are in fact complex plans and tentative proposals contingent on strict environmental stipulations. Arguably such sensationalism highlights an increasing similarity in the way academia and activism communicates with the wider community, especially in relation to environmental issues.

Indeed, *The Conversation* may be doing the community a service by helping to publicly expose the close alignment that now exists between the philosophical position of many conservation scientists and the cultural mindset of environmental activism. Furthermore, it has highlighted the yawning gulf that can often exist between the understanding of arms-length researchers and the practical realities of land management.

This latter point is evident in much of the academic opposition to forest fuel reduction burning which is largely founded on dubious fears of biodiversity impacts that exemplify a lack of appreciation of the rigorous bureaucratic planning and conduct of burning programs, including the logistical capabilities of forest management agencies to conduct successive burns at anywhere near the frequencies that some scientists are mooting as ecologically-damaging.[98]

The Conversation has at times also highlighted some fascinating incidences of disquiet within academia. One example stems from the unified chorus of academic condemnation which accompanied the Victorian Government's limited re-introduction of mountain cattle grazing in the Alpine National Park in 2010. This was achieved under the guise of a scientific research trial which almost certainly involved an element of political reward for disaffected rural voters, but was nevertheless based on a very real knowledge gap of whether or not grazing reduces fuel hazard in Victoria's foothill and montane forests, and sub-alpine woodlands.

Outraged academics and environmental activists alike missed this fundamental point by presuming that the trial was taking place in alpine grasslands in which the effects of grazing had already been extensively studied. In fact, less than 100 hectares of the trial's approximately 26,000 hectare study area was comprised of alpine grassland as it was focussed on the lower altitude forests and semi-cleared valley woodlands and (non-alpine) grasslands on and around the former Wonnangatta Station which had been farmed for several generations up to the 1950s.

In the wake of this confected outrage, one senior academic pondered why those scientists vehemently opposed to the trial had been unwilling to acknowledge its reality and whether this was a case of strategically ignoring inconvenient truths to pursue a political/ideological agenda in relation to national park management. In an article on *The Conversation*, he decried such misplaced academic activism as having considerable potential to damage Australia's scientific reputation.[99]

Generally though, *The Conversation* operates as a vehicle for strident academic views about environmental issues typically from a 'pro-environment' perspective. Its articles on forest conservation and the majority of reader comments which they attract, regularly display the dominant themes of eco-activism in which: 1) natural resource use is derided as an anathema to a healthy environment; 2) resource management and environmental regulations are routinely decried as inadequate; 3) resource use is guilty of severely damaging the environment until proven innocent (rather than the other way around); 4) responses to such criticism emanating from within resource use industries is dismissed as self-serving and untrustworthy; and 5) provoking the ire of resource use industries is seen as a badge of honour that simply affirms the veracity of any accusations levelled against them.

This latter point was amply illustrated when the ANU announced in

October 2014 that it would divest its shares in seven resource use companies after receiving independent advice that they were ranked low in terms of 'social responsibility' and because, according to a university spokesperson: "Their track record hasn't been as good as some others in terms of their immediate impact on their local environment."[100]

When those targeted companies became outraged at being unfairly maligned, several articles appeared on *The Conversation* implying that this confirmed the righteousness of the divestment decision.[101,102] This betrayed the unstated sub-text of those supporting the divestment decision which appeared to decree that resource use corporations couldn't possibly be anything other than socially irresponsible agents of environmental destruction. As it turned out, those advising the ANU on corporate socio-environmental responsibility were found to have undertaken poor research and to have "strong connections to other left-wing organisations" presumed to be ideologically-opposed to mining and other resource uses.[103]

This episode also exemplified the ANU's penchant for defining its public image and arguably revealed who it regards as its major stakeholders. Share divestment is a private matter that must be constantly happening somewhere, every minute of every hour of every day. That the ANU would announce its own share divestment decisions by media release seems to be indicative of a need to publicly embellish its reputation amongst environmental stakeholders, including vocal elements of the student body who had been lobbying the university to take such action. Presumably these groups are considered to be more important than the resource sector industries even though they reportedly fund a considerable slice of the university's research budget and employ many of its graduates.

Misusing academic credibility

Another concern about *The Conversation* is that by restricting article contributors to only practising academics (including advanced students), it elevates the observations of resource management concepts and practices by arms-length academic researchers above the in-depth knowledge of academically-trained industry practitioners who have often spent long careers at the 'coal face'. Those practitioners may make comments in response to an article, but it is the content of an article with what may well be only imagined, poorly understood, or misinformed insights that ultimately goes onto the public record.

It is also apparent that in allowing some authors to contribute articles on forestry issues, *The Conversation* has contravened its own charter in relation to providing an independent forum free from commercial or political bias.[104] These authors include a former Greens political candidate working as a lecturer in an unrelated discipline; and a formerly prominent environmental activist who was at the time studying an aspect of international forestry.[105] Undoubtedly both have knowledge about forestry gleaned from their non-academic endeavours – but presenting them as though they are impartial experts on Australian forestry anoints them with an undeserved academic credibility.

Another example has been *The Conversation*'s publication in May and July 2013 of two forest carbon articles written by an environmental law academic who was also Deputy Director of the 'progressive' think-tank, The Australia Institute (TAI).[106] These articles had formerly been published as 'opinion pieces' on the TAI website. During 2012-13, the TAI had produced five 'research papers' and nine 'opinion pieces' about native forestry and the economics of timber production, typically advocating its cessation.[107]

TAI's linkages to environmental activism were laid bare in a leaked 2011 Greenpeace-Australia strategy document outlining a proposed anti-coal industry campaign. This revealed that the TAI would be paid around $400,000 to "gradually undermine the social licence of the industry" by "challenging the economics of coal".[108] As the tone and frequency of the afore-mentioned TAI publications about forestry could also be described as 'challenging the economics' of the native forest timber industry, it would be unsurprising if this was also a campaign funded by environmental activist benefactors.

The two TAI 'opinion pieces' re-published by *The Conversation* had been accompanied by a disclosure that their author was receiving funding from a forest carbon offset company, Forests Alive, which was being managed by former Wilderness Society forests campaigner, Virginia Young. Forests Alive facilitates the payment of carbon credits to landowners for "avoided emissions by stopping native forest logging". Think tank 'opinion pieces' hardly constitute new research, and republishing them on an ostensibly scholarly website, arguably bestows them with a questionable academic credibility.

Academic credibility has been a vexed concept in relation to Australian forestry for some time now. This was exemplified just prior to the 2004 Tasmanian Election when an 'open letter' signed by 100 university academics and scientists calling for the state's old forests to be 'saved' was published in *The Australian*.[109] Most of the letter's contributors had either highly questionable or no expertise in what it was espousing.

More recently, just four-days before the 2010 Tasmanian election, Hobart's *Mercury* newspaper published an 'open letter' to the state's political leaders from 'an eminent group of professionals' calling for the reform of Tasmania's forest industry.[110] This group included 26 of the University of Tasmania's 'most senior professors and lecturers' from a variety of disciplines including philosophy, accounting and corporate governance, economics and finance, law, student governance, geography and environmental studies. Perhaps only the six from the latter discipline could be presumed to have much knowledge of what was being advocated.

Given that the credibility afforded to Australian academia is rarely questioned, the broader community deserves better than scientists embellishing their personal views by linking them to their academic qualifications and/or employment at an academic institution to distinguish them as being supposedly knowledgeable, objective and trustworthy.

Episodes such as these should be a serious concern for Australian universities if they hope to retain community confidence in their integrity and objectivity. They simply reinforce a growing perception of Australian academe as a bastion of left-leaning ideology which, with regards to environmental issues, is largely characterised by a preservationist philosophy that prioritises nature over resource use and views the two as competing rather than complementary values.

Historically, there appears to have been little effort devoted to quantifying the extent to which Australia's academic institutions are dominated by 'left-wing' thinking. One very dated study from 1970 found that just over 50% of academics at the University of Sydney identified their political ideology as 'left' compared with only 12% of the general population at that time.[111]

However, more recent North American studies suggest that the lop-sided domination of 'left-wing' thinking may now be far more pronounced. For example, a 2005 survey by the University of Toronto of over 1600 (mostly) humanities and social sciences professors in four US universities, found that 72% were liberal (i.e. left-leaning in American parlance) while just 15% identified themselves as conservative. Amongst professors at the more prestigious Ivy League universities, the split was an even more lopsided 87:13 in favour of left-leaning liberals.[112]

While the political leanings of employees of higher education institutions may not be important of itself, the blinkered view of environmental issues that is inherent to 'left-wing' ideology, especially amongst Greens political parties and their cohorts and supporters, does raise considerable concern about what the future generation of land and environmental managers is being taught.

While the distinction between academia and activism is already blurred in relation to forestry, it seems likely to become more so given the cross-over of prominent activists into academia. For example, the former long-serving CEO of the Australian Conservation Foundation, Don Henry, is now a Professor of Environmentalism (the first ever) on the staff of the University of Melbourne's Melbourne Sustainable Society Institute;[113] and former leading forest activist Aiden Ricketts has been teaching a course on activism and constitutional rights at Lismore's Southern Cross University.[114]

Ricketts' former work for the North East Forest Alliance's anti-logging campaigns included helping to produce the group's *Intercontinental Deluxe Guide to Blockading* which purported to provide "everything you wanted to know about stopping the Earth-raping juggernaut but had nobody to ask".[115] In an article in *The Age Good Weekend* magazine in June 2003, Ricketts was quoted as saying "The only way to make non-violent action value-added is to place people in direct life-threatening situations that are so difficult that it takes police hours to get the person out safely".[116]

Although high profile activists entering academia is at least a case of knowing where they've come from, the same can't always be said of academics or academic institutions harbouring ideological biases that can go undetected in ostensibly credible, peer reviewed research prepared in partnership with environmentalist benefactors.

While the relationship between the ANU Fenner School and the Wilderness Society has already been discussed, a more recent partnership has been the Zero Carbon Australia 2020 joint initiative between the University of Melbourne's Melbourne Sustainable Society Institute (MSSI) and the climate-focussed environmental group, Beyond Zero Emissions. In October 2014, they released a jointly-prepared report under the Zero Carbon Australia banner, entitled *"Land Use: Agriculture and Forestry Discussion Paper"*. This was a superficially impressive 172-page document citing hundreds of scientific references.[117]

According to Beyond Zero Emission's website, "the project was made possible by the generous support of a private donor and significant additional assistance from MSSI". The report itself acknowledges three 'Major Supporters', including the first-listed 'Graeme Wood Foundation'. Mr Wood, a travel booking entrepreneur and well known environmental philanthropist, made the then largest political donation in Australian history to the Australian Greens in 2010.[118] In relation to forestry, Wood has been loudly disdainful of Tasmania's forest industry and, in 2011, bought the state's Triabunna woodchip mill in conjunction with a financial partner. Under his ownership, the mill was deliberately decommissioned to thwart the potential revival of the local timber industry.[119]

The report's authors were drawn from the University of Melbourne's Melbourne Sustainable Society Institute. The author of the sections on forestry and revegetation – Dr Chris Taylor – had at one time been employed by a consortium of environmental groups to produce a major report about Victoria's 2009 'Black Saturday' bushfires;[120] is a lead researcher and public advocate of the contentious notion that timber harvesting exacerbates bushfire risk (which is regularly used to denigrate the timber industry); and is a self-professed 'architect' of the 'Great Forest National Park' proposal.[121]

The major findings of Zero Carbon Australia's "*Land Use: Agriculture and Forestry Discussion Paper*" matched the central tenets of environmental activism (and that of recent contentious research papers about Victoria's forests), in concluding that:

- "Clearfell logging ... prevents ongoing carbon sequestration in living forests."
- "Disturbance to our tall forests increases the likelihood that fires will be more severe and more frequent."
- "Rehabilitated and older forests are more resilient to the impact of fire."
- "An expanded reserve network recognising the importance of forests for carbon stocks and sequestration potential is recommended, as well as stopping clearfell logging."

Arguably the blurred distinction between academia and activism over the environmental impacts of forestry has led to the emergence of two separate tribes of forestry 'experts':

- one comprised of conservation scientists who shun forest and bushfire science and practical forestry expertise, typically ignore or misrepresent wider context and perspective, selectively draw from a body of science largely produced by themselves and their associates, demonstrably supports environmental activist agendas, and actively promotes their work through the media including as part of environmental campaigns; and
- another comprised of forest and bushfire scientists and practitioners who grapple with forest management issues on a daily basis, are fully conversant with the contextual balance between forest conservation and use, know the operational

and silvicultural practices being already used to safeguard environmental values, but rarely rate a mention in the media.

Due to its inherent association with environmental activism and the unquestioning support this gets from the majority of the media, the former tribe is far more influential in shaping community and political attitudes. This is effectively giving unwarranted respectability to research largely prepared with minimal or no input from the most important body of scientific expertise on forests.

Forest science has its own academics, but they are fewer in number and have been demonstrably less willing or able to publicly defend Australian forest management. This may be indicative of an unwillingness to depart from conventional expectations of the behaviour of objective scientists.

Lack of appreciation for land management

Forest and fire scientists and practitioners have always appreciated from their training and experience that effective conservation is about land management rather than land tenure. However, the recent ANU Fenner School forests conservation papers suggest that the same can't be said for most conservation scientists given their obvious prioritisation of more national parks. For them it seems that land/environmental management is a secondary consideration only to be applied once resource use industries have been forced out by a change in land tenure.

This is an immature attitude that may be largely rooted in misunderstanding of what resource use actually means and how extensive it is; as well as a refusal to consider conservation as a landscape-scale concept where the whole is more important than what happens on any particular hectare. This may in-part reflect the progressive shift of environmental research away from in-house applied research formerly undertaken by scientists employed by Government land management agencies, to institutionally-based scientists working more remotely from their study areas. Under this arrangement scientists may have less appreciation of the local land management concepts and operational practices being employed in the field.

Research was formerly a key role of Australia's State forest management agencies for decades. Using Victoria as an example, as far back as the late 1970s the Forests Commission was permanently employing two ecologists in monitoring the ecological impacts of contemporary integrated timber production.[122] By the mid-1980s, forestry researchers in Victoria's Department of Conservation, Forests and Lands were working on nine major programs including the ecological impacts of fire, soil and tree nutrition, insect pests, and plant ecology, in addition to silvicultural and forest productivity issues.[123]

More recently in 2000-01, the equivalent Department still had 58 staff employed in forest research working on 67 projects concerning biodiversity, ecosystem health, and vitality. This included the release of 26 peer-reviewed scientific papers, although much research undertaken at government-funded facilities was for clients and so was not formally peer-reviewed for journal publication.[124] This ongoing work and that of others arguably underpins the substantial reservation of Victorian State forests within special protection zones (in addition to national park expansion) from the early 1990s, as well as more stringent operational measures, such as the retention of 'habitat trees' adopted to reduce harvesting impacts and complement the biodiversity conservation being mostly achieved in the majority of forests that are not used.

Between 2001 and 2006, "… the Victorian Government invested over $37 million in forest management research, development and education. Most of this investment was allocated to major service providers including the School of Forest and Ecosystem Science (University of Melbourne), the Arthur Rylah Institute of Ecological Research …"[125] This is indicative of the shift to contracted research partners, although it is likely that most of the Department's formerly employed researchers were still present within those partner institutions. For example, most of Victoria's former state forestry researchers had departed to the University of Melbourne but still retained state contracts for mainly ecological research.

By 2014, it had become difficult to accurately identify what was being spent on forests-related research in Victoria (and presumably also in other states) given that forestry was now only a minor part of a mega-department overseeing primary industries, including agriculture, drawing on an annual research budget of $100 million and employing over 600 researchers. The former Victorian Department of Environment and Primary Industries' *DEPI Science Strategy 2014-19* confirms the perception of research being by then mostly undertaken more remotely by contract partners and collaborators:

> Science in DEPI is supported and delivered by a broad spectrum of research funders, collaborators, partners and providers, and overseen by scientifically literate staff. These relationships take time to establish and develop, and a long-term commitment is required to realise their benefits.[126]

By 2014, Forestry Tasmania was the nation's only dedicated forestry agency which still retained a (small) internal research capacity. WA and SA had very little forest industry research capacity, whereas some forestry research continued to be funded by the Queensland and NSW governments but was not undertaken by land management agencies.[127]

Overall, there are serious concerns about the sharply declining investment in forestry research and development at a national scale. This has been due to a range of factors including the declining relative importance of the forest industry to the national economy, reduced government involvement in the industry due to the privatisation of former state-owned plantation resources, low industry profitability, cost-cutting, and inadequate recognition of the importance of research and development. Given that most forests are no longer used for wood production, the majority of forest research now has a conservation focus.[128]

Environmental issues in general and forestry issues in particular, seem to provoke strong personal feelings amongst scientists from a wide range of disciplines. Undoubtedly amongst the ranks of Australia's conservation scientists there are some who overtly support environmental activist campaigns and are willing to focus their efforts on achieving a pre-conceived agenda of more national parks and reserves so as to consign traditional forest use or forestry activities (such as most fuel reduction burning) to the annals of history.

The prevailing attitude of conservation science in relation to forests arguably reflects the success of decades of environmental activism in creating a populist conventional wisdom, coupled with the more recent threat of climate change. The former has arguably created a generation of environmental researchers and students motivated to save the environment from the excesses of capitalism; while the latter has arguably engendered a sense of urgency that dictates that no less than the future of the planet rests on their efforts to identify new ways of living.

With these underlying motivations, it is hardly surprising if some conservation scientists simply dismiss traditional forestry activities as damaging and outdated and target their research to confirming this without even considering the evolution of these activities themselves, or assessing whether they may in fact offer superior ecological outcomes compared to an alternate national parks expansion. This has skewed the research effort and is indicative of why, for example, far greater effort is directed towards trying to find endangered wildlife in State forests which may be potentially available for future timber harvesting, rather than studying what is living in national parks which (in the minds of the researchers) have already been 'saved'. Similarly there is an unwarranted focus on quantifying subtle ecological damage potentially caused by fuel reduction burning, compared to measuring the obvious catastrophic ecological damage caused when heavy accumulations of fuel in long unburnt forests is obliterated in hot summer bushfires.

In addition, the perceived urgency of avoiding climate change 'tipping points' has undoubtedly emboldened some scientists to break the shackles of academic standards to become spruikers for societal change, and on the evidence in regard to forests, this has spread to other disciplines. Perhaps, as they see it, if the world ends there will be no environment and no need for academic standards.

Unfortunately these days, the wise response to headlines, such as – "Researchers propose new strategies to manage forests" – must be to firstly question the motives of the messenger, and the backgrounds of the researchers and who they are associating with. As objective conservation science becomes increasingly diluted by the apparent willingness of its researchers to support ideological eco-activist causes, it is arguably inevitable that there will be an erosion of public trust in the integrity of environmental research and the institutions that oversee it. Ultimately, the environment will be the loser as our capability to effectively manage forests and its undisputed major threat – fire – is progressively weakened.

Chapter 4 Endnotes

1 Richard Horton, 2015, Comment: "Offline: What is medicine's 5 sigma?", *The Lancet*, Volume 385: 1380, April 11 2015. Note: Richard Horton was at the time the editor of *The Lancet*.

2 *Early Learning Centre project wins international recognition*, Melbourne University Staff/Students E-news, August 15/165.

3 In reality, old growth mountain ash forest is not logged and hasn't been for over 30-years; Leadbeater's possum appears to be thriving rather than on the brink of extinction; trees being harvested are not 150-years old but mostly advanced regrowth from the 1939 fires and some 1926 regrowth; and the wood from the harvesting produces a mixture of solid timber and paper products.

4 My Environment 2014, 'Great Forest National Park' website created by the environmental group, My Environment Inc.

5 *Forest Management Plan for the Central Highlands*, Department of Natural Resources and Environment, Victoria (May 1998). Appendix B defines the 'Wet Forest' ecological vegetation class as being dominated by mountain ash (*Eucalyptus regnans*), and Appendix C delineates the area of 'Wet Forest' and assigns its proportional extent within various conservation and use public land tenures. It states that the net harvestable area of 'Wet Forest' (i.e., mountain ash) comprises 39% of the total 'Wet Forest' area. Therefore 61% of mountain ash forest in the Central Highlands was reserved at that time (i.e., 1998).

6 DELWP 2017, *A review of the effectiveness and impact of establishing timber harvesting exclusion zones around Leadbeater's possum colonies*, Department of Environment, Land, Water and Planning with contribution from VicForests, July 2017. The additional reservations have been in accordance with the recommendations of the Victorian Government's Leadbeater's Possum Advisory Group which produced its Final Report in January 2014.

7 Conservation Advice, *Gymnobelideus leadbeateri*, Leadbeater's possum, provided by the Threatened Species Scientific Committee to Federal Environment Minister, Greg Hunt, and approved by him on 22 April 2015, p. 6 of 53 pp.

8 Blair 2016, *First after the fire: Leadbeater's possum sighted at historical location*, by Dave Blair, Long Term Ecology Group web blog, 11 August 2016.

9 DELWP 2017, op. cit.

10 Another research focus has been on the value of fuel reduction burning as a bushfire

management tool, but this is part of a wider forests agenda rather than the campaign for the Great Forest National Park.

11 "Sending Leadbeater's possum down the road to extinction", by David Lindenmayer, *The Conversation*, 14 December 2012.

12 Chris Taylor described himself as one of the architects of the Great Forest National Park proposal during an address to a Victorian Greens public meeting about the proposed park where he spoke alongside Bob Brown, 19 June 2016, just 13 days before the Federal Election.

13 "Why Victoria needs a Giant Forest National Park", by David Lindenmayer, *The Conversation*, 30 September 2013.

14 "Victorian forestry is definitely not ecologically sustainable" 17 January 2013.

"Victoria's logged landscapes are at increased risk of bushfire", 25 August 2014.

"A job for Victoria's next leaders: save the Central Highlands", 28 November 2014.

"Forestry agreements need a full overhaul, not just a tick and flick", 1 April 2015.

"Victoria must stop clearfelling to save Leadbeater's possum", 23 April 2015.

"Native forests can help hit emissions targets if we leave them alone", 23 July 2015.

"Ashes to ashes: Logging and fires have left Victoria's magnificent forests in tatters", 25 November 2015.

15 My Environment 2014, op. cit.

16 There have been many public events where Professsor Lindenmayer and/or his conservation research associates have promoted the Great Forest National Park proposal. Just a couple of examples are: A Key Note Speech at the (so-called) Launch of the Great Forest National Park at Federation Square, Melbourne: 'The Science behind the Great Forest National Park' 25 November 2014 (four days before the Victorian State Election); A speech to ANU Alumni: 'Professor David Lindenmayer presents the Great Forest National Park', Melbourne, 20 June 2016 (two weeks before the 2016 Federal Election); Greens Public Meeting to provide information about the Great Forest National Park proposal, Melbourne, 19 June 2016 (speaker was Chris Taylor), just two weeks before the Federal Election.

17 Google search: 'David Lindenmayer on the ABC' brings up many relevant radio and television interviews. Professor Lindenmayer's views on timber production have been often published in newspaper articles since 2008. These include:

"Logged forests a 'fiercer fire risk': Black Saturday warning sounds warning over Tasmanian logging", David Beniuk, *The Mercury*, 7 September 2014.

"Like a voice in the wilderness", Rosslyn Beeby, *Canberra Times*, 17 September 2011.

"Just 1% of Central Highlands old growth survives", Adam Morton, *The Age* and *Sydney Morning Herald*, 12 September 2011.

"Old-forest loss catastrophic: study", Rosslyn Beeby, *Canberra Times*, 13 September 2011).

"The decline and fall of the forest's grand old masters", Bridie Smith, *Sydney Morning Herald*, 10 December 2012.

"Central Highlands carbon storage worth more than logging", Josh Gordon, *The Age*, 20 June 2016.

18 Lindenmayer et al (2009) – *Effects of logging on fire regimes in moist forests*, by D.B. Lindenmayer, M.L. Hunter, P.J. Burton and P. Gibbons, *Conservation Letters 2: 271-277* (October 2009).

The paper's findings were promoted through a series of media reports which mostly coincided with the 2009 Victorian Bushfires Royal Commission's coverage of the effect of land management policies and practices on bushfire risk, and the Tasmanian State election in March 2010:

8 December 2009: *ABC News* report, "Scientist links forest logging to bushfires";

11 February 2010: *ABC Science* report "Logging makes forests more flammable: study";

1 March 2010: ANU Media release and article in *Australasian Science* entitled "Logging creates fire traps".

2 March 2010: report in Tasmania's *The Mercury*: "Logging legacy labelled greater fire risk"; and

5 March 2010: five-minute interview of Professor Lindenmayer: "Industrial logging linked to frequency and severity of fire".

19 Pozible 2014, Pozible Great Forest National Park project.

20 Lindenmayer, Blair, McBurney and Banks (2015), "Ashes to ashes: Logging and fires have left Victoria's magnificent forests in tatters", *The Conversation*, 25 November 2015.

21 Burns et al, 2015, "Ecosystem assessment of mountain ash forest in the Central Highlands of Victoria, south-eastern Australia" by E.L. Burns, D.B. Lindenmayer, J. Stein, W. Blanchard, L. McBurney, D. Blair and S.C. Banks, *Austral Ecology*, Vol 40:4 (386-399), June 2015.[Note: The paper was accepted for publication in August 2014].

22 At the time of writing, the science within this paper is being reviewed by forest scientists from the Institute of Foresters of Australia with a view to challenging the IUCN's listing of these forests as 'critically endangered'. The paper's most obvious error is in misrepresenting the proportion of the forest reserved from timber production as being only 20%, whereas 61% had been reserved in 1998 and this had increased by 2015.

The full list of noted problems is:

Asserting that 80% of the forest is designated for paper pulp and timber production.

Implying that all burnt forests are subject to salvage logging.

Labelling mountain ash forest as the most carbon-dense forest on earth.

Asserting that the forests will be consigned to a 'landscape trap'.

Assuming that regenerating logged forests will significantly exacerbate bushfire risk. Most of these claims and assertions are derived from earlier ANU research which has been actively contested (i.e., the claim that logged forests exacerbate bushfire risk – see Attiwill et al, 2014); considered to be highly contentious (i.e. the claim of a supposed landscape trap – see Ferguson and Cheney 2011); or found to be wrong (i.e. the claim that mountain ash is the most carbon-dense forest in the world – see Sillet et al 2015).

The notion that the mountain ash ecosystem could be in danger of disappearing needs to be considered in the context of it occupying over 320,000ha in Victoria and Tasmania, most of it reserved for conservation. Statewide across Victoria there is 250,000ha of mountain ash of which 75-80% is reserved for conservation.

23 *Forest Management Plan for the Central Highlands*, Department of Natural Resources and Environment, Victoria, May 1998.

24 G.J.W. Webb, 2008, "The dilemma of accuracy in IUCN Red List categories, as exemplified by hawksbill turtles (Eretmochelys imbricata)" *Endangered Species Research*, Volume 6:161-172, December 2008. On the question of exaggerating the risks of extinction, this paper in-part noted that:

> Yet scientific precaution often clashes with the goals of advocacy, with which many IUCN member organisations are involved. Exaggerating risks of extinction, whether ethical or not (Shrader-Frechette & McCoy 1999), is an effective strategy for achieving conservation outcomes and the end justifies the means in the eyes of many (Mrosovsky 1983, 2002, 2003).

25 Scientific concerns about the Conservation Advice which informed the critically endangered listing of Leadbeater's possum were that it: 1) it grossly understated the area of suitable habitat; 2) its insufficient consideration of the known capacity of the possum to recover from disturbances such as severe wildfire and timber harvesting; and 3) its misconception that habitat area is directly analogous to animal population which is at odds with the clumped distribution of the species. These concerns raise doubts over the magnitude of the TSSC's assumed extent of Leadbeater's possum population decline over the previous three generations (18 years), which is the key trigger for a 'critically endangered' listing. This was articulated in a submission by the Institute of Foresters of Australia to the Threatened Species Scientific Committee in 2017.

26 D. Lindenmayer, D. Blair, L. McBurney and S. Banks, 2015, *Mountain ash: Fire, Logging and the Future of Victoria's Giant Forests*, CSIRO Publishing.

27 P. Fagg and M. Poynter, 2016, *Book Review: Mountain ash: Fire, Logging and the Future of Victoria's Giant Forests*, by Peter Fagg and Mark Poynter, *Australian Forestry* Vol 79:3 (September 2016). This review concluded that:

> The content in this particular book ... with its many misleading, one-sided, or erroneous statements and serious omissions of highly relevant facts is disturbing to these reviewers who have specialised in silviculture, forest ecology, and forest and fire management.

While the book is richly illustrated (although with some bias), and contains some interesting and useful information, it cannot be recommended as a balanced treatment of the ecological knowledge of the montane ash forests or of their management.

28 *You Tube*: Great Forest National Park Cinema Advertisement, 2016.

29 Victorian Government nest box installation and monitoring work undertaken since 2014 in response to recommendations by its Leadbeater's Possum Advisory Group was, by 2016, showing a 53% possum occupancy rate of the 496 recently monitored nest boxes. Similarly, of the 72 artificial tree hollows created at 18 sites (i.e. four per site) as part of a collaborative project between the Arthur Rylah Institute and Vic-Forests, Leadbeater's possums had been found nesting in 37 (52%) of them at 78% of the sites. Source: *Supporting the Recovery of Leadbeater's possum: Progress Report December 2016*, Department of Environment, Land, Water and Planning, State of Victoria.

30 LTEG (2016), *Fact checking a fact sheet*, by unnamed author/s, Long Term Ecology Group web blog, 28 May 2016.

31 McBurney (2016), *Are nest boxes an effective management tool for Leadbeater's possum?* by Lachie McBurney, ANU Fenner School Long Term Ecology Group web blog, 9 May 2016.

32 Keith et al 2017, *Experimental Ecosystem Accounts for the Central Highlands of Victoria: Final Report*, by H. Keith, M. Vardon, J.L. Stein, J.A. Stein, and D. Lindenmayer, Threatened Species Recovery Hub, ANU Fenner School of Environment and Society (July 2017).

33 Deloitte Access Economics (2015), *Economic Assessment of the native timber industry in the Central Highlands RFA area – Report 1: Economic and financial impact*, by Deloitte Access Economics for VicForests, October 2015.

34 R. Harris, 2017, "Logging industry wins battle for review into Leadbeater's possum endangered status", by Rob Harris, *Herald Sun*, 11 September 2017.

35 *The Leadbeater's possum Review*, by D. Blair, D. Lindenmayer, L. McBurney, S. Banks and W. Blanchard, ANU Fenner School of Environment and Society, August 2015.

36 "Leadbeater's possum population crashes by two-thirds in past 20 years: report", *ABC Radio National Breakfast*, 11 October 2017.

37 "More sightings of an endangered species don't always mean its recovering", David Lindenmayer, *The Conversation*, 16 October 2017.

38 ANU Long Term Ecology Group 2016, "Fact Checking a Fact Sheet" 28 May 2016.

39 Lindenmayer, Burton and Franklin 2008, *Salvage logging and its ecological consequences*, by D.B. Lindenmayer, P.J. Burton, and J.F. Franklin, Island Press (Washington, USA), 2008.

40 "Salvage logging in the montane ash eucalypt forests of Central Highlands of Victoria and its potential impacts on biodiversity", D.B. Lindenmayer and K. Ough, *Conservation Biology*, Vol. 20:4, 2006.

41 Most of the ANU's Central Highlands forest research contains no acknowledgement

that the majority of mountain ash forests are not used for timber production. However, the following publications by ANU environmental scientists have acknowledged the existence of a balance between conservation and use in the Central Highlands mountain ash forests, but misrepresented it as 20% reserved and 80% usable. In reality the proportion reserved for conservation has grown from 61% in 1998 to around 65-70% in 2017:

Life in the Tall Eucalypt Forests, by Lindenmayer and Beaton, Reed New Holland (2000).

Forest Pattern and Ecological Process: A Synthesis of 25 years of Research, by Lindenmayer, CSIRO Publishing (2009).

Burns et al, 2015, "Ecosystem assessment of mountain ash forest in the Central Highlands of Victoria, south-eastern Australia" by E.L. Burns, D.B. Lindenmayer, J. Stein, W. Blanchard, L. McBurney, D. Blair, and S.C. Banks, *Austral Ecology*, Vol 40:4 (386-399), June 2015.

Lindenmayer and Sato (2018), "Hidden collapse is driven by fire and logging in a socioecological forest ecosystem", by D.B. Lindenmayer and C. Sato, Proceedings of the National Academy of Sciences of the United States of America (pre-print publication), 30 April 2018.

42 "Silvicultural recovery in ash forests following three recent large bushfires in Victoria", P. Fagg, M. Lutz, C. Slijkerman, M. Ryan and O. Bassett, *Australian Forestry*, Vol. 76: 3-4, October 2013.

43 Routley, R., Routley, V. 1974, *The Fight for the Forests – the take-over of Australian forests for pines, woodchips and intensive forestry*, Research School of Social Sciences, Australian National University, Falcon Press.

44 The Wilderness Society 2013, *WildCountry – Victoria*: The US-based Wildlands Network (formerly Project).

45 Wildlands Project 2013, Agenda 21 Course: Lesson 3: This online lecture explains that the Wildlands Project aims to put 50% of the USA's rural lands under reservation for conservation purposes.

46 The Wilderness Society 2008a, *Annual Review for the Year ended 30 June 2008*, p. 4.

47 Long Term Ecology Group website: www.longtermecology.com/great-forest-national-park.

48 Greenpeace Australia-Pacific Ltd 2008, *Annual Financial Report*, 31 December 2007, p. 7.

49 The Zero Emissions Network is now known as Beyond Zero Emissions. The record of its blog from the 2007 Bali Climate Conference no longer exists online.

50 H. Keith, *Wilderness News*, June 2008, *Green Carbon*. This article, written by one of the co-authors of the Mackey et al *Green Carbon* report, included a text box stating: "The Wilderness Society made an organisational submission [to the Garnaut Review] that spells out the compelling science about forests and carbon. And we co-ordinated thousands of Australians to have their say on this critical issue by making their own submissions."

51 Garnaut Climate Change Review 2008, *Final Report*, Chapter 7, p. 165.
52 B. Mackey, H. Keith, S. Berry and D. Lindenmayer, 2008, *Green Carbon – the Role of Native Forests in Carbon Storage – Part 1: A green carbon account of Australia's south eastern eucalypt forests, and policy implications*, ANU E-Press, p. 41: "We are grateful to The Wilderness Society Australia for a research grant that supported the analyses presented in this report."
53 The Wilderness Society no longer lists its formal policies on its website, but refers to *Our Vision* which does not articulate detailed policy aims.
54 Wilderness Society 2009, Media Release (10 September).
55 The Intergovernmental Panel on Climate Change (IPCC) Fourth Assessment Report 2007, noted that "wood products can displace more fossil-fuel intensive construction materials such as concrete, steel, aluminium and plastics, which can result in significant emission reductions."
56 F. Ximenes, H. Bi, N. Cameron, R. Coburn, M. Maclean, D. Sargeant, M. Mo, S. Roxburgh, M. Ryan, J. Williams and K. Boer, 2016, *Carbon stocks and flows in native forests and harvested wood products in SE Australia*, Forest and Wood Products Australia, Project Number: PNC285-1112. Research funded by the Department of Agriculture, Fisheries and Forestry.
57 D.B. Lindenmayer, M.L. Hunter, P.J. Burton and P. Gibbons, 2009. "Effects of logging on fire regimes in moist forests", *Conservation Letters 2: 271-277*.
58 "Scientist links forest logging to bushfires", *ABC News*, 8 December 2009.
59 P. Attiwill, M. Ryan, N. Burrows, P. Cheney, L. McCaw, M. Neyland and S. Read, 2013, "Timber harvesting does not increase fire risk and severity in wet eucalypt forests in southern Australia", *Conservation Letters*, 2013, 1-14. Cheney is a former Head of the CSIRO Bushfire Research Unit, and McCaw is a prominent WA bushfire researcher.
60 H. Keith, D. Lindenmayer, B. Mackey, D. Blair, L. Carter, L. McBurney, S. Okada and T. Konishi-Nagano, 2014, "Managing temperate forests for carbon storage: impacts of logging versus forest protection on carbon stocks", *Ecosphere*, Vol 5, Issue 6, Article 75.
61 The mistake is attributable to the paper's reliance on data from the source reference – Raison, J. and Squire, R. 2008, *Forest Management in Australia: Implications for Carbon Budgets*, National Carbon Accounting System Technical Report No. 32, Australian Government, Department of Climate Change, December 2008.

Table 9 (p. 140) of this reference gives a breakdown of the distribution of the above-ground biomass in a harvested 50-year-old Victorian mountain ash forest. This data has been inappropriately used as the basis for conclusions about the current harvesting of 75-year-old ash forests in which trees are substantially larger thereby resulting in a different breakdown between sawlogs and pulp, as well as between log removals and residue.

Further exacerbating this error is that in the study area examined by the paper – Victoria's Central Highlands' ash forests – VicForests exclusively harvests regrowth forests in which forest product removals often comprise 80-85% of the above-ground biomass. Accordingly, it is not uncommon for as little as 15-20% of the above-ground biomass of clearfell harvested ash forests to be left on-site as residue.

62 On 21 February 2014, the paper's lead author, Heather Keith, was part of an *ABC Radio National* discussion which argued that timber production should cease in Australia's forests so that they can be fully reserved for carbon: "Forests more valuable for carbon: ex-Treasury official."

63 Personal comment, Mike Ryan, forester Vicforests', April 2014: Two of the Keith et al 2014 authors had previously been briefed by VicForests that the harvesting area and number of coupes nominated in the Timber Release Plan (TRP) included contingencies and that it therefore greatly overstates the actual amount of harvesting planned over the nominated five-year period. Despite this warning, Keith et al 2014 cited an area to be harvested from 2011-16 taken straight from the TRP which accordingly overstated the actual harvesting during this period by approximately a factor of three.

64 *Response to the Institute of Foresters of Australia*, by Heather Keith, David Lindenmayer and Brendan Mackey (undated two-page document) – received by email 11 March 2016.

65 The most recent publications incorporating these mistakes were in 1) "Native forests can help hit emissions targets – if we leave them alone", by David Lindenmayer and Brendan Mackey, *The Conversation*, 23 July 2015; and 2) *Mountain ash: fire, logging, and the future of Victoria's giant forests*, by David Lindenmayer, David Blair, Lachlan McBurney, and Sam Banks. CSIRO Publishing, November 2015.

66 The 3rd Australian Forests and Climate Forum held at the Australian National University, Canberra, 22 February 2014 http://www.eventbrite.com.au/e/3rd-australian-forests-climate-forum-real-direct-action-tickets-10040707025#m_1_100

67 H. Keith, D. Lindenmayer, B. Mackey, D. Blair, L. Carter, L. McBurney, S. Okada and T. Konishi-Nagano, 2014, op. cit.

68 W. Grant and R. Lamberts, 2014, "The 10 stuff-ups we all make when interpreting research", *The Conversation*, 3 October 2014.

69 R. Smith, 2006, "Peer review: a flawed process at the heart of science and journals", by Richard Smith, *Journal of the Royal Society of Medicine*, 99(4): 178-182.

70 M.J.I. Brown, 2015, "Vanity and predatory academic publishers are corrupting the pursuit of knowledge", *The Conversation*, 3 August 2015.

71 OMICS International website.

72 Stated on the first page of the paper when viewed online at the Journal of Biodiversity and Endangered Species website.

73 P.E. Smaldino and R. McElreath, "The natural selection of bad science", *Royal Society Open Science*, 21 September 2016

74 S. Gandevia, 2016, "We need to talk about the bad science being published", *The Conversation*, 19 July 2016.

75 The 3rd Australian Forests and Climate Forum, op. cit.

76 The Australian Greens 2013, Extinction Emergency: Save our Forests Seminar, Melbourne, 19 August 2013.

77 Australian National University Policy 007403 – Code of Research Conduct.

78 Australian National University Policy 000359 – Academic Expertise and Public Debate.

79 Lindenmayer, Hunter, Burton, Gibbons, op. cit.

80 Australian National University 2010, Media Release (1 March): *Forest logging creates fire traps – academic*.

This also coincided with an article in *Australian Science* by the paper's lead author, David Lindenmayer, entitled "Logging creates fire traps".

81 "Logging legacy labelled greater fire risk", *Hobart Mercury*, 2 March 2010; and "Industrial logging linked to frequency and severity of fire", featuring a 5-minute interview of Professor David Lindenmayer, *ABC Radio*, 5 March 2010.

82 D.B. Lindenmayer, R.J. Hobbs, G.E. Likens, C.J. Krebs and S.C. Banks, 2011, "Newly discovered landscape traps produce regime shifts in wet forests". *Proceedings of the National Academy of Sciences of the United States of America* (PNAS), Vol 108 No 38, pp. 15887-15891.

83 1) Australian National University 2011, Media Release (12 September): *Forest logging increases risk of mega fires*.

2) Adam Morton, 2011, "Just 1% of Central Highlands' old growth survives", *The Age* and *Sydney Morning Herald*, 12 September 2011.

3) Roslyn Beeby, 2011a, "Old-forest loss catastrophic: study", *Canberra Times*, 13 September 2011.

4) Roslyn Beeby, 2011b, "Like a voice in the wilderness", *Canberra Times*, 17 September 2011.

84 David Beniuk, 2014, "Logged forests a 'fiercer fire risk': Black Saturday study sounds warning over Tasmanian logging". *The Mercury*, 7 September 2014. The study in question was "Non-linear effects of stand age on fire severity" by Chris Taylor, Michael McCarthy and David Lindenmayer, *Conservation Letters*, Volume 7 Issue 4, 355-370 (July/August 2014).

85 Keith, Lindenmayer, Mackey, Blair, Carter, McBurney, Okada, Konishi-Nagano, 2014, op. cit.

86 ABC 2014a, "Forests more valuable for carbon – former Treasury official". *ABC RN Breakfast*, 21 February 2014. Guests: Dr Frances Perkins – Economist; Senator Richard Colbeck; Ross Hampton – CEO Australian Forest Products Association; ANU Vice Chancellor Professor Ian Chubb; Dr Heather Keith – ANU forest catbon researcher.

87 Beeby 2011b, op. cit. – quote attributed to David Lindenmayer.

88 D. Lindenmayer, 2012, "Sending Leadbeater's possum down the road to extinction", *The Conversation*, 14 December 2012.

89 ABC 2014b, "Logging near towns helped fuel the worst fires in Australia's history". ABC Radio National, *Bush Telegraph,* 13 August 2014. Guests: Professor David Lindenmayer, Fenner School, Australian National University; and Chris Taylor, Research Fellow, University of Melbourne. Quote from David Lindenmayer.

90 Beniuk 2014, op. cit. – quote attributed to David Lindenmayer.

91 "Possible logging bail-out to save Leadbeater's possum", *ABC RN Breakfast*, 23 April 2015. Guests: Victorian Environment Minister, Lisa Neville; Professor David Lindenmayer, Fenner School, Australian National University. Quote from David Lindenmayer.

92 Australian National University, Office of the Vice Chancellor, Media Release, 15 October 2008: *ANU Research: The Facts*, Contact: Jane O'Dwyer, Director, Communications and External Liaison Office.

93 ANU Media and Outreach Awards, 11 December 2015.

94 Australian National University Media Release, 26 November 2008: *Media work leads to public impact: Winner.*

95 Australian National University Strategic Communications and Public Affairs Unit: Get Media Coverage.

96 *The Conversation* website.

97 E. Ritchie, B. Laurance, C. Bradshaw, D.M. Watson, E. Johnston, H. Possingham, I. Lunt and M. McCarthy, "Our national parks must be more than playgrounds or paddocks", *The Conversation*, 24 May 2013.

98 AFAC 2015, *Overview of prescribed burning in Australasia.* National Burning Project: Sub-Project 1, Australian Fire and Emergency Service Authorities Council (AFAC) and Forest Fire Management Group (FFMG), March 2015, Section 5.2, pp. 44-51

99 M. Adams, 2011, "Science and alpine grazing: politics and responsibility," *The Conversation*, 18 April 2011.

100 Emma Kelly, 2014, "ANU divests in seven resource and mining companies". *Canberra Times*, 3 October 2014.

101 Jotzo 2014, "Outrage at ANU divestments shows the power of its idea", *The Conversation*, 13 October 2014.

102 C. Adams, "Divestment backlash shows companies need to improve sustainability reporting", *The Conversation*, 17 October 2014.

103 Judith Sloan, 2014, "ANU divestment decision based on flawed, careless methodology", *The Australian*, 18 November 2014.

104 *The Conversation*, op. cit.

105 *The Conversation* op. cit., biographies of contributing authors Tom Baxter and Russell Warman.

106 A. Macintosh, "Tasmanian Forests Agreement: Liberal society needs an alternative", *The Conversation*, 10 May 2013.

A. Macintosh, 2013a, "Profits from forests? Leave the trees standing". *The Conversation*, 9 July 2013.

Both articles disclose that its author receives funding from a carbon offset company Forests Alive. This compay's website describes its projects as "the first within Australia to be accredited under the Verified Carbon Standard, and among the first VCS projects in the world in the field of avoided emissions by stopping native forest logging". The Managing Director of Forests Alive is Virginia Young, a well-known former Wilderness Society forests campaigner, who is described on the website as being instrumental in the *Green Carbon* research (by Mackey et al, 2008), which is described as underpinning the science of forest carbon offsets.

107 The Australia Institute describes itself as the country's 'most influential progressive think tank'. Its staff comprises a mixture of economists, former (mostly) Greens political advisors, and environmental activist group operatives. During Andrew Macintosh's time at the Australia Institute, Richard Dennis was its Executive Director. Dennis was formerly a political advisor to Greens Leader, Bob Brown, and the Institute was under his direction strongly focussed on climate change, mining and forestry. During that time, Andrew Macintosh and Richard Dennis either seperately or jointly authored up to a dozen 'research papers' and 'opinion pieces' about native forestry for the Institute – in most cases advocating the reduction or closure of the timber industry.

108 J. Hepburn, (Greenpeace Australia-Pacific), B. Burton (Coalswarm) and S. Hardy (Graeme Wood Foundation), *Stopping the Australian Coal Export Boom – Funding Proposal for the Australian Anti-Coal Movement* (unpublished), November 2011.

109 *Tasmania's Old Forests* – a half-page advertisement in *The Australian*, 15 September, 2004.

110 Sue Neales, 2010, "Push for new forestry rules", *The Mercury*, 16 March 2010.

111 Rowan D'Souza, 2006, "Academics keep left", *Institute of Public Affairs Review*, Vol. 58, No 2, pp. 14-15.

112 Howard Kurtz, 2005, "College faculties a most liberal lot, study finds", *The Washington Post*, 29 March 2005.

113 Melbourne Sustainable Society Institute, 2014, website (accessed December 2014)

114 Chris Dobney, 2014, "SCU public advocacy course under attack", *The Byron Shire Echo*, 21 October 2014.

115 North East Forest Alliance website.

116 Janet Hawley, 2003, "Trees amigos", *The Age Good Weekend*, 28 June 2003.

117 ZCA/BZE/MSSI 2014, *Land Use: Agriculture and Forestry Discussion Paper*, Zero Carbon Australia, Beyond Zero Emissions/Melbourne Sustainable Society Institute, The University of Melbourne, October 2014, http://bze.org.au/landuse

118 Wikipedia 2015, *Graeme Wood (businessman)* – accessed March 2015. Wood gave a $1.6 million donation to the Greens prior to the 2010 Federal election, and was more recently reported to have given a further $0.6 million to the Greens prior to the 2016 Federal election. Further to this, the Graham Wood Foundation was a funding partner of the joint strategy proposed to undermine the economics and social licence of the coal industry in 2011. The other funding partners were Greenpeace Australia-Pacific and Coalswarm.

119 J. van Tiggledon, 2014, "The destruction of the Triabunna mill and the fall of Tasmania's woodchip industry", *The Monthly*, July, 2014.

120 Dr Chris Taylor was formerly employed by The Wilderness Society, the Australian Conservation Foundation, and the Victorian National Parks Association to produce a substantial joint report about the 2009 Victorian Black Saturday bushfires. Subsequently in 2014, he was the lead author of a disputed research paper which alleged that timber harvesting exacerbated those bushfires, "Non-linear effects of stand age on fire severity" by Chris Taylor, Michael McCarthy and David Lindenmayer, *Conservation Letters*, Volume 7 Issue 4, 355-370 (July/August 2014). Since then he has promoted his view that logging increases bushfire severity and community risk in the media, on *The Conversation* website, and speaking at a Greens public meeting.

121 Declared by Dr Taylor at a Greens Public Meeting advocating the so-called 'Great Forest National Park' proposal at which he spoke alongside Bob Brown, 19 June 2016, several weeks before the Federal Election. As heard by Peter Fagg who attended the event.

122 R. Loyn, M. McFarlane, E. Chesterfield and J. Harris, 1980, "Forest Utilisation and the Flora and Fauna of the Boola Boola State Forest in South Eastern Victoria", Forests Commission Victoria, *Bulletin No. 28*.

123 F.R. Moulds, 1991, *The Dynamic Forest*, Lynedoch Publications, pp. 169-170.

124 State of Victoria, Department of Sustainability and Environment 2005, *Victoria's State of the Forests Report 2003*.

125 State of Victoria, Department of Sustainability and Environment 2009, *Victoria's State of the Forests Report 2008*, p. 33.

126 Victorian Department of Environment and Primary Industries 2014, *DEPI Science Strategy 2014-19*, p. 4

127 G.A. Kile, E.K.S. Nambier and A.G. Brown, 2014, "The rise and fall of research and development for the forest industry in Australia", *Australian Forestry*, Vol 77, Nos 3-4, pp. 142-152.

128 Ibid.

5

Politics and the bureaucracy: Implementing a cultural agenda

"The power and significance of the Ministers' private offices is too great. The political advice stream has grown at the expense of the traditional public service"

John Cain[1]

One afternoon in mid-April 2015, close to a hundred politicians and forest industry identities crowded into a meeting room in Victoria's Parliament House to mark the launch of a bi-partisan Forest and Wood Products Support Group. According to the launch invitation, the group would assist state parliamentarians "… who wish to work with industry and learn more about the importance of forest and wood products to their electorates and the Victorian economy".[2]

This was a reconstitution of a similar 'Friends of Forestry' parliamentary group which had operated under Victoria's previous Liberal Government. With a new Labor Government having been elected just six-months earlier, the state's timber industry leaders believed that reconvening this support group was critically important.

In short addresses to the gathering, both the responsible Government Minister and the Opposition's Shadow Minister lauded the socio-economic importance of a sustainable wood products industry. Both also acknowledged the threat to its future being posed by incessant lobbying for a proposed 'Great Forest National Park' that would supposedly save the endangered Leadbeater's possum. While most of the forests amd woodlands in which the possum occurs are already reserved in various ways, the new national park would 'lock-up' the remaining unreserved mountain ash forest, north east of Melbourne, upon which the state's hardwood industry was now largely reliant.

While the right things were said to reassure industry attendees that their contribution mattered and would be front and centre in any future government consideration of new national parks, few would have left the gathering without a gnawing apprehension. Memories of Labor's previous period in State government were still raw given its alarming predisposition to elevate political needs above practical forest management and socio-economic realities.

Within a week this apprehension would be heightened by the tenor of the Victorian Government's response to a highly questionable Federal Government decision to suddenly elevate Leadbeater's possum onto the 'critically endangered' list. Speaking on the ABC's RN Breakfast program, the Victorian Minister for Environment, Climate Change and Water, Lisa Neville, refused to rule out the possibility of closing most of Victoria's native hardwood timber industry to create the new national park being touted as the possums' saviour.[3]

Months later it would be revealed by Australian National University ecologists advocating an end to timber production supposedly to save Leadbeater's possum, that, immediately prior to the 2014 Victorian election, the then Labor Opposition had privately committed to declaring the 'Great Forest National Park' if elected. This was at odds with the formal policy they had taken to the election.

> In fact, immmediately before the last state election, Labor made a written commitment to one of us (David Lindenmayer) via formal email to establish the Great Forest National Park to secure populations of Leadbeater's possum. They have even set-up a Task Force to do just that.[4]

Doing backroom deals and making private promises is hardly uncommon in modern politics. However, it surely reflects badly on the state of a democracy when a political party campaigning for government hides a contentious policy rather than testing its public acceptance by openly taking it to an election. Further to this, privately committing to a policy largely to appease the views of some scientists and their research associates who mostly don't even reside in the State, have never managed forests, and have never publicly acknowledged that most of the forests in question are already in some way reserved, is surely a slap in the face to Victoria's land management agencies and the several thousand workers in the state's forestry and timber sector who would lose their jobs if such a policy was actually implemented.[5]

After taking office, the Andrews Labor Government made good on its promise to establish a 'Task Force'. Publicly at least, its Forest Industry Taskforce was somewhat less focussed on establishing the 'Great Forest National Park' than its supporters would have hoped.[6] However, two Government actions lend weight to the contention that the Taskforce was specifically established to manufacture an aura of legitimacy to support the pre-determined intention to establish the new national park. Firstly, the appointment of one of the nation's formerly most powerful environmental activists – Don Henry, recently retired leader of the Australian Conservation Foundation – to chair the Taskforce; and secondly, the prolonged delay in publicly releasing a commissioned socio-economic assessment of the regional timber industry reportedly because of concern that it would 'undermine' the Taskforce. That assessment by Deloitte Access Economics showed that the regional timber industry was a strong contributor to the economy and a direct employer of over 2,100 workers mostly in regional areas where alternative employment is scarce.[7]

Just eighteen months before the election of the Andrews Labor Government, an advisory group of government-employed scientists and industry representatives conceived under the previous Napthine Liberal Government had found that maintaining a sustainable timber industry was not incompatible with the conservation of Leadbeater's possum.[8] During the interim period before the incoming Labor Government established its Taskforce, a work program stemming from the advisory group's recommendations had made great strides in showing that the possum was far less rare and far more resilient than had been previously assumed.[9] As the results from this work were clearly undermining the campaign for a new national park, the Greens and their environmental group associates had tried to discredit it while redoubling their efforts to reassure the Labor Opposition of the need for the new park despite the increasing evidence that it was unnecessary. That an intention to declare the new park still became an (albeit undisclosed) Labor Party policy attests to the success of their efforts.

While the past treatment of forestry issues has ostensibly been shaped by pragmatic economic and/or environmental needs, the enthusiasm for creating national parks exhibited by most state Labor Governments since the mid-1990s is suggestive of a deeper conservation ideology hardwired into the Labor Party's DNA. Arguably, if not for the moderating influence of trade unions keen to maintain jobs, Labor's record on forests could well be barely distinguishable from the wish-lists of the environmentally-zealous Greens and their associates within the environmental movement.

This is not so surprising given that the environmental movement has spent decades fostering strong links with the Australian Labor Party (the ALP). This strategy had been instigated in 1983 following interventions by the Federal and NSW Labor Governments respectively in relation to Tasmania's Franklin River and the rainforests of northern NSW. To the environmental movement, these actions had tagged the ALP as the preferred political vehicle for realising its various conservation agendas.[10]

More than thirty years later, the continued influence of environmental activism on ALP thinking is exemplified by the existence of a Labor Environment Action Network (LEAN) which reportedly mounts aggressive grassroots campaigns within the party to shape its environmental policies. Established in 2004 as an internal cross-factional party organisation, it was (in 2016) being run by Felicity Wade, a former Wilderness Society campaigner and partner of Wilderness Society national campaigns director, Lyndon Schnieders. The organisation reportedly enjoys heavyweight patronage, including former ALP National President turned Senator, Jenny McAllister, and former Federal Environment Minister, Tony Burke.[11]

LEAN has undoubtedly played a leading role in shaping Labor's attitude towards forestry, and it is instructive that former Victorian Labor Premier, Steve Bracks, revealed in an ABC Radio interview promoting his 2012 autobiography that, in his view, Victorian Labor had an unofficial policy to end all timber production in Victoria's public native forests. He went on to say that more should be done in this area to add to his Government's legacy of creating "more national forest and marine parks ... than any other Government in Victoria's history".[12]

This admission fits with the track record of successive Victorian Labor Governments led by Bracks and John Brumby from 1998 to 2010. During this period, the area of Victorian State forest available for timber harvesting was substantially reduced and tens of millions of dollars was spent on buying-back sawlog licences to facilitate the declaration of new national parks in the state's south west, north and east. Over roughly the same period, Labor Governments in NSW, Queensland, WA, and more latterly Tasmania, also presided over a collectively huge expansion of forested national parks and other conservation reserves that has decimated multiple-use forestry and severely curtailed native forest wood production in southern and eastern Australia.

Labor's particular fondness for declaring new national parks is suggestive of an unshakable belief that 1) new parks are very popular amongst the electorate; 2) they are the best form of conservation; 3) that there are few conservation values in non-park public land tenures, such as State forests; and 4) that the community benefit of national parks outweighs the costs of closing or downsizing rural industries. While there is little doubt that national park expansion is popular – particularly amongst the inner-urban demographic – the latter three presumptions are either wrong or highly contentious. Yet their adoption by politicians (including some of all persuasions) arguably shows the extent to which hearts and minds have been influenced by decades of rarely challenged environmental propaganda. This is an uncomfortable reality lost on political luminaries such as former NSW Premier, Bob Carr, who recently nominated the 350 national parks created during his 10-year reign as amongst his greatest achievements.[13]

As a result of recent national park expansion in Australia's southern and eastern states, thousands of forestry and timber industry workers have lost their livelihoods. Some of this can be justified by the need to address genuine conservation concerns or to conduct necessary industry reforms. In some cases in Victoria, NSW and Tasmania, annual timber harvests needed to be wound-back to compensate for faster-than-sustainable harvesting of designated wood production forests (often diminished by reservation), or to counter the loss of available timber resources in huge mega-fires. Often these downward harvest corrections have then been seized upon by Governments as opportunities for national park expansion that has then either closed the timber industry outright, or squeezed it into a much reduced area of State forest thereby recreating the conditions for a repeat of the initial problem.

However, much of the recent national park expansion can't be as easily justified by rational imperatives and can largely only be categorised as politically expedient acquiescence to incessant campaigning against resource uses by the environmental lobby and its Greens party political arm. In these cases, Labor's treatment of forests as a veritable cache of political capital to be periodically cashed-in for 'green' votes, has ensured that often little or no serious consideration has been given to relevant facts about the actual environmental impact of activities such as timber production, or the consequences of ending them.

The most significant consequence has been losing a revenue generating activity along with its associated workforce and expertise which has unquestionably weakened the capability to effectively manage the much broader forest estate that exists beyond the wood production zones – especially with regards to fire. While national park expansion is routinely credited with huge conservation benefits, these have in reality been either minimised or at least counteracted by a consequentially weakened fire management capacity.

Irrespective of actual conservation outcomes, the garnering of crucial 'green' votes from Australia's inner urban demographic was integral to the lengthy reign of Labor State Governments in most of the country's southern and eastern states for the approximately fifteen-year period from the late 1990s to the early 2010s. In contrast to Labor, Liberal or Coalition Governments have never been as focussed on 'saving' forests. What forest policy reforms they have enacted have tended to be underpinned by the central proviso of maintaining a reasonable balance between commercial resource use and conservation. While Labor has at times articulated a similar aim, their actions have typically facilitated an unbalanced position whereby most or all targetted forests are reserved for conservation with only minimal or no commercial uses being maintained.

The dual aims of Liberal/Coalition Governments are perhaps best exemplified by the development of Regional Forest Agreements (RFAs) under the Howard Government during the mid to late 1990s. The RFA process was defined as producing "20-year plans for the conservation and sustainable management of Australia's native forests" with the broad aim to "... provide certainty for forest-based industries, forest-dependent communities and conservation".[14] A more recent example was the Leadbeater's Possum Advisory Group process overseen by Victoria's Napthine Government in 2013 in accordance with terms of reference requiring it to recommend "medium and longer-term actions focused on ensuring the persistence of the species and its coexistence with a sustainable timber industry".[15]

Despite such processes having delivered considerable conservation gains, they've been routinely vilified by environmentalists and the Greens for giving legitimacy to the continuance of native forest wood production. While increased reservations resulting from these processes have also undoubtedly taken some toll on the livelihoods of forestry and timber workers, they have at least been founded upon a belief in the sanctity of multiple-use forestry and science-based initiatives such as nationally-agreed conservation criteria, as opposed to being instituted for purely political motives.[16]

Unlike Labor, the Liberal/Coalition has rarely surrendered to the temptation to close timber industries to create more national parks simply to improve their electoral prospects. The contrasting propensity for Labor State governments to take electoral advantage of anti-forestry populism has simply encouraged environmental activists to intensify their 'whatever-it-takes' mode of lobbying to further engender adverse public sentiment about forest uses, particularly during the lead-up to State elections.

The critical role of this in shaping public opinion and (subsequently) driving forest policy change was noted by Greg L'Estrange, then CEO of Tasmanian timber giant, Gunns Ltd, in his controversial keynote address to a Forestworks industry conference in September 2010. In using this address to announce his company's decision to cease harvesting native forests and divest its hardwood sawmilling assets, L'Estrange acknowledged that the timber industry and the environmental movement were "pitted in a fierce battle for the support of the Australian people, who in turn balance the political debate", and admitted that "we (*the forestry sector*) have lost the public debate and the support of the broader community". He went on to say that:

> ... the industry has been out-thought and out-played, with the ENGO's [environmental non-government organisations] using three key leverage points: public emotion, multi-level government involvement; and certification – market action. Whilst the industry has maintained a stance that science will prevail ...[17]

While few in the forestry sector would dispute the broad thrust of his summation, most would be troubled by its inference that the environmental movement has "won" while the forestry sector has "lost". While this infers a fair debate, the reality is that environmental activists have run deceitful campaigns misrepresenting the scale and extent of forest use and its impacts, while strategically ignoring the already favourable balance between it and conservation. They've also been substantially assisted by much of the city-based media which effectively promotes 'green' causes almost without question, whilst treating opposing views with suspicion thereby granting them comparatively little oxygen.

The forestry sector has been inclined to beat-itself-up over its inability to get equal media treatment. However, it has been pitted against an environmental group collective that includes wealthy, professional corporations whose very existence relies on promoting environmental conflict through the media to maintain a flow of funding from outraged donors. Accordingly they have an imperative and the resources to devote far more time and effort to fostering relationships with the media. By comparison, forestry and timber sector associations or companies are either reliant on voluntary efforts or able to devote only relatively small budgets to public relations.

Most importantly, the activists have a far simpler and more marketable message of threatened eco-destruction that appeals to both the media's need for sensationalism and it's craving to be a popularly acclaimed agent for public good. Effectively counteracting this is almost impossible in an era where the media is more pre-disposed to superficial and simplistic sloganeering rather than complex scientific explanations of why and how forests are managed.

That environmental campaigns are continuing to push timber production out of native forests is hardly something that Australian society should be proud of given that they are typically small-scale, renewable activities which produce essential materials while posing a negligible environmental threat at the landscape scale. Nevertheless, the celebrations which greet each announced sawmill closure or new national park betrays a dominant but disconnected urban community that has been exposed to one-sided portrayals for so long that it has lost perspective on rural issues.

Notwithstanding the value of initiatives like Victoria's bi-partisan Forestry and Wood Products Parliamentary Support Group, the nation's political class, dominated as it is by urban interests, also collectively lacks perspective on rural issues. In particular, the Labor Party has been demonstrably less concerned about how the community comes to embrace popular environmental causes than by the electoral implications of supporting or opposing them.

Politicising the bureaucracy

In the now distant past, government policy was based on the wise counsel of senior bureaucrats supported by government agency scientists with a thorough understanding of rational resource use and land management. There was also a clear separation between politics and bureaucracy that enabled the purity of expert advice to be maintained through the inevitable cycle of political change as governments of a different ideological hue came and went.

Over time, the bureaucracy has been politicised to an extent that it is now commonplace for senior bureaucrats heading-up public service departments to be immediately dismissed upon a change of government. The appointment of new bureaucrats to realign those departments to the ideology of the incoming government has been typically accompanied by departmental rebranding and reoganisations with associated staff purges that have progressively taken their toll. There are now real questions about whether state and federal bureaucracies still have the requisite experience and knowledge to guide sensible policy development, notwithstanding concern about the extent to which their advice is even required given the propensity for today's parliamentarians to rely on political advisers more concerned with short-term electoral implications than technical needs.[18]

With regard to public land management, Victoria may well have been the worst affected state judging by the reflections of recently retired senior government forester, Peter McHugh:

> I can count seven major name changes after the Forests Commission was gobbled-up into Conservation Forests and Lands back in 1983 ... Since then we've had DCE, CNR, NRE, DPI, DSE, DEPI and finally DELWP. I can't remember exactly how many other minor internal changes and convoluted job titles have been wrought but I seem to have accumulated a large collection of name tags and business cards.
>
> It was sometimes very difficult to remain buoyant while getting used to yet another tongue twister acronym. We became the butt of so many tiresome jokes like CNR = Constant Name Review, DSE = Dept. of Sparks & Embers while NRE became No Rational Explanation. But for most bushies all the corporate branding was lost on them because as far as they were concerned I still worked for *"The Forestry"*.
>
> And in my experience nothing upsets staff and consumes more emotional energy than endless tea room chatter, speculation and worry over which box in the latest restructure diagram they may occupy. Needlessly shuffling office accommodation, reducing vehicles, new corporate uniforms, stationary and logos all had the same effect.

> The loss of corporate memory and reduced depth of practical field experience in combination with high managerial turnover gives me reason for concern for the future of forestry and fire. There now seems to be an abundance of data and information but is there enough knowledge and wisdom?[19,20]

The staff disruption associated with these frequent reorganisation and rebranding exercises has been cited as a significant reason for the steady decline in the annual rate of fuel reduction burning in Victoria from a mid-1980s peak to the low depths plumbed immediately prior to the devastating 'Black Saturday' bushfires of 2009.[21]

Further to this, the regularity of agency reorganisations has invariably changed career pathways within the bureaucracy. It was once almost a prerequisite for senior bureaucrats to have risen through the ranks gaining along the way an intimate understanding of what they were managing. This has now been supplanted by a system of inter-changeable senior administrators with broad governance experience but far less technical understanding of what they are responsible for.

This was all-too-evident in mid-2016 when Victoria's Department of Environment, Land, Water and Planning announced the 'Executive Leadership Team' of its newly created Forest Fire and Regions Group. Amongst the five newly appointed Regional Directors with responsibility for forests and fire were several with extensive local government experience, including CEO and senior executive roles in up to five different municipalities. One who had overseen major projects such as the redevelopment of a major tourist resort, establishment of a wind farm, and duplication of a highway; and another who was responsible for a business plan which had developed 'the most livable regional city in Australia'. Another appointee was a qualified social worker with a Bachelor of Environmental Horticulture who had formerly worked in such diverse roles as a State Director of the Australian Agency for International Development (AusAid) and as the Assistant Director of Settlement and Ethnic Affairs for the Department of Immigration and Ethnic Affairs.[22]

That the brief biographies of these new appointees trumpeted their lack of knowledge about forests and fire as though it were a virtue was exemplified by the brief bio of the only trained forester amongst the group which did not even include this fact. Instead, she was lauded for "working collaboratively with a wide range of stakeholders and local communities on priority projects such as removal of asbestos from the Cheetham Saltworks site and the Regional Coastal Plans."[23]

All may be experienced administrators and leaders, but their collective lack of knowledge about what they are now responsible for gives little confidence in their capability to provide meaningful input or oversight over forest policy development. Furthermore, the appointment of leaders with so little technical knowledge surely has potential for serious public embarrassment, such as for example, the spectacle of the then Secretary of the former Department of Sustainability and Environment being unable to explain the rationale for his department's planned burning targets during the 2009 Victorian Bushfires Royal Commission.[24]

Presumably in-part because of the lack of expertise amongst senior bureaucrats, forest policy has for some time been formulated with an over-reliance on the advice of 'spin doctors', pollsters, and unknowing and unaccountable political advisers more concerned with the implications for gaining or retaining power. Under Labor, this may have been further compounded in more recent times by the need to be seen to be doing even more for the environment or risk losing a substantial part of its voter base to the Greens, which has made no secret of its intention to eventually usurp Labor as one of the nation's 'big two' political parties.

Reportedly, the dominant influence of political apparatchiks stems back to the Whitlam era when Government Ministers were granted leave to engage a larger personal entourage of advisors and media managers. This reform was also enthusiastically embraced by the subsequent Fraser Government and has been supported and expanded by both sides of politics ever since.[25]

The numbers of public servants now employed in such roles under Governments of either political persuasion has jumped to extraordinary levels. In early 2016, it was reported that Victoria's Andrews Labor Government was employing 942 personnel in communications and media roles within the bureaucracy, as well as a further 164 political advisors and media managers specifically assigned to Government MPs at an overall cost of $110 million per annum. Reportedly, this was twice the number of ambulance paramedics and three times the number of highway patrol police being employed in Victoria at the time.[26]

The primary role of these advisors and communicators has evolved into maintaining and reinforcing a positive Government image, including a particular determination to avoid or minimise any adverse publicity adjudged as having potential to inflict political damage. Their consequent determination to control the Government message has led to a greater involvement of parliamentarians and their unqualified advisors in decision-making that was once far more appropriately dealt with by Government agencies and their trained and experienced staff.

This increased reliance on political advice has somewhat displaced scientific knowledge, considered planning, and practical local experience as the driver of public land management. It may also have created a lightning rod for minority protest groups which have come to realise that directly complaining to the responsible Minister or his/her political advisor can often get their demands far more quickly and satisfactorily addressed than having to negotiate with government agencies as was once more typical of land management disputes.

In reality, direct Ministerial interference in forest management (or other areas for that matter) is nothing new and there are many past examples where government field foresters have felt sensible forest management to have been somewhat compromised by, for example, political directives spawned from complaints by disgruntled sawmillers regarding sawlog supplies. The pendulum has long ago swung the other way, and it is now environmental activist groups which are more likely to gain by appealing directly to the highest political authority.

For example, during the last two years, complaints by local environmental activists have led Victoria's Minister for the Environment to order the suspension of scheduled regeneration burns in harvested coupes in the Toolangi forests, north of Melbourne. On one of these occasions, this Ministerial intervention reportedly led to tens of thousands of dollars in additional government expenditure as the opportunity to burn was lost thereby necessitating the hire of earth-moving machinery to create a suitably disturbed receptive seed bed. Another notable instance was the postponing of planned fuel reduction burns in the Otway Ranges in 2015 (and reportedly also on earlier occasions) apparently due to concerns about the safety of resident koalas. This came to light following the devastating Wye River bushfire which started in late 2015 and ultimately burnt through the same country.[27]

Direct political intervention is problematic for informed decision-making because it is motivated by extraneous factors and personal biases. However, the now routinely close relationship between Government Ministers and public service department heads somewhat obscures the level of political input into bureaucratic decision-making. An example is the propensity for government agencies to give-in to the often exaggerated claims of environmental activists whenever there is a whiff of adverse publicity. To what extent is such a decision arrived at by the bureaucracy or directed by the responsible Minister?

A Victorian example from 2015 was the campaign by the Euroa Environment Group (EEG) that ultimately led to the Department of Environment, Land, Water and Planning (DELWP) abandoning a planned 3,300 hectare fuel reduction burn in the Strathbogie Ranges, north of Melbourne. Invariably, surrendering to unreasonable demands solves nothing and only emboldens the protestors to embark on further campaigns. In the wake of the burn's cancellation, the EEG embarked on a new campaign against a small 27 hectare selective timber harvesting operation planned by VicForests for the same area. As EEG spokesman, Bert Lobert, exclaimed:

> This is dogged, determined, organised people power. We live, work and play in and around the forest and we aim to see that forest get a better deal than the one hatched out in the CBD somewhere. We took on DELWP over planned burning and reason won out; watch out VicForests.[28]

In Victoria, the political determination to avoid controversy was also evident as far back as 2006 when the Department of Sustainability and Environment was prevented from allowing three of its officers to present the Norman Wettenhall Memorial Lecture on fire management on the eve of a State election.[29] Similarly, the government's commercial forestry agency, VicForests, did not respond to regular criticism of native forest timber production during the 2006 Victorian election campaign and even elected to avoid the potential for unwanted attention by scheduling its major symposium on the 'big issues facing Australian forestry and forest industries' for the week after the election. For keeping its head down to avoid controversy that (it had been led to believe) could damage the re-election prospects of its political masters, VicForests gained nothing except perhaps some relief that the government's cuts to timber resource accessibility in East Gippsland could have been far worse.

A more recent example of politically-motivated controversy avoidance is the Victorian Government's behaviour during and since the 2015 Wye River bushfire. Soon after the fire had destroyed 116 houses, the Institute of Foresters of Australia's Victorian Division (the IFA) called for a public inquiry into the fire suppression strategy.[30] Most experienced fire-fighters supported this call because the fire – which began as a small lightning strike in relatively accessible country – had been unable to be controlled during its first five days despite burning only slowly under benign weather conditions. Eventually hot, windy conditions on the sixth day carried it across poorly secured containment lines into the small coastal settlements of Wye River and Separation Creek where it did all the property damage.

After the IFA's call for an inquiry appeared on the front page of *The Age* newspaper, senior Victorian emergency services bureaucrat/s lobbied the Australasian Fire and Emergency Service Authorities Council (AFAC) to cancel its contract with the IFA's Victorian Division Chairman (a voluntary position) who they were employing in a permanent part-time role.[31] This appears to have been an attempt to silence a critic (of which there were many others) presumably as part of a political directive, and perhaps also to protect the personal standing of the responsible bureaucrat/s.

Ultimately in the face of mounting public concerns about the fire, the Victorian Government was compelled to direct its Inspector-General for

Emergency Management (IEGM) to conduct a review of the fire even as it continued to burn out-of-control. Again though, this review seemed to be deliberately designed to avoid any exposure of problems with the fire-fighting strategy through assigning terms of reference which did not meet the test of a truly independent review; did not include any contribution from an independent expert in forest fire management; and did not consider other circumstances within the IGEM's charter which may have affected the conduct of the fire-fight.[32]

These deficiencies were in-part justified by deferring to a future coronial inquiry which the Government had also agreed to in January 2016. However, this inquiry never eventuated and the Victorian Coroner ultimately released a 30-page determination in September 2017 advising that it was not necessary to investigate the fire.[33] One of the reasons given for this ruling was "that there is no legitimate coronial purpose that is likely to be served by holding a public hearing in this matter".

This ruling means that critically important allegations about the conduct of the fire will never be formally examined. Reportedly, these include the initial appointment of a fire controller with no expertise in forest fire control, and a Ministerial intervention which delayed a critically important back-burn thereby virtually consigning it to fail due to the onset of dangerous weather conditions. The failure of this back-burn led to the fire decimating the Wye River and Separation Creek hamlets. Far from being unworthy of formal investigation, most experienced fire management practitioners fear that if the Wye River fire-fight is allowed to exemplify a new normal for forest fire-fighting standards, the Coroner will be required to inquire into fire far more in the future than ever before.

The significance of the Wye River fire is that (excluding the value of house losses) it arguably represents Australia's most expensive ever forest fire-fight (on a per hectare basis) given that just 2,500 hectares was burnt over a 33-day suppression campaign typified by substantial use of costly water bombing aircraft in support of large numbers of ground-based fire-fighters. As the fire was not in a remote location and was mostly subject to mild weather, the damage that it wrought – the loss of valuable private holiday real estate and a substantial economic hit to seasonal tourism along the iconic Great Ocean Road – appears to have been eminently preventable. After one bad day occurred six days after it had started, the fire continued to burn only very slowly for the further 27 days that it took to contain it.[34]

Despite the shortcomings associated with the Wye River fire-fighting campaign, the Victorian Government has continued to insist that the fire was a "success" based on the effective evacuation of the two affected holiday settlements without any loss of life. While that was undoubtedly a positive outcome, the continual reference to it seems to have been orchestrated to obscure the problems inherent to the fire-fighting strategy which necessitated it. This has only added weight to the contention that there is much to hide.

In Tasmania, the prominent role played by then Federal Labor Environment Minister, Tony Burke, in cajoling and ultimately forcing the 'Tasmanian forests peace deal' in 2013 also raised eyebrows. In particular a 'midnight dash' from Canberra to Hobart to offer a hundred million dollars of Federal Government money to force a deal just as negotiators had agreed to walk away with any prospect of a result thought to have been lost.[35] Minister Burke would subsequently play the leading role in lobbying the United Nation's World Heritage Committee to add 170,000 hectares of Tasmanian forests to the already existing 1.4 million hectares of Tasmanian Wilderness World Heritage Area in 2013. This included a last minute commitment of $500,000 of taxpayer's money to fund a cultural heritage study whose absence would normally invalidate such an application.[36]

Burke's behaviour in these processes suggests that he was motivated by far more than an obligation to party political agendas. Instead, his willingness to go above and beyond was suggestive of a personal conviction that befits his reported background as an environmental activist in his younger years, albeit that the extent of his activism and in what capacity he acted has never been disclosed. Certainly his zeal and apparent determination to leave a legacy by facilitating the end of Tasmania's decades-long forests' conflict, as well as creating huge new marine national parks, is indicative of such a background.

Further to this he would later gain notoriety amongst fisheries management experts by mishandling the so-called 'super trawler' episode where he initially approved of an application to allow one of the world's largest commercial fishing boats to catch the allocated east coast quota of jack mackerel, only to subsequently change his mind in the face of largely irrational hysteria whipped-up by environmental activists and recreational fisherman.[37]

Indeed, Burke's performance under the Gillard Government from 2010 – 13 created some speculation that he had redefined the role of Federal Environment Minister from vetting and potentially amending development proposals, to being a far more black-and-white arbiter of whether proposals should or shouldn't proceed, and a proactive enabler of environmental activist agendas. However to put Burke's role into context it should be remembered that he was also intensely vilified by the Greens and their environmentalist cohorts for refusing to nominate Tasmania's so-called 'Tarkine' region for National Heritage listing, reportedly due to the importance of the area for mining.[38]

Politically expedient forest policy

The politically expedient nature of most proposed and actual changes to forest policy over the past 15-years is evident from their election-eve timing. Prominent examples have been competing Tasmanian forests policies announced by Labor and the Coalition just days before the 2004 Federal election; the Victorian government's commitment to end timber production in the Otway Ranges announced three weeks before the 2002 state election, and its commitment to create new forested national parks in East Gippsland just a week before the 2006 election. In NSW there was the promise to incorporate 15 'icon' wood production forests totalling 65,000 ha into national parks in north eastern NSW made prior to the 2003 election.[39] In Western Australia's 2001 election, the Labor opposition promised to create 30 new national parks, including the 12 which had already been agreed to under the Regional

Forests Agreement process, and immediately acted on these commitments once elected.[40]

More recently during the 2015 NSW election campaign, Labor Opposition leader, Luke Foley, promised to create a 315,000 hectare koala preserve in the state's mid-north coast forests should he be elected as premier.[41] This promise was made without consulting local communities and would reportedly have caused a significant loss of forestry and timber industry jobs whilst doing little to help the koala which is principally under threat from habitat loss for urban development, disease, wildfire, and dog attacks.[42]

Indeed, some of the most blatant examples of science and common sense being overridden by political imperatives have involved public announcements of new national parks which have either preceded or subverted the bureaucracy's due process of objectively assessing the potential ramifications of changing public land tenures. For example in 2002, the Victorian Government only directed its environmental assessment agency – the Victorian Environmental Assessment Council (VEAC) – to undertake an investigation of public land use in the Otway Ranges after it had already announced the end of timber harvesting to create a huge new national park. Similarly in 2006, the Victorian Government announced that forests in East Gippsland would be reserved in new national parks even though VEAC was only part-way through an investigation into that very question.[43]

Even where government environmental assessment bodies – such as VEAC and the Natural Resources Commission (NRC) in NSW – have been properly directed to investigate proposed changes to public land tenure, their work has often been tainted by a perception of political interference. As their investigations have invariably been initiated in response to environmental activism there is undoubtedly a covert pressure to produce outcomes that reward it.

It has also become apparent that State governments are unwilling to undertake studies that may cost millions of dollars without them recommending changes to the status quo. Almost without exception these public land use investigations have recommended substantially more State forests becoming new national parks irrespective of whether there is any objective environmental justification for it, or whether it will unfairly damage the socio-economic fabric of nearby communities.

In NSW, a recent Parliamentary inquiry into public land management noted that many participants had "raised concerns that the [national park] conversion process had been politicised" and questioned whether "the economic and social values of an area have been equally considered in this process" This view was even shared by the NSW National Parks and Wildlife Advisory Council which called for "greater consideration of economic and social issues … during the assessment process when new lands are being added to the reserve system".[44]

Of the above-listed examples of politically-expedient forest policy appropriations, only the 2004 Federal election defied the trend of electoral advantage for Labor by touting for 'green votes'. Just days before the election, then Labor Opposition leader, Mark Latham, adopted the Greens' policy of closing Tasmania's native forest timber industry. This outraged many Tasmanians who saw it as pandering to mainland city voters and the unwarranted trashing of an industry crucial to the state economy. In response, the Coalition under incumbent Prime Minister, John Howard, theatrically committed to virtually maintaining the status quo on timber resource access in accordance with the 1997 Tasmanian Regional Forest Agreement. The resultant Liberal gain of two of Tasmania's five Labor-held seats was widely credited as a critical determinant of the Coalition's subsequent election victory.

In the aftermath of the 2004 Federal election, a short few years of relative stability ensued under a Coalition Government. During this time there was a concerted push within the Federal Labor caucus to re-establish connections with its Tasmanian blue-collar base, including the native forest timber industry. This persisted until the lead-up to the 2007 Federal election when new Labor leader, Kevin Rudd, committed to a new Tasmanian forests policy that would supposedly protect jobs whilst conserving more 'environmentally-sensitive forest areas' presumably as a nod to the party's inner-city constituency.[45]

The difficulty of achieving both aims raised suspicions amongst the timber industry and its supporters, although the influential Construction, Forestry and Mining and Energy Union (CFMEU) supported Rudd's new policy after gaining assurances as to its intent.[46] While Tasmanian Labor Premier, Paul Lennon, acknowledged that these dual aims could be achievable if the Federal Government could fund appropriate value-adding timber projects, he strongly reiterated his opposition to any further forest preservation promises which put forward Tasmanian timber workers as 'election bait' as had happened in the past.[47]

The determination of the Tasmanian Labor Government to maintain its timber industry arguably highlights a traditional difference between it and the attitude to forests of its mainland State Labor counterparts presumably because Tasmania's timber industry has historically made a comparatively greater socio-economic contribution to the state. However, this changed after the Tasmanian election in March 2010, when Labor was forced into a governing alliance with the Greens. In return for their support in enabling a minority Labor government to be formed, the Greens made political demands which undoubtably included renewed efforts to close or substantially reduce Tasmania's native forest timber industry.

Just five months later at the Federal level, Labor's Julia Gillard also acquiesced to the demands of the Greens and three independents in order to form a minority government after the August 2010 election had resulted in a 'hung parliament'. This involved a formal signed agreement between Gillard and then Greens Leader, Bob Brown, committing to legislative undertakings on a range of environmental issues. Although its contents have never been publicly disclosed, it would almost certainly have included provisions to 'preserve' Tasmania's forests given Brown's career-long determination to 'save' trees in his home state.

In late 2010, a process which became popularly known as the 'Tasmanian forests peace deal' was instigated. Although this was a protracted and painstaking series of negotiations between timber industry and environmental group representatives, it was effectively cheered-on by both the Tasmanian and Federal Labor governments in accord with their commitments to Greens political allies. The ensuing two-and-a-half year process effectively outsourced a government policy function to two self-interested and highly polarised groups and thereby ignored a wide range of other community stakeholders.[48]

The rising influence of the Greens has also been noted at local government level which, particularly in the inner-city and suburbs, has effectively become a training ground and platform for aspiring Greens politicians seeking to advance a range of conservation and social justice agendas which have minimal or no direct relevance to municipal plans and services. Well publicised causes advocated by some local governments include changing the date of Australia Day, advocating same-sex marriage, and calling for improved refugee welfare.

In some cases urban-based municipal councils have misdirected efforts towards trying to influence the management of public forests for which they have zero responsibility. In an attempt to justify this, some municipalities have developed policies on State forest timber harvesting and fuel reduction burning on the grounds that these activities may create health issues for their rate-payers by supposedly damaging water supplies or polluting the air.[49] Others have instigated bans on the procurement of wood products sourced from native forests for use in municipal works in a further attempt to damage hardwood industries and reduce or eliminate the need for forest harvesting.[50] For example in late 2016, the Banyule City Council which occupies part of Melbourne's mid-north eastern suburbs voted to sign the Wilderness Society's Ethical Paper Pledge and passed a motion supporting the 'Great Forest National Park' proposal.[51] Both initiatives are intended to cripple a regional Victorian timber industry in which Banyule's rate-payers have no stake.

While the proposed national park has no direct relevance to a municipality located in Melbourne's inner suburbs, the Banyule council's signing of the Ethical Paper Pledge committed the municipality to sourcing its paper needs wholly from recycled fibre. In due course it selected a brand of paper imported approximately 16,000 km from Austria.[52] Media coverage of Banyule's decision to boycott Australia's only domestic paper manufacturer (despite it also producing a recycled paper product) was damning, and prompted a public plea by the local MP – Victoria's Opposition Leader, Matthew Guy – for Banyule City Council to display more common sense.[53]

The reduced influence of professional forestry with the bureaucracy

Australia-wide concerns about the capability of forest management bureaucracies are rooted in the decline of forest industries generally, including the recent creation of narrowly-focussed government commercial forestry agencies. The creation of these agencies such as the Forests Products Commission in WA, the Forestry Corporation of NSW, and VicForests in Victoria, has precipitated a transfer or loss of practical forestry experience away from significant involvement in non-commercial public land management functions, including the critically important landscape fire management function.

There is a perception that state land management bureaucracies have largely replaced this lost forestry knowledge and practical experience with a new generation of conservation scientists, administrators and field managers with less experience and lower enthusiasm for key forest management practices which had evolved from hard-won experience over the previous century. This includes reduced support for public forest timber harvesting and fuel reduction burning than was evident in the past. When considered in conjunction with other changed societal and demographic factors, this cultural change is a significant factor in why we are experiencing worsening environmental outcomes due to a weakened capability to control bushfires.

Arguably, the problem is best exemplified by the Victorian experience where the public land management agency (formerly the Department of Sustainability and Environment) traditionally had overall responsibility for all forest management activities, including commercial uses. However in 2004, the Government separated commercial timber production from broad public land management functions by creating a separate agency, VicForests, specifically to manage only timber production coupes for the several years needed to harvest and then successfully regenerate them. As the Department's most experienced operational forestry practitioners were traditionally involved in timber harvesting and associated road construction/maintenance and silvicultural activities (such as regeneration burning), they mostly ended-up working for VicForests.

This left the Department still retaining overall responsibility for forest management across the whole landscape (including oversight of VicForests environmental performance through regular audits) but lacking its formerly most experienced operational practitioners whose shift to VicForests had taken them away from broader landscape management functions.

The problems this has created are still evident in the current Department of Environment, Lands, Water and Planning's (DELWP) lack of forestry expertise and only lukewarm enthuasism for timber harvesting despite being required to oversee the compliance of VicForests' operations to the regulatory framework. Reportedly in late 2017, only one member of DELWP's forestry compliance unit has any practical timber production experience. This apparently manifests itself in a confrontational attitude to VicForests over alleged minor transgressions of timber harvesting regulations. Whereas in the past such incidences would have been resolved by on-site discussion and mutually agreed actions, there is now the perverse prospect of a government suing itself whenever DELWP threatens legal action against VicForests.

The reduced influence of professional foresters in forest management bureaucracies is also a reflection of the extent to which the environmental lobby has progressively achieved various conservation agendas through infiltrating land use policy processes at both state and local government levels. This has arguably been assisted by legislated commitments to community consultation that have enabled some environmental groups to develop close, influential affiliations with State Ministers and their advisors, or in some cases, government departments that best represent their interests.

For example in Victoria, the Bracks Labor Government established an Environment Liaison Office (ELO) in 1999 specifically to enable five environmental groups to speak directly to the government and opposition with a single unified voice.[54] In NSW, the Environmental Liaison

Office represents an alliance of nine leading environmental groups with the aim of ensuring their shared concerns and priorities are communicated consistently to Members of Parliament. The ELO monitors parliamentary debates, inquiries and legislation, and produces a regular summary of parliamentary debates relating to environmental and natural resource management issues.[55]

There has been some disquiet that the direct access given to such unaccountable groups unfairly exceeds that afforded to other stakeholders, such as the timber industry, whose aspirations to similar levels of State government influence have become more problematic since forestry issues have become politicised.[56] It may also at times have elevated the concerns of environmental activists, which are often ill-informed or misguided, above the knowledge of the trained and far better informed forest management personnel within the bureaucracy.

In addition, policy processes that invite public input have provided opportunities for environmental lobby groups to mobilise mass support for their agendas through standard pro-forma submissions that make it easy for their members and supporters to proffer their views without the hard-work and research usually required to prepare an informed submission. This is a worldwide phenomenon that has reportedly reached unprecedented levels in Australia raising concerns that it enables unelected and unaccountable city-based groups to distort the proportional diversity of public interest in rural issues and create pressure to override the scientific and practical basis of land management.

There are many examples of the forestry profession being marginalised or excluded from government land management processes in which they have special expertise. In Western Australia, the 2001 election of a Labor government coincided with the increased influence of the environmental movement on government land management policy. Those trained and experienced in managing native forests were marginalised as forest planning was consigned to a Conservation Commission comprising a Chairman, who was an environmental law academic, and citizen representatives lacking any experience in practical forest management. For several years the main source of forest management advice to the Minister was a 'Round Table on Forestry' which did not include a forester or anyone with forestry experience.[57]

In NSW, professional forestry expertise was excluded from a Community Reference Panel convened to examine how sustainable harvesting and regeneration of private native forests should be treated by the *Native Vegetation Act 2003*. Without forestry input, the panel incredibly advised the relevant Minister that cyclical timber harvesting and regeneration should be classed as 'clearing', thereby undermining the legitimacy of private landowners practicing renewable forestry on properties that may have been managed that way for generations.[58]

In a related move, the NSW Government gave the task of developing a private native forestry Code of Practice to a Natural Resources Advisory Council without any input from forest owners amidst a suspicion that it would be overly influenced by environmental activists with no direct stake in its outcome.[59] The Draft Code of Practice for Private Native Forestry subsequently released for public comment in mid-2006 created outrage amidst claims that its overly zealous bureaucratic and environmental protection requirements would place 60% of NSW private native forests in reserves causing sawmill closures and substantial job losses.[60] Eventually, in the face of sustained protest, the NSW government shelved it in late August 2006 in what was then a rare set-back for 'green' ideology.[61]

In Victoria, there has been a history of Government disregard for forestry expertise since the early 1980s when the Cain Labor Government dismantled the 65-year old Forests Commission. This included, the abolition of the 'Forester' public service employment classification thereby downgrading forest scientists into a body of non-technical employees that included clerical and administrative staff. However, after pressure from the public service union, it was eventually agreed to reclassify foresters as 'Scientists'.[62] Again in the early 1990s, the need for professional forestry expertise was further questioned when Victoria's newly created Department of Conservation and Natural Resources began advertising for senior forest management positions without stipulating the need for

Forests, fire and a flawed conservation culture

professional forestry qualifications. This was questioned by a group of 'Concerned Foresters' without translating into any change.[63]

Later, the Victorian Government would decline to include any forestry expertise on its Board of Inquiry convened by the then Commissioner of Emergency Services to investigate the 2003 Alpine bushfires.[64] As foresters are the leading exponents of public land fire management and forest fire control, their exclusion is as ridiculously analogous to excluding medical consultants and surgeons from an inquiry into medical malpractice.

Similarly, the Victorian Government has in recent years taken advice about the value of its fuel reduction burning program from a Bushfires Royal Commission Implementation Monitor (BRCIM) who is a former policeman with no forest fire management qualifications or experience. Yet in his 2012, 2013, and 2014 Final Reports, the BRCIM articulated a firm view that the 5% annual fuel reduction burning target implemented after recommendation by the 2009 Victorian Bushfires Royal Commission would not work, without providing any rationale for such an assertion. A similar view was articulated by Victoria's Commissioner for Environmental Sustainability [Dr Gillian Sparkes] in the 2013 *State of the Forests Report*.[65]

The classic example in Tasmania was the convening of a forestry roundtable in 2010 to thrash out a so-called 'peace deal' negotiated by environment group and timber industry representatives. Forestry Tasmania, which managed the forests in question, was able to provide input when asked but was not invited to be part of the negotiations despite its in-depth knowledge, experience and role in making forest management plans and strategies that were central to the negotiations. A similar Taskforce established with a similar aim by the Victorian Government in 2014 also included no forest management practitioners, and was chaired by one of the nation's former leaders of environmental activism.[66]

In addition to this, almost every Australian forest policy review since the Regional Forest Agreements of the mid-late 1990s has attracted allegations of covert or direct political interference to favour a particular conservation model of increased national parks by reducing multiple-use State forests. Unfortunately in only a few cases have those who have been most disadvantaged by these reviews or forest use investigations made the effort to document why and how they were conducted. However, the following two chapters (6 and 7) are based on in-depth observations and records of two such processes that give a powerful picture of decision-making unduly shaped by political agendas and 'greened' bureaucracies in lieu of the moderating influence of professional forestry.

The progressive eschewing of professional forestry experience from its own field of endeavour by State government bureaucracies, is arguably in-part a product of decades of relentless public misportrayals of foresters as 'loggers' (or an integral part of the so-called 'logging industry') by environmental campaigners. This is a sad distortion of the reality that State forestry agencies and their personnel have always occupied a planning and regulatory role that has sat between the timber industry and those concerned with environmental protection. Although State government foresters had an effectively adversarial relationship with the timber industry, they also enjoyed high levels of cooperation and common purpose, particularly in relation to confronting the forest fire threat. Unfortunately, the extreme nature of this threat to the environment and the joint efforts to confront and manage it has never been understood or appreciated by the environmental movement.

Reversing the trend?

In contrast to Labor, the Liberal and National Parties have traditionally provided greater support for rural industries, including a greater inclination to favour environmental protection through resource management and regulation rather than simplistic strategies of shutting-out industries by creating more national parks. While the election of Coalition Governments has invariably offered a temporary respite to a seemingly inexorable progression towards total forest 'lock-up' they had until recently made few attempts to actually reverse this trend.

This changed from 2012 when Queensland's Newman Coalition Government re-opened some State forests to timber production and grazing after they'd been closed by previous Labor Governments. In Tasmania, a Liberal Government elected in 2014 sought to reverse the previous Labor Government's commitment to turn 400,000 hectares of State forest into new national parks by repealing the legislated provisions of the Tasmanian 'forests peace agreement'. In addition in 2014, the Abbott Federal Coalition Government tried but failed to partially-revoke an extension to Tasmania's Wilderness World Heritage Area which had been engineered by the Gillard Labor Government just a year earlier.

The Abbott Government also recognised that costly and unnecessary environmental conflict in Australia is in-part being driven by cashed-up environmental groups maintaining an illusion of eco-crisis to keep the donations rolling-in. The subsequent conduct of a parliamentary review into the activities of registered environmental groups in contravention of operating obligations required to maintain tax deductible gift recipient status, unsurprisingly drew the ire of those groups and their supporters keen to portray it as an attack on free speech when in reality it was an attempt to confer some accountability to their activities.[67]

Despite the findings of the inquiry, no Government action has been forthcoming at the time of writing.[68] However, the propensity for environmental groups to portray the removal of tax deductibility from their donations as a threat to their very existence underscores just how reliant they must be on sensationally misrepresenting reality to maintain their relevance. Arguably though, the damage has been wrought long ago given the extent to which their brand of environmental ideology has been accepted largely without question throughout society, including in political and bureaucratic ranks. With regard to forests, this is evident in how conflict, often inspired by barely rational activists' claims, is mostly being resolved by eschewing multiple-use management for largely unmanaged landscape-scale reservations.

Efforts to halt or reverse the march towards total forest 'preservation' typically generates negative publicity from media reports that primarily articulate the perspectives of environmental groups and the Greens. Partly due to this, Liberal and Coalition Governments are now routinely chastised by most of the media and much of the wider community as 'environmental vandals'. From this it has become glaringly obvious that any government wishing to earn an 'environmentally responsible' tick of approval from the Greens and their activist cohorts must be actively engaged in closing-down resource use industries and/or creating more national parks on land or in the sea.

Irrespective of the positions of the major parties at the Federal level, on-ground forest policies are primarily the province of State governments. In the mainland states, there is a clear expectation that further national park expansion will be routinely promised by Labor and/or the Greens at each state election, largely irrespective of the socio-economic and/or environmental merit of such proposals.

While there is frustration and despondency about whether this purely political treatment of forest policy will ever cease, the decline of local manufacturing and the consequent loss of regional and rural jobs is heightening socio-economic concerns to a degree that now should make it harder to justify the blatant misuse of forest policy for political purposes. In Victoria at the time of writing, the recent closure of the Hazelwood power station with a loss of 750 jobs has occurred in the same Latrobe Valley region where over 1,000 jobs would be lost by closing the Central Highlands timber industry to declare a 'Great Forest National Park' proposed by environmental activists and supported by the Greens.

This appears to have made the Victorian Labor Government rethink its obvious preference for declaring the 'Great Forest National Park'. However, rather than put aside its national parks agenda, the government may have re-set its sights on other forests where a new park could be declared without creating such a dramatic loss of timber industry employment. In 2016, the Government directed its environmental assessment agency, VEAC, to investigate conservation values in State forests. Given past experience this is a likely precursor to yet another forested national park being promised by the Government in the lead-up to 2018 State election.

Over many years, Australian regional and rural communities have watched with growing cynicism as what they regard as a city-centric enviro-political agenda has been enacted, whilst far more pressing concerns such as the state of country roads and rural health are neglected. There has been considerable disquiet about government-sanctioned declines of traditional rural industries such as timber and firewood production, honey production, and even agriculture which in some jurisdictions has been adversely impacted by bureaucratic requirements surrounding native vegetation protection on private lands.

In Victoria, this disquiet has at times erupted in mass demonstrations such as took place in June 2005, and again in November 2006 when 'Push for the Bush' rallies took to Melbourne's streets. Indeed, this rural angst was credited with driving the greatly improved performance of the National Party in the 2006 Victorian election and was at the time regarded as a concern for the Bracks Labor Government. Ironically, Bracks had gained power largely on the basis of rural dissatisfaction with the performance of the previous Kennett Liberal government.[69] More recently in March 2017, an estimated 1,000 residents of the small town of Heyfield marched through Melbourne's streets to protest against a government cut to log supply (ostensibly to better protect Leadbeater's possum) that was threatening 270 jobs at its local sawmill.

Further to this pre-existing concern, the recent US Presidential election victory by rank outsider, Donald Trump, has magnified questions about the tolerance of voters towards the political establishment, particularly in areas long disadvantaged by loss of rural industries and associated manufacturing. Under the present political climate, jobs unnecessarily sacrificed for the sake of a political ideology would seem unlikely to be tolerated by a substantial bulk of voters now seemingly looking for excuses to step away from conventional politics and its embrace of inner-city causes, including 'saving' forests.

It remains to be seen whether this changes anything in the Australian political context. However, forestry is fertile ground for voter malcontent after two decades of largely political decisions on forests that have in some cases needlessly trashed the livelihoods and lifestyles of rural people to appease a vocal minority with only a superficial understanding of what they are protesting about and no material stake in the consequences.

In Victoria, the decision to phase-out local timber production to create a new national park in the Otway Ranges stands as the classic example of an out-of touch polity outraging rural voters. When Victorian Premier, Steve Bracks, announced the decision to create a Great Otway National Park on the eve of the 2002 Victorian election, he justified it with words indicating an appalling lack of understanding of what he was doing. In his 2012 autobiography, Bracks repeated this justification in relation to the state's native forest timber industry as a whole:

> It is now possible to stop all native forest logging in Victoria. The plantation forests of the State's south west are now mature and could, with transitional transport assistance from the Government, provide a stable source of timber for the existing mills. That would be a logical next step after the native forest reforms during my period in office. More can be done in this area – indeed more should be done.[70]

In reality, the native forest timber industry is based on producing slow-grown sawn eucalypt (hardwood) timber, whereas the 'plantation forests of the State's south west' are comprised of pine (softwood), which produces a range of sawn and unsawn products; and eucalypt (hardwood) plantations that are being grown only over a 12-15-year rotation to produce woodchips for export. Those plantations cannot produce the logs required by the native hardwood sawmilling industry. Furthermore, an industry with infrastructure designed to saw native hardwood logs cannot suddenly saw softwood logs, even if the Government was to provide 'transitional transport assistance' to cart logs ridiculous distances [500-700 km] from south-west Victoria to 'existing mills' as far away as East Gippsland.

It defies belief that Bracks still isn't aware of these realities given that they would have been conveyed to his advisers and Ministers loudly and often by timber industry representations prior to, and in response to, his decision to create the Great Otway National Park. Indeed, his refusal to acknowledge the truth even 10-years later only adds weight to cynical local speculation that the decision was based more on the fact that Bracks owned a holiday home on the Otways coast at the time and didn't much enjoy sharing the roads with occasional log trucks. Arguably, it is such perceptions of elitism fuelled by political decisions that devastate lives and livelihoods for little practical reason, that reportedly lie at the heart of the rise of the likes of Trump in the US, as well as One Nation and an increasing cadre of independent politicians in Australia.

After a series of similarly unnecessary political decisions over the past 15-20 years the forest sector's depth of distrust of the political establishment (particularly left-of-centre parties) is exemplified by the NSW Forests Products Association submission to a recent Parliamentary Inquiry which featured the following statement in large, bold type within a highlighted text box on the first page:

> The greatest constraint upon forest industries in NSW is the lack of confidence that governments may deal fairly with integrity and honesty, with the supply commitments that they have placed into legislation, contracted to industry, and with the faith on which industry has invested.[71]

Chapter 5 Endnotes

1 John Cain, former Victorian premier, in a Letter to the Editor published in *The Age*, 8 December 2015.

2 Launch of the Forest and Wood Products Support Group 2015, 14 April 2015.

3 ABC 2015, *Possible logging bail-out to save Leadbeater's possum*, RN Breakfast, ABC Radio National, interview with ANU scientist Professor David Lindenmayer and Victorian Minister for Environment, Climate Change and Water, Lisa Neville, 23 April 2015.

4 Lindenmayer, Blair, McBurney and Banks (2015), "Ashes to ashes: Logging and fires have left Victoria's magnificent forests in tatters", *The Conversation*, 25 November 2015.

5 Deloitte Access Economics 2015, *Economic Assessment of the native timber industry in the Central Highlands RFA Area – Report 1: Economic and financial impact*, prepared for VicForests, October 2015. This report found that the regional timber industry that would be forced to close by the declaration of a 'Great Forest National Park' annually generates over $570 million in regional economic activity and directly employed 2117 workers, not including secondary wood products manufacturing jobs.

6 The Forest Industry Taskforce was administered by the Victorian Department of Environment, Land Water and Planning. Its purpose articulated on a Taskforce website was "for the major stakeholders to reach common ground on a durable, long-term set of

recommendations and proposals to government, about future issues facing the industry, job protection, economic activity, protection of our unique native flora and fauna and threatened species, such as the Leadbeater's possum".

7 Deloitte Access Economics 2015, op. cit.

8 Victorian Government 2014, *Leadbeater's possum Recommendations: Report to the Minister for Environment and Climate Change and the Minister for Agriculture and Food Security*, Leadbeater's Possum Advisory Group, 20 January 2014.

The Advisory Group was co-chaired by the Jenny Grey, CEO Zoos Victoria and Lisa Marty, CEO Victorian Association of Forest Industries. The Terms of Reference for the Advisory Group were to develop recommendations for government focused on supporting the recovery of the Leadbeater's possum while maintaining a sustainable timber industry. It did not examine the option of closing the industry.

9 DELWP 2015, *Supporting the Recovery of the Leadbeater's possum: Progress Report October 2015*, Department of Environment, Land, Water and Planning, Government of Victoria.

10 Drew Hutton and Libby Connors, 1999, *A History of the Australian Environment Movement*, Cambridge University Press, p. 165.

11 Graham Lloyd, 2016, "How Hardcore greens trumped unions on RET" *The Australian*, 11 October 2016.

12 Steve Bracks with Ellen Whinnett, 2012, *A Premier's State*, Melbourne University Press p. 279

13 Troy Bramston, 2015, "Carr's financial lesson for Labor, 20 years on", *The Weekend Australian*, 4-5 April 2015.

14 Department of Agriculture 2015, Regional Forest Agreements

15 Leadbeater's Possum Advisory Group, Terms of Reference, Department of Environment and Primary Industries, June 2013.

16 Department of Agriculture 2015, op. cit.

17 "New look Gunns ready for Forest Wars truce", *ABC Radio National*, 17 September, 2010.

18 Ross Gittens, 2015, 'Broken public service leads to broken governance", *The Age*, 7 December 2015

19 Peter McHugh – A Fraternity of Foresters, August 2016 (unpublished recollections).

20 The various incarnations of Victoria's forested public land management agencies since the abolition of the Forests Commission Victoria in 1983 are: Department of Conservation Forests and Lands (1983-90), Department of Conservation and Environment (1990-92), Department of Conservation and Natural Resources (1992-96), Department of Natural Resources and Environment (1996-2002), Department of Primary Industries (2003-13), Department of Sustainability and Environment (2003-13), Department of Environment and Primary Industries (2013-14), and finally the Department of Environment Land Water and Planning (2014-current time). Note that from 2003-13 non-commercial land management responsibilities were split between two departments. Further complicating

this picture is that responsibility for most (but not all) commercial timber production was vested in a seperate new agency, VicForests, created in 2004. In addition to this, Parks Victoria was established in 1996 to manage national parks and other reserves declared under the *Parks Victoria Act 1998*.

21 Bushfires Royal Commission 2009, *2009 Victorian Bushfires Royal Commission Final Report*, Volume 2, Chapter 2.

22 *Forests, fire and regions update*, Department of Environment Land Water and Planning, June 2016.

23 Ibid.

24 Bushfires Royal Commission 2009, op. cit.

25 Ross Gittens, 2015, op. cit.

26 Richard Mickelburough, 2016, "State of spin", *Herald Sun*, 11 January 2016.

27 B.D. Dexter, A. Hodgson and R. Incoll, 2016, An evaluation of the Inspector-General for Emergency Management review of the Wye River – Jamieson Track fire and related matters: What are the real lessons learnt? (ISBN: 9780994253125).

28 Will Murray, "Logging postponed in Strathbogie forest, for now", *Euroa Gazette*, 2 November 2016.

29 Norman Wettenhall Memorial Lecture 2006, *Can we prevent disaster fires in one of the most fire-prone landscapes in the world*, Melbourne 23 November 2006. Advertised speakers Richard Loyn, Gordon Friend, and Liam Fogarty of the Department of Sustainablity and Environment were replaced at short notice due to the Victorian Government's determination to avoid any controversial media coverage just days before the 2006 State election.

30 Richard Willingham, "Forestry industry demands inquiry into the devastating Wye River fire", *The Age*, 7 January 2016.

31 Personal comments, Gary Featherston, former Chairman of the Victorian Division of the Institute of Foresters of Australia, February 2016.

32 B.D. Dexter, A. Hodgson and R. Incoll, 2016, op. cit.

33 Coroners Court of Victoria: Decision by Coroner whether or not to hold an inquest into fire, by Judge Sara Hinchey, State Coroner, Court Reference 2016 0308, 20 September 2017.

34 Denis O'Bryan, 2016, *The Jamieson Track bushfire and its escape into Wye River and Separation Creek townships*, August 2016 (unpublished report by Red Eagle Bushfire Protection Services).

35 "Governments intervene in the forest peace talks", *ABC News*, 10 August 2012.

Matthew Denholm, 2012, "Tasmanian MLCs told $300 million hinges on forest peace deal", *The Australian*, 6 December 2012.

Matt Maloney, 2012, "Forest peace: talks fail to reach a deal", *The Examiner*, 11 August 2012.

36 Geoff Law, 2015, *The Tasmanian Wilderness – A Case for Long-lasting Civil Society Involvement in Protecting World Heritage*, paper presented to the World Heritage Watch Conference: UNESCO World Heritage and the Role of Civil Society, Bonn, Germany, 26-27 June 2015.

37 Bob Kearney, Emeritus Professor of Fisheries, "Opposition to Margaris super trawler not evidence based", *The Conversation*, 15 August 2012.

38 "Burke rejects national heritage listing for the Tarkine", *The World Today*, ABC Radio National, 8 February 2013.

39 Janet Hawley, "Tree Amigos", *The Age Good Weekend Magazine*, 28 June 2003.

40 Attiwill et al 2001, *The Environmental Credentials of Production, Manufacture and Re-Use of Wood Fibre in Australia*, prepared for the Australian Department of Agriculture Fisheries & Forestry by the University of Melbourne, p. 128.

41 Sean Nicholls, "'We have to act': Luke Foley promises Australia's first koala national park", *Sydney Morning Herald*, 19 January 2015.

42 Alice Burnet, "The great koala debate", *The Belligen Shire Courier Sun*, 27 January 2015.

43 Mark Poynter, 2007, *Saving Australia's Forests and its Implications*, Coretext ISBN 978-0-9775029-1-2, p. 54.

44 NSW Legislative Council 2013, *Management of public land in New South Wales*, Legislative Council General Purpose Standing Committee No. 5, Report 36 , May 2013. Chapter 4: *The Conversion Process* (pp. 54-58).

45 "Labor forest policy brawl", *The Hobart Mercury*, 30 April 2007.

46 ABC 2007a, "Industry, conservationists slam ALP's forestry policy", *ABC News Online*, 30 April 2007.

47 "Forestry workers won't be used as bait: Tasmanian Premier", *ABC Radio AM*, interview with Tasmanian Premier Paul Lennon, 30 April 2007.

48 Bruce Montgomery, 2012, "In Tassie politics, its hard to see the sense from the trees", *Crikey*, 6 March 2012.

49 In western Victoria, the Surf Coast Shire formally adopted a policy supporting the closure of the local native forest timber industry in the Otway Ranges in 2001. Similarly, the Shire of Yarra Ranges cited concerns about harvesting in minor parts of Melbourne's water supply catchments as the catalyst for developing a forest policy despite having no responsibility for timber harvesting on public land.

50 The Wilderness Society was for a time actively encouraging municipal councils to institute procurement policies that prohibited the purchase of wood products sourced from Australian native forests.

51 Roger Fyfe, former professional forester and Banyule resident who attended the Council's meetings, November 2016.

52 Ibid.

53 "MP pleads for sense", *Heidelberg Leader*, 20 December 2016.

54 Environment Victoria 2005, *About the Environmental Liaison Office*, July 2005.
55 Nature Conservation Council of NSW 2015, *Parliamentary Liaison*, March 2015.
56 Gary Johns and John Roskam, 2004, *The Protocol: Managing Relations with NGOs*. Report prepared by the Institute of Public Affairs to the Prime Minister's Community Business Partnership.
57 Roger Underwood, 2003, *Community forestry and public involvement in forestry* (unpublished).
58 Garsden et al 2002, "Vegetation Plans and Forestry on Private Land", by Garsden, Coombe and Dyason, *The Forester*, Vol. 43, No. 1, March 2002.
59 Warwick Ragg, "One step back on PNF Code in NSW, one step forward in QLD", *Australian Forest Grower*, Spring 2004.
60 "Timber reserves could cost jobs", *Australian Financial Review*, 8 August 2006.
61 "Private native forestry regulation back to the drawing board", *ABC News Online*, 28 August 2006.
62 J. Gillespie and J. Wright, 1993, *A Fraternity of Foresters*, Jim Crow Press.
63 B.D. Dexter, 2017, personal recollections.
64 Stephen Pyne, 2006, *The Still-Burning Bush*, Scribe Short Books.
65 B.D. Dexter and A. Hodgson, 2015, *Submission to the Inspector-General Emergency Management's Review of Performance Targets for Bushfire Fuel Management on Public Land*, March 2015.
66 Don Henry, former long time leader of the Australian Conservation Foundation.
67 ABC 2015b, "Parliamentary inquiry into environmental organisation's donation spending", *ABC RN Breakfast*, 1 April 2015.
68 Australian Government 2016, *Inquiry into the Register of Environmental Organisations*, House of Representatives Standing Committee on the Environment, April 2016.
69 Farrah Tomazin and Matthew Murphy, 2005, "Focus: Has Bracks lost the bush", *The Age*, 9 June 2005.
70 Steve Bracks with Ellen Whinnett, 2012, *A Premier's State*, Melbourne University Press p. 278-279.
71 NSW Forest Products Association 2012, Submission to the NSW Parliamentary Inquiry into the Management of Public Land in NSW, 15 August 2012.

6

Politics and bureaucracy 1:

Reserving Victoria's river red gum forests (2005-08)

"For every complex problem there is an answer that is clear, simple, and wrong"
H. L. Mencken

Victoria's river red gum (*Eucalyptus camaldulensis*) forests and woodlands occupy uncleared parts of the floodplains of the Murray River and its major tributaries in the state's north. They include the iconic Barmah Forest which, together with the Millewa Forest on the opposite (NSW) side of the Murray, comprises over 60,000 hectares and is the world's largest natural river red gum forest.

By 2005, these floodplain river red gum forests and associated woodlands had been used for cattle grazing and selective timber and firewood harvesting for up to 150-years. They also supported significant wetlands. Indeed, Victoria's Barmah and Gunbower forests had been listed as sites of international importance under the Ramsar Convention on Wetlands since 1982, while NSW red gum forests on the northern side of the Murray River had been Ramsar-listed since 2003. This high level conservation status had been maintained largely under a multiple-use State forest management regime that included periodic timber harvesting and seasonal cattle grazing.[1]

These Ramsar-listings are also highly significant because they take account of and have been maintained despite the red gum forests and their wetlands comprising one of the nation's most modified landscapes.[2] Due to demands for irrigated agriculture and flood mitigation, Murray basin river flows have been regulated by upstream dams since the 1930s. This has completely overturned the natural flooding regime which had sustained the unique ecology of these forests for possibly tens of thousands of years. That water availability is the overwhelming determinant of their health has long been recognised, as is exemplified by the title of a 1989 conference on River Murray wetland and forest management – *The Barmah Forest: Dying for a Drink*.[3]

By the turn of the millennium, these floodplain red gum forests were in very poor health, as were many other forests in the midst of southern Australia's worst recorded drought. Compounded as this was by having to compete for water with irrigated agriculture, the situation could only have been significantly alleviated by reforms to water management.

Despite this reality, an environmental group, the Victorian National Parks Association (VNPA), was claiming that the dire health of these red gum forests was due to supposedly rampant logging and cattle grazing. In 2001, the VNPA was awarded a $69,000 grant from the Myer Foundation specifically to fund a campaign to 'save' these forests by ending these activities.[4] Almost immediately their campaign was supported by other mainstream environmental groups, most notably the Australian Conservation Foundation (ACF) and the Wilderness Society (TWS), with a collective focus on creating a large red gum forests national park.

In fact, selective timber harvesting and grazing were already restricted by public land tenures and other management regulations to less than half of Victoria's floodplain river red gum forests and woodlands.[5] These forests were contained in a mix of public land tenures (but mostly State forest) that already provided for high levels of conservation while supporting a variety of commercial and recreational uses. This [inconvenient] reality was unsurprisingly ignored by environmental campaigners well practised in pushing a simplistic national parks agenda in response to almost any forestry issue.

In stark contrast to what these environmental campaigners were implying, simply rebadging all these forests and woodlands as national park would, on its own, do nothing to improve poor health caused by a lack of water. Indeed, the adoption of a passive national park management regime was arguably more likely to make matters worse by sidelining the active land and water management tools and flexibility that could otherwise assist in maintaining forest health, for example, through regrowth thinning to reduce between-tree competition for moisture.

Of particular significance was that the local communities, which lived cheek-by-jowl with these forests and arguably appreciated and used them more than anyone else, were happily accepting of this pre-existing mix of public land tenures. These communities recognised the need to improve forest health and vitality and, in many cases, had been participating in important forest and water management decisions for up to twenty years through bodies such as the Barmah-Millewa Forum, which had been created under the Murray Darling Basin Agreement.[6]

Despite this, in early 2005 Victoria's Bracks Labor Government responded to the incessant environmental campaigning for a large red gum national park by directing its Victorian Environmental Assessment Council (VEAC) to investigate public land use in the floodplain river red gum forests. In announcing this directive, the Government noted that it would be "receptive" to the creation of new national parks.

VEAC and the current situation, 'green' politics and local fears

VEAC investigations are typically a lengthy and expensive process running across several years. They include various opportunities for community groups or concerned individuals to make public submissions that can potentially help to shape their outcome. With regard to northern Victoria's river red gum forests and woodlands, there was considerable local disquiet about the prospect of the existing mix of public land tenures being overturned to create new national parks. Arguably, this local opposition was the strongest and most coordinated which VEAC or its predecessor agencies (such as the Land Conservation Council) had ever encountered in any public lands investigation undertaken before or since.[7]

The most prominent opponent of the mooted expansion of national parks was a consortium of 25 local and statewide bodies known as the Rivers and Red Gum Environment Alliance (the RRGEA). It included three local Shire councils, an indigenous group (the Bangerang People), a range of timber industry and agricultural groups, six local community and environmental groups, and nine recreational user groups, including hunters and fishermen. Collectively, the RRGEA was highly knowledgeable and experienced in the science and practicalities of forest and water management

One of the RRGEA's central arguments was that the existing mix of public land tenures was already adequately protecting conservation values whilst enabling a suite of highly valued recreational and commercial uses subject to appropriate environmental safeguards. They could see no justification for changing this to facilitate an expansion of national parks and believed that, if enacted, such a move would inflict significant socio-economic damage to local communities.[8]

According to the RRGEA, biodiversity conservation was already the primary or major management focus on an estimated 73% of the public lands in the nominated study area prior to the VEAC investigation (see Table 1). Only on around 10% of the public lands was conservation a lesser focus, mostly in areas used to supply community services or to house essential utilities. Their analysis invalidated the environmental campaigners' claim that more national parks were needed because biodiversity was not being sufficiently protected, and highlighted the spurious rationale upon which the Victorian Government had commissioned VEAC to undertake its River Red Gum Forests Investigation.

By and large the public expects government agencies to act with integrity. They expect that an independent investigative body, such as VEAC, will conduct a scientific process that is fair, objective, free of any pre-meditated personal or group bias, and uninfected by ideological agendas. Unfortunately the outcomes of previous public land use investigations undertaken by VEAC and its immediate predecessor, the Environmental Conservation Council (ECC), had already raised doubts about the integrity of the process. Each of these past VEAC (Otways and Goolengook) and ECC investigations (Box-Ironbark) had also been commissioned by the Bracks Labor Government in response to environmental activism. Each had resulted in the declaration of substantial new national parks and conservation reserves, with consequent complete (or

Table 1: Levels of effective biodiversity conservation on public lands in the designated River Red Gum Forests Investigation Study Area in 2005, prior to the VEAC investigation

Management of Biodiversity Conservation	Public Land Category	Area (ha)	Proportion of Public Lands in Study Area
Primary Focus	National Parks	52,120	31.2%
	State Parks	9,925	
	Nature Conservation Reserves	11,895	
	State Forest Special Protection Zones (SPZ)	10,035[A]	
Major Focus	Regional Parks and other parks	7,775	41.8%
	Natural Features Reserves	32,605	
	River Murray Reserve	16,060	
	State Forest – unproductive (Mid Murray)	3,793[C]	
	Wildlife Co-operative Management Area	2,565	
	State Forest GMZ & SMZ (Mildura)	38,081[D]	
	Other unused State Forest	11,144[E]	
Joint Focus	State Forest GMZ & SMZ (Mid Murray)	43,857[F]	16.3%
Minor Focus	Earth Resources	125	10.7%
	Plantations	175	
	Services and Utility	5,880	
	Community Use Areas	2,690	
	Water production, supply, and drainage	12,665	
	Uncategorised Public Land	6,620	
	Historic / Cultural Features Reserve	705	
TOTALS	STUDY AREA	268,715 ha	100%

Notes: The specified areas are taken from VEAC's Draft Proposals Paper (July 2007), Executive Summary, and other sources listed below. It should be noted that the total public land area statement in the Draft Proposals Paper is 5,800 ha higher than that in VEAC's earlier Discussion Paper (Table 9.3, p. 129).

State Forest SMZ = Special Management Zone GMZ = General Management Zone

A – from the VEAC Discussion Paper, Table 9.3, p. 129.

C– from the *Estimate of Sawlog Resources, Mid Murray FMA*, Department of Sustainability & Environment (March 2002).

D– from the *Forest Management Plan for the floodplain State Forests of the Mildura Forest Management Area*, Department of Sustainability & Environment *(July 2004)*. No timber harvesting occurs in the Mildura FMA, but firewood is collected.

E– refers to other State forests outside the Mildura and Mid Murray FMAs. No timber harvesting occurs, but firewood is presumably collected.

F– sustainable timber harvesting, some grazing, and firewood collection permitted.

near-complete) removal of some traditional activities and substantially reduced opportunities for other community recreational and commercial uses.

Amongst those local communities which were likely to be most affected, this past experience had engendered a perception of VEAC investigations as a government mechanism designed to justify already pre-determined policy changes to public land tenure that would severely disrupt country livelihoods and lifestyles.

On the other hand, the advantage of a lengthy VEAC investigation to the Victorian Government was that the passage of time would soften the impact of contentious decisions (particularly about closing the native hardwood timber industry) by enabling them to be announced under a guise of supposed scientific rigour and extensive community consultation.

To be fair to VEAC, it must be noted that it's earlier Otways investigation was not required to consider all options for future land use because the Victorian government had already made a pre-election committment to close the local timber industry and declare a new national park before it had even commissioned VEAC's investigation. Also, VEAC's Goolengook Forest investigation had been abandoned part-way through when the government promised to end timber production during the 2006 state election campaign.[9] These two episodes only further reinforced the perception that public land policies were being primarily based on political (ie. electoral) imperatives, with VEAC being appropriated mainly to create an illusion of considered and objective decision-making.

A major concern with this past public land tenure change was that it couldn't be justified in terms of significantly improved environmental outcomes. For example, in the Otways prior to the Government's announcement of a new national park, the supposed threat posed by timber production was grossly overstated. In reality it was already a minor forest use restricted to within just a net 22% portion of the region's 160,000 hectares of public forest. Only around 300 hectares was being annually harvested and regenerated, which equated to just 0.15% of the total forest area. Nevertheless, it supported a sustainable local industry that employed around 125 people and annually added $20 million to the regional economy.[10]

Accordingly, the forced closure of the Otways timber industry has done little to improve already good conservation outcomes while causing some significant socio-economic damage to local communities. Understandably amongst those communities most likely to be adversely affected by VEAC's River Red Gum Forests investigation, there were serious misgivings about any investigation of forest use instigated at the behest of strident environmental activism.

Questionable political and bureaucratic behaviour

At various times during VEAC's River Red Gum Forests Investigation, the behavior of the Victorian land management bureaucracy was suggestive of a pre-awareness of the investigation's final outcome. In particular, the Department of Sustainability and Environment (DSE) abandoned its partially-completed State Forest Resource Inventory (SFRI) project in the river red gum forests as soon as the VEAC investigation was announced in early 2005. While the official explanation for this was a lack of funding, informal sources within DSE regarded it as an acceptance that the VEAC process would create new national parks thereby ending red gum timber production and removing any further need for forest resource assessment.

DSE's failure to complete the SFRI process ultimately would prove to be short-sighted as it forced VEAC to estimate future forest growth and sustainable timber yield from outdated growth data that had already been regarded as inadequate for estimating long term sawlog supplies.[11] In effect, the lack of updated data made it more likely that timber production would not be allowed to continue.

Part-way through VEAC's River Red Gum Forests Investigation, the Victorian Government faced an election. During the campaign for the November 2006 state election the Premier, Steve Bracks, and Minister for the Environment, John Thwaites, launched a 'National Parks and Biodiversity Policy'. In-part it stated that the government would "… create new Red Gum National and Forest Parks if recommended by VEAC".[12]

It was later revealed that this statement was included in the policy so as to secure second preference votes from the Australian Greens.[13]

This announcement at about the halfway point of the River Red Gum Forests Investigation effectively signalled to VEAC that a recommendation to expand national parks was expected by the government thereby putting covert pressure on the agency to deliver such an outcome. The announcement also effectively pre-empted the outcome of the critical last step of the process whereby, in accordance with the *Victorian Environmental Assessment Council Act 2001*, the government has six-months to decide whether or not to implement VEAC's final recommendations.

The government's red gum national parks pre-election pledge was interpreted by the regional media as a disincentive for local residents to participate in the VEAC process if they disagreed with the notion of expanded national parks. Accordingly, it can be viewed as distorting the subsequent public consultation process.[14]

From the start, the capability of VEAC to conduct an informed investigation of forested public lands had been questionable. None of its four nominated Councillors had any specific forest management expertise and there was little practical land management expertise amongst VEAC's support staff which was overwhelmingly populated by personnel with ecological and conservation backgrounds.[15]

Although a Melbourne-based DSE forest resource specialist would later be seconded to VEAC to assist with part of the process, he also had no specific experience of the river red gum forests. Disturbingly, no local forester with a working knowledge and understanding of the unique silvicultural characteristics of these forests played a significant role in the VEAC process.

This only further reinforced the strong perception amongst regional Victorians that VEAC was a city-based bureaucracy ideologically predisposed to view forests as fragile preserves in which there is little or no place for human use. Over the course of earlier VEAC investigations, plenty of anecdotal evidence had arisen suggesting that VEAC did not undertake its role objectively and that it was staffed by personnel who were in some cases openly biased towards an agenda of national park expansion. Indeed at least one VEAC staffer had reportedly glorified her role in private as one of saving the environment by 'making national parks'.[16] Whilst this anecdote may be hard to verify, if true, it is a disturbing revelation for a body charged with responsibility for objectively advising on public land use and natural resource management.

Arguably also indicative of this preconception, was the work of a VEAC staffer in writing a paper subsequently published by the Royal Society of Victoria in October 2006, at about the half-way point of the agency's River Red Gum Forests Investigation. It provided some insights into the workings of VEAC in striving to create an impression of an objective process at a time when doubts were emerging about whether this was indeed the case. The paper also attempted to foster an impression that an outcome recommending more national parks had majority community support. At best, this seemed to be irregular behaviour for a supposedly objective government scientist involved in an ongoing public lands investigation.[17]

A critically important part of VEAC's River Red Gum Forests Investigation was the establishment of a Community Reference Group (CRG) comprised of key local stakeholder representatives. While the CRG was meant to be a significant component of the investigations's public consultation process, VEAC did not strongly direct it and appeared to actively prevent it from functioning as its delegates had expected. Consequently, many became disillusioned and cynical about a process that appeared to be merely going through the motions.

Several CRG delegates later recalled that at the start of the process they were collectively informed by VEAC Chairman Duncan Malcolm, that the group *would not* function in a democratic manner which would have allowed issues to be discussed and debated with a view to determining a consensus position. Accordingly, the CRG operated simply as a forum where VEAC disseminated information to the delegates who could then make a comment. VEAC Councillors attending CRG meetings apparently simply listened to delegate concerns and answered questions where required, but there was never any suggestion that these concerns would be seriously considered beyond the meeting room.[18]

In addition, CRG delegates were only provided with selected draft chapters of VEAC's initial *Discussion Paper* to consider and were forbidden from discussing their content with constituents of the stakeholder groups that they were representing. Furthermore, CRG meetings were invariably so short that the delegates were often not given sufficient time to fully voice their concerns. At the CRG's final meeting in Shepparton, the VEAC Councillors refused to answer many of the questions directed at them and one Councillor sat through the entire meeting without providing any input at all.[19]

VEAC's CRG process has been unfavourably compared with a similar process undertaken several years earlier during the development of the Department of Sustainability & Environment's (DSE) Mid-Murray Forest Management Plan. A CRG delegate who had also participated in the earlier DSE process noted that it had been a far more transparent and participatory exercise, and included a more balanced government representation, including local foresters with a good working knowledge of the river red gum forests.[20]

The CRG process further strengthened the local perception that VEAC's investigation was working to meet a pre-determined outcome of substantial national park expansion. This impression was further magnified by the demonstrably close relationship between VEAC and the Victorian National Parks Association (VNPA) during the course of the investigation.

The VNPA's national parks expansion campaign had long overstated the threat to the forests from past management, and had continually articulated the misconception that VEAC had a community mandate to recommend new parks in the investigation area. Their close relationship with VEAC was exemplified by them effectively acting as the public defender of VEAC whenever concerns about the investigation were raised in the media; and by being party to VEAC's intentions over and above other stakeholder groups. For example, at the CRG meeting in July 2007 during which VEAC released its confidential *Draft Proposals Paper*, it was patently obvious that the VNPA delegate was already aware of its contents. This was reportedly later confirmed when a VEAC staffer admitted that he had told one person of its contents prior to its release.

Defenders of VEAC would probably dismiss the above observations as merely the subjective and not unexpected reflections of disaffected locals made in the hindsight of unwanted change foisted upon their communities. However, there is other more compelling evidence of the shortcomings of the VEAC process which cannot be so easily brushed aside. This includes:

- VEAC's failure to meet the Terms of Reference for its River Red Gum Forests Investigation which in-part required it to recommend outcomes which accorded with the principles of 'ecologically sustainable development';
- VEAC's deliberate manufacturing of a case for change by:
  1. understating the environmental benefits already attributable to the pre-existing mix of public land tenures, use, and management;
  2. downplaying the adverse socio-economic impacts of its recommendations for changing public land tenure, use, and management; and
  3. constructing a hypothetical community demand for more national parks through the use of a contentious Choice modelling exercise.
- VEAC's tendency to ignore or summarily dismiss inconvenient truths and to manipulate evidence that could otherwise challenge its preferred outcomes, as was evident in:
  1. its insistence on developing an unrealistic environmental water proposal despite evidence that it could never be achievable;
  2. the misuse of Ecological Vegetation Classes (EVCs) as a surrogate for 'ecosystems' and the inappropriate conflation of woodlands lost to agriculture with true floodplain forests so as to fashion a stronger than warranted case for more land reservation;
  3. its determination to recommend the complete removal of cattle grazing and firewood collection, and foster the doubling of coarse (fallen) woody debris for ecological purposes, de-

spite grave misgivings amongst local communities about the increased threat of fire that these changes would cause;
4. exaggerated portrayals of the impact of cattle grazing to justify a determination to remove it from the forests;
5. its reluctance to learn about or acknowledge the success of long-standing community efforts in improving conservation outcomes in wetlands, on Crown water frontages, and in relation to the Superb Parrot;
6. its apparent willingness to make recommendations that would knowingly cause significant hardships to local communities, and to effectively dismiss these local concerns as being of lesser significance than the alarmist demands of arms-length environmental ideologues;
7. vague and deceptive conduct in revising sustainable timber yield – particularly their unprecedented use of the slowest-ever recorded tree growth as the basis for expectations of future forest growth – that largely ignored the likelihood of a return to higher rainfall and the potential for improved water management due to greater environmental flows; and
8. its refusal to research or even acknowledge the climate change mitigation benefits of sustainable wood production.[21]

Departing from due process – VEAC's failure to meet its own Terms of Reference

VEAC's deviation from the Terms of Reference (TOR) announced for its River Red Gum Forests Investigation by the Minister for Environment in April 2005, represents a serious departure from due process that neither they nor the Victorian Government has ever acknowledged. These TOR included a requirement for VEAC to "make recommendations relating to the conservation, protection, and ecologically sustainable use of public land as specified in Section 18 of the *Victorian Environmental Assessment Council Act 2001*".

The first requirement listed in Section 18 of the *Victorian Environmen-*

tal Assessment Council Act [2001] is 'the principles of ecologically sustainable development'.[22] These principles are defined in Part 1, Section 4 of the later *Commissioner for Environmental Sustainability Act 2003*.[23] They effectively describe ecologically sustainable development as a concept that strongly integrates the conservation of biodiversity with economic development, and community well-being and welfare.

In VEAC's River Red Gum Forests Investigation Study Area, it is obvious that a workable balance between conservation, economic and social needs was already being met under the pre-existing mix of public land tenures, and its management and use (see Table 1). By comparison, it can be argued that VEAC's recommended changes to land tenure contravened the requirements of ecologically sustainable development by substantially reducing opportunities for human use and community well-being by giving an unbalanced weighting to biodiversity conservation above all else (see Table 2).

Whilst one of the guiding principles of ecologically sustainable development (in the *Commissioner for Environmental Sustainability Act 2003*) does allow for major change to be implemented where there are "threats of serious or irreversible environmental damage", this could never be considered to be relevant to the pre-VEAC management of public lands in the investigation's study area, given that:

- biodiversity conservation was already either the singular primary or major focus of management on 73% of the public lands in the investigation study area which was already contained in national parks, other conservation reserves, or State forests where timber production and/or grazing were not permitted (as per Table 1);

- the two activities portrayed by pro-park environmental campaigns as being the most threatening (i.e. timber harvesting and cattle grazing), were not even specifically listed as a 'potentially threatening process' in VEAC's initial *Discussion Paper* (Appendix 8)[24] and were in any case, already excluded from most of the public lands in the investigation's study area; and

Table 2: The approximate impact of the VEAC River Red Gum Forests Investigation on human use of the river red gum forests and woodlands in the defined Study Area

Activity	Public land management prior to VEAC Investigation (2005)		Public land management after VEAC final recommendations (2008)	
	Area where permitted	Proportion of study area public lands	Area where permitted	Proportion of study area public lands
Timber harvesting	41,160 ha	15.3%	9,980 ha	3.7%
Cattle grazing	~98,000 ha	36.5%	unused road reserves only	<1%
Hunting	~140,000 ha	52.1%	~30,000 ha	11.2%
Horse & dog access	196,105 ha	72.9%	~53,000 ha	19.7%
Firewood collection	106,910 ha	39.8%	~15,000 ha	5.6%

Notes: Total area of public land in the VEAC River Red Gum Forests Investigation study area was 268,715 ha.

Timber harvesting – the pre-VEAC net harvestable area was outlined in the Department of Sustainability and Environment's Mid-Murray FMA Plan (April 2002). VEAC's final recommendations created new National parks and other reserves which reduced the area of State forest by around 90% and this in turn reduced the harvestable area by around 75%.

Cattle grazing – was previously permitted in 86,000 ha of State forest, as well as 12,000 ha of Crown water frontage and unused road reserves. After the VEAC process and a phase-out period, grazing was only permitted to continue in unused road reserves comprising something less than 1% of the region's public lands.

Hunting – the change in available area was due to proposals to exclude hunting from 23 wildlife reserves and wetlands where it was previously permitted. As a trade-off, some accredited hunters were promised limited involvement in feral animal control on public lands.

Horse and dog access – it was presumed that the only places where horses and dogs were previously prohibited was the 62,045 ha that is National and State park, plus water production areas (2120 ha), wildlife cooperative areas (2565 ha) and service and utility areas (5880 ha). It was difficult to accurately gauge where horses and dogs would be permitted after the VEAC process because it was not specified for some land tenures. However, there will be a substantial reduction in available area because of the loss of State forest to National park expansion.

Firewood collection – it was presumed that firewood collection was previously permitted only in State forests. As a result of the VEAC proposals it substantially declined because State forest area was reduced by 88%. However, VEAC allowed for firewood to be also collected in some small areas (of unspecified size) within the Murray River Park near Mildura, Robinvale and Nathalia.

- these activities were already being regulated in a manner which had been verified as environmentally appropriate according to international standards being applied under the Ramsar Convention to protect wetlands of international importance.[25]

Further to this, VEAC openly acknowledged that most of the perceived benefits from their recommendations were of an intangible 'feel-good' nature that would accrue to people residing outside the investigation study area, mostly in Melbourne. On the other hand, the tangible socio-economic pain arising from its recommendations would be borne by those residing locally within the investigation study area. This admitted inequity was also strongly at odds with the general thrust of ecologically sustainable development which recognises the importance of a strong, diversified local economy in enhancing the capacity for environmental protection.

Dubious justification for change

Prior to VEAC's investigation, the prevailing mix of public land tenures was primarily based on the Land Conservation Council's (LCC) recommendations for the Murray Valley Study Area which had been adopted by the Victorian Parliament in 1986. While conducting its River Red Gum Forests Investigation from 2005-08, VEAC never sufficiently explained why this prevailing mix of land tenures should change, except to say in general terms, that it was necessary "to meet nationally-agreed [*JANIS*] criteria for a comprehensive, adequate and representative reserve system." And that it "reflects the shifting priorities for public land use since the last systematic assessments in the investigation area, the majority of which were more than 20 years ago".[26] While these 'shifting priorities' were never defined, they undoubtedly reflect urban-based attitudes shaped by decades of environmental campaigning largely founded on substantial exaggerations of ecological threat.

Arguably, VEAC supported its determination to meet the nationally-agreed JANIS criteria by misusing the Ecological Vegetation Class (EVC) concept. Delineated, modelled and mapped EVCs are meant to reflect the original pre-European vegetation (nominally as at 1750). They are assigned a conservation status based largely on their current extent as a proportion of their modelled former extent prior to European settlement However, especially in such a highly modified landscape, EVCs delineated and mapped in these forests today are unlikely to accurately reflect what was present over 200-years ago primarily due to the dramatically changed forest flooding regimes.

VEAC's sub-division of 34 distinct, broad vegetation communities into 168 smaller, seasonally variable, and often barely discernible EVCs, made it easier to create a stronger impression of conservation vulnerability based on modelled area-depletion since European settlement. This was arguably further enhanced by VEAC's use of two additional conservation status categories. Given the rules associated with the JANIS criteria, this automatically translated into an unwarranted need for greater conservation action.[27]

Further to this, VEAC's justification for national park expansion based on supposed gaps in the conservation reserve network was encapsulated in the following statement drawn from its *Draft Proposals Paper* (July 2007):

> Public land occupies some 269,000 hectares of the total Investigation area (1,220,000 hectares) and comprises some 22 per cent of the former extent of River Red Gum forests and related ecosystems. As these ecosystems are poorly represented on public land and under significant threat from damaging processes, VEAC proposes that a substantial area be protected within the conservation reserve system.[28]

However, this contradicted VEAC's own earlier *Discussion Paper* (2006) which had explained that most of the original extent of floodplain river red gum forests was still in existence:

> While most of the original ecosystems of the broad alluvial plains of northern Victoria have been cleared or extensively altered since European settlement, most of the original river red gum forests ..._ remain across the floodplains. Regularly flooded land was generally considered unsuitable for conventional agriculture.[29]

In reality there is no clear agreement on what the pre-European distribution of the river red gum floodplain forests and woodlands actually was. Some scientists and historians contend that it only occupied narrow strips and lower-lying areas along rivers and other watercourses and that, therefore, a considerable portion of today's floodplain did not naturally support heavily-stocked river red gum forests. They believe that it was not until river regulation for irrigated agriculture that changed flooding regimes created conditions conducive to widespread tree colonisation. This view contradicts VEAC's justification for more reservations based upon current red gum forests being only remnants of a much larger area now largely lost.[30]

Veteran forests researcher, Vic Jurskis, believes that much of today's floodplain river red gum forests are neither natural nor very old. After examining the diaries and journals of early settlers such as Edward Curr, explorers such as Charles Sturt, and pioneering foresters such as Manton, he contends that today's extensive areas of dense regrowth red gum forest – such as the combined Barmah and Millewa Forests which straddle the mid-Murray River near Echuca – are occupying large areas that were formerly reed beds or very open grassy woodlands. These areas have apparently been invaded by trees mainly as a result of the cessation of aboriginal burning around 150-years ago, and the substantial reduction of regular flooding due to river regulation.[31] The nation's foremost expert on river red gum silviculture, Barrie Dexter, partly agrees that there has been tree colonisation of the floodplains since European settlement, but also believes that there must have been extensive natural stands of red gum forest on parts of the floodplain prior to settlement, given the recorded wood volumes attributed to early timber cutting.[32]

Accepting that today's floodplain river red gum forests are in-part 'man-made' puts a different context on their management given that the dedication of native forests in national parks is usually predicated on providing the strongest form of protection to natural, high quality biodiversity and heritage.[33] That much of today's forested floodplain did not exist in its current form 150-years ago, suggests that their existing values have been shaped by generations of active human interference and management, thereby significantly diluting any compulsion to conserve them as remnant natural areas in national parks.

Despite this, there is little to suggest that VEAC paused to consider the worthiness of these forest values as they strove to justify the need for more national parks and reserves. A key to this was manufacturing a justification for more reserves by using the technique of Choice modelling to quantify the supposedly high community demand for new reserves through a theoretical cost-benefit analysis.

Choice modelling involves randomly surveying citizens by telephone to ascertain what they would be hypothetically willing to pay for a standard range of change options. According to many economists, the usefulness of Choice modelling is limited by inherent biases due to:

- its integrity being completely reliant on how the current situation and various change options are presented to the survey participants;
- it often reflecting an inflated willingness to pay because no real payments are required;
- the likelihood that survey outcomes are influenced by the budgetary constraints of the participants; and
- because it is difficult to obtain a truly random sample.[34]

Arguably, it was the use of Choice modelling which confirmed the suspicions of most sceptical observers that VEAC's River Red Gum Forests Investigation was less of an objective land use study to inform policy, than an exercise in justifying an already determined (but undisclosed) policy direction – namely, to drastically expand the area of national parks. Particular aspects of the Choice modelling exercise which raised concerns were:

- Survey participants were not fully informed of the current management of the forests, including the conservation works already being undertaken under the *Living Murray Initiative* or other government policies and practices. This implied that there was no conservation management being undertaken in the red gum forests, thereby creating an exaggerated imperative for changes that would address this failing.

- Survey participants were not fully informed about the current state of the forests. For example, information supplied to participants about 'healthy Red Gum trees' nominated 67,000 hectares of forest as being healthy – an area that approximated the area that was already contained in national, State, regional, and other parks. This implied that forests in other public land tenures outside parks (such as State forests) were unhealthy, thereby manufacturing a false imperative for participants to support the creation of more parks to improve forest health.[35]
- The survey simplistically presented change options as an either/or choice, when in reality the interaction between human use and biodiversity conservation is highly complex with many grey areas. For example, the selective harvesting of river red gum forests for timber is not incompatible with forest health as the questions implied. Indeed, such harvesting has long been credited with maintaining or improving the health of these forests by reducing between-tree competition for scarce moisture during drought conditions.
- That almost 50% of survey participants located outside of the VEAC investigation area displayed a substantially greater willingness to pay presumably because they have no direct association with the area and would be unaffected by any adverse socio-economic implications arising from changes to public land tenure, use, and management.[36] This skewed the results considerably in favour of creating more national parks and reserves.
- The survey process did not fully inform participants of the potential socio-economic implications of their choices. Different results may have been forthcoming if respondents were made aware that particular choice options would virtually eliminate local industries.

While VEAC relied on the Choice modelling results to claim there was very strong community demand for a new approach to river red gum forest management, it is worth noting that 42% of those Melbournians approached to participate in the exercise actually declined, presumably because they didn't think the issue warranted their attention. In addition, 34% of Melbourne respondents and 48% of respondents from towns outside the study area had reportedly never visited the Murray River red gum forests.[37] This raises considerable doubts about the credibility of the exercise as a justification for change.

Nevertheless, the hypothetical payments which the survey respondents had nominated for arbitrary environmental improvement options were fed into a Benefit-Cost Analysis which compared them to the estimated actual costs of the market-based impacts of the public land changes supposedly required to facilitate those environmental improvements (such as a closed timber industry). From this hypothetical comparison, VEAC determined "... that the [*value of*] environmental benefits of the VEAC draft recommendations dominate the costs in terms of lost timber, grazing and duck hunting opportunities"[38]

That such an analysis was used to justify VEAC's recommendation of expanded national parks and reserves understandably angered those small local communities required to bear the brunt of actual costs (such as lost jobs and business closures) incurred by changing public land management purely on the basis of what largely far-away people would be hypothetically willing to pay for supposed environmental benefits. This local anger was articulated by the Nationals MP for Northern Victoria, Damien Drum, in a speech to the Victorian Parliament soon after VEAC's final recommendations had been adopted:

> The fact that we will lose a couple of hundred jobs in the middle of an economic crisis... does not worry the government. ... It has choice modelling which is directed at the latte sippers of metropolitan Melbourne, who will always back up the supposed benefits outlined in a VEAC report.[39]

Environmental activism, timber and cattle

VEAC also sought to justify its determination to considerably change public land use by arguing that timber harvesting and cattle grazing were overtly damaging the environment and needed to be either stopped or substantially reduced. This reflected the environmental campaigning which had convinced the government to commission the VEAC to investigate in the first place.

While environmental activists have vilified logging for decades, managed timber harvesting conducted in Victoria's floodplain river red gum forests since the early 1900s has never featured the clearfelling, high intensity slash burning, and export woodchipping that has provoked anti-logging campaigns elsewhere in southern Australia. Instead, it was always based on selective single tree or small patch tree removal with nil or only minimal burning for fire protection purposes. Over its history it produced sawlogs, marine piles and wharf timbers, sleepers, fencing timbers, charcoal and firewood rather than woodchips, latterly with sawmill offcuts chipped and sold as garden mulch. Immediately prior to the VEAC process, it generally supported small, family-run sawmills based on a relatively tiny annual sawlog allocation of around 6,000 m^3 (equating to less than 1% of the annual Victorian native sawlog harvest in 2005-06).[40]

This gentler form of seasonal timber production dictated by forest flooding, was appropriate to the silvicultural characteristics of river red gum forests and, as far as could be remembered, had rarely if ever been subject to vociferous anti-logging protests. Indeed, less extreme environmental activists had at times nominated such small scale, low-impact operations supplying small family-run sawmills, as their acceptable alternative to the more intensive, so-called 'industrial logging' of southern Australia's more productive mountain and foothill forests.

The relatively low environmental impact associated with river red gum harvesting was largely ignored by VEAC, and their major justification for substantially ending it seems to have been an assertion that even without their national park expansion recommendations, the current growth of the forest had slowed to such an extent that the calculated sustainable sawlog yield would no longer be achievable into the future.[41]

VEAC's discussion of current and future timber production was highly contentious. They used undefined factors that made it difficult to determine the underlying basis of their estimates of future forest growth and sustainable yield. These included:

- the use of 'High' and 'Low' site quality classes without explanation of how they were reconciled against the three traditional site qualities (SQI, II, and III) into which the forest had always been classified. This was significant because most of the forest was classified as SQII and therefore could conceivably fall into either of VEAC's defined 'high' or 'low' classes;

- using 'original' and 'recent' growth rates without explaining how they related to historical growth data strongly influenced by flood and drought sequences; and
- referring to 'frequent' and 'reduced' flooding without explanation of what this means in terms of average numbers of flood years (per ten year period) and how this related to 'original' and 'recent' growth.

Compounding these uncertainties was VEAC's reliance on basic growth data supplied by the Department of Sustainability and Environment (DSE) which had previously been acknowledged as an inadequate basis for decisions about future timber production.[42] Consequently, despite VEAC's determinations of future forest growth and sustainable yield being highly questionable, they nevertheless articulated upper and lower limits of sustainable yield contingent on their recommendations for reduced timber industry access into the forest and recommendations for additional environmental water releases being accepted or rejected.

According to VEAC's calculations, even if the previously available area of productive State forest was maintained, the sustainable sawlog harvest would drop by 38%. Critics regarded this as overly pessimistic because it was based upon recent very low, severely drought-affected growth rates continuing in perpetuity, and ignored the likelihood of improvements to environmental watering. Even under the worst climate change projections rainfall was expected to be significantly higher than had occurred during the then severe drought, and the issue of greater environmental water releases was already being addressed by other government initiatives – such as the Northern Region Sustainable Water Strategy and the Food Bowl Modernisation Project – which were expected to increase the frequency of forest flooding.

It is also worth noting that overly restrictive habitat tree prescriptions had, since their introduction in the mid-1990s, been reducing per hectare timber yields and depressing regrowth.[43] Since 2002, the Department of Sustainability & Environment had acknowledged that these prescriptions were having a negative impact on regrowth health and vigour but had yet to act on a stated commitment to revise them. Such a revision could also have made more timber available, boosted the sustainable yield, and better conserved stand health during drought years.

It is also likely that VEAC's investigation substantially understated the socio-economic significance of the region's red gum timber industry. A socio-economic assessment undertaken for VEAC calculated that the industry had a gross value of $9.6 million per annum with employment of 102 jobs, including multiplier or flow-on effects.[44]

The municipal councils in which the red gum timber industry was located firmly believed that this had understated its actual socio-economic value. In challenging VEAC's timber industry assessment, they referred to an alternative analysis from the Latrobe University's Regional Economic Modelling and Planning System (REMPLAN 2.0) which, in 2001, had shown that the total gross value of direct output from timber-related sectors in the North-Central Murray region was $98 million per annum. This was generated from forestry and logging ($9.27M); sawmill products ($19M); furniture ($29.42M); and other wood products ($40.6M). REMPLAN further estimated that 477 persons were employed in the four associated sectors – forestry and logging (59), sawmill products (59), furniture (163) and other wood products (196).[45] These estimates dwarfed VEAC's assessment of the red gum industry's worth and employment.

The Gannawarra Shire Council which encompassed much of the region in which the red gum timber industry was still operating, was particularly vocal in support of the REMPLAN analysis and suggested that VEAC's figures may be deliberately understating the industry's socio-economic contribution so as to downplay the adverse community impacts of its expanded national parks recommendations.[46]

VEAC's socio-economic assessment prepared by Gillespie Economics effectively dismissed the 2001 REMPLAN analysis citing concerns about its reliance on modelling of secondary data (rather than primary data); the inclusion of a much wider industry under the furniture sector category; and potential overlap and consequent double-counting between related sectors.[47] Nevertheless, the size of the discrepancy between REMPLAN and Gillespie Economics was so great as to raise considerable conjecture about which approach more accurately measured the state of the industry and what would be lost as a result of VEAC's national park expansion recommendations.

A subsequent 2009 estimate of the socio-economic value of the NSW red gum timber industry on the northern side of the Murray River, lends weight to the contention that VEAC had significantly understated the socio-economic value of Victoria's floodplain red gum timber industry. In a submission to the NSW Natural Resource Commission's investigation of that state's public land red gum forests, the NSW Forest Products Association estimated the gross value of its red gum timber industry at $72 million per annum supporting 550 jobs. This was based on a bigger NSW industry with access to approximately three-times the total log volume (of all classes) compared to the Victorian red gum industry.[48]

The other forest use targeted by VEAC was cattle grazing which had become entrenched in local heritage over the previous 120-years. Presumably to emphasise its destructive impacts, VEAC routinely described it as 'intensive grazing' despite grazing pressure in red gum forests being typically light. In 2005, Jansen and Robertson had classified "all forest sites as low grazing (less than 5 dry sheep equivalent/hectare/yr)" when they studied the ecological condition of riparian river red gum forests in a 620 km belt on the NSW side of the Murray.[49]

On the Victorian side of the river, cattle and some horse grazing in the then-named 'Barmah Common' (now the Barmah Forest) began around the mid-1880s following an approximately 40-year period of spring/summer sheep grazing. The forest was initially divided into two sections, each under a manager who had been nominated by the stock owners to tend their animals on the Common. Gradually, Government control was increased, and in 1909 the then Forests Department appointed a herdsman to look after the stock.

By 1951, the Forests Commission's concerns about lack of tree regeneration led to the exclusion of grazing from 6,400 hectares in the eastern section of the forest. Observations in red gum regrowth less than 6 years old provided some direct evidence of cattle damage – in sparsely stocked areas nearly every seedling showed signs of butting, trampling and browsing damage. Low seedling stocking and multi-stemmed regrowth suggested that regular and heavy grazing was the cause.[50] Jacobs (1955), in commenting on grazing of red gum forests in NSW, had also noted that seedlings were eaten by sheep, cattle, horses, rabbits and kangaroos.[51]

From the 1960s, cattle were dehorned to limit structural damage to open-grown young regrowth stands and summer and winter stocking levels were adjusted throughout the year in accordance with winter/spring flooding and the availability of summer feed. Grazing management was determined in consultation with the Barmah Grazing Advisory Committee and was under the full time supervision of a herdsman.

A 1955 Barmah Forest tourist guide describes 3,000 cattle grazing in the forest each summer.[52] However by the early 2000s, grazing in the forest was being strongly regulated by the Department of Sustainability & Environment in conjunction with the Barmah Forest Cattlemen's Association. During summer, 1500-2000 head of cattle were normally permitted to graze across 29,660 ha of public land – equivalent to at most just one beast per 15 hectares. In winter, the herd was reduced to around 800 head equating to just one beast per 37 hectares.

There is no doubt that cattle can damage soils and waterways under wet conditions. In order to mitigate damage to waterways and river banks, four dams had been constructed in the middle of the forest to draw the cattle away from vulnerable areas. In addition, stock numbers were reviewed in response to seasonal conditions that could affect the vulnerability of the environment.[53] This enabled grazing pressure to be better regulated in accordance with conditions to minimise environmental damage.

Research and experience in NSW state forests had also found that grazing could be actively utilised as a management tool to improve ecological outcomes.[54] This confirmed long-standing observations that carefully managed cattle grazing could be beneficial in controlling exotic weeds such as Patterson's curse that otherwise detrimentally competes with native flora.[55] An earlier study had also shown that cattle could be beneficial to the early survival of red gum seedlings by reducing competition for moisture by grass and weeds.[56] There was also plenty of anecdotal evidence that grazing could reduce fuel loads and thereby assist fire management, although this may have been more apparent in the past when more intensive 'set grazing' had been used.

In Victoria, the positive environmental benefits of well managed grazing had been recognised by the Department of Sustainability & Environment (DSE), which had already developed a set of 'ecological grazing principles' acknowledging its value as " ... an ecological management tool to achieve biodiversity benefits and to restrict adverse changes in the floodplain forest environment" and in assisting "... in the consolidation and recruitment of native plant species by helping to maintain or shift the vegetation composition to the EVC benchmark".[57]

The DSE had learnt much from the NSW experience and since 2003 had trialled various grazing regimes in the Lower Goulburn State Forest. These had confirmed that ecological benefits could be realised where grazing was carefully controlled in both timing and stock numbers targeted to meet nominated objectives. The success of these trials had led DSE to produce a draft *River Red Gum Forest Ecological Grazing Strategy* (prepared in June 2005), which had demonstrated a commitment to continued, but more closely regulated cattle grazing.

Within a matter of months VEAC's River Red Gum Forests Investigation would begin to tear down this committment in deference to the popular but questionable notion that completely removing stock grazing from the forests would automatically improve conservation outcomes.

In reality, assessing the environmental impact of cattle grazing in river red gum forests is difficult because their long history of disturbance means that there are few (if any) areas of original (ungrazed pre-European) vegetation that could benchmark a baseline study. This was acknowledged by Jansen and Robertson in their 2005 study of grazing in NSW red gum forests which found "... little clear evidence of significant impacts of grazing on biodiversity at low stocking rates ... most likely due to a lack of reference sites in near pristine condition."[58]

Further compounding the capability to quantify the impacts of grazing was the added complexity of effects caused by alterations to natural frequencies of flooding and fire, and changed faunal composition stemming from European settlement. These difficulties were concisely summarised by researcher Amy Jansen when presenting a paper to the Royal Society Meeting on the Barmah Forest in June 2005, during which she pragmatically noted:

> We don't know what the forest was like before due to confounding factors of fire, flood and native fauna, seasonal condition, and pre and post-European land management practices. ... We can't go back to what it was.... Under circumstances of low stocking it is very hard to say exactly what specific damage grazing does due to factors such as variable stocking rates & grazing intensity; variable timing of grazing either periodic or year round; and flood regimes and variable seasonal conditions which combine to influence the effects.[59]

In view of this uncertainty, the question of whether to suddenly cease all cattle grazing was far from the 'no-brainer' which VEAC and its supporters regarded it as. Arguably there was little justification for a complete cessation of cattle grazing and considerable doubt over whether removing it would provide any significant environmental benefit. On the contrary, there was a strong case that its cessation would worsen conservation outcomes by removing a useful ecological management tool.

Fire management

Together with the removal of cattle grazing, two other VEAC recommendations were criticised as having significantly increased the bushfire threat:

- reducing opportunities for domestic firewood collection; and especially
- deliberately encouraging a doubling of coarse woody debris on the forest floor in the public lands closest to the Murray River ostensibly to improve wildlife habitat.

Deliberately fostering a substantial increase in woody fuel loads whilst removing activities, such as grazing and firewood collection, that help keep the forest open was meant to benefit native wildlife by recreating pre-European forest habitat. However, it raised local concerns that prolific regeneration of exotic weeds and the root parasite, dwarf cherry (*Exocarpos strictus*), would further lessen the health of river red gums while creating dense cover that would both add to the fire risk and favour populations of foxes and feral cats, thereby counteracting any supposed wildlife benefits.[60]

Of further concern was the means by which the need for increased woody debris was apparently determined and how natural this actually was. The suggestion that current levels of coarse woody debris are substantially less than what naturally occurred in the forest is derived from a 1997 BSc Honours Thesis by Reuben Robinson based on measurements taken in the NSW Millewa State Forest. A reading of the thesis shows Robinson's claim that the natural forest structure included 125 tonnes/ha of woody debris, had come from the measurement of just three elongated 0.1 hectare quadrats within a single 3.5 hectare area of what he considered to be 'old growth' forest.[61]

According to Jurskis, this 'old growth' forest was in fact 60-year old regrowth growing in a former reedbed swamp that had dried out due to the changed flooding regime created by the building of the Hume Reservoir in the 1930s. Furthermore, at least one of Robinson's quadrats included a windrow of woody debris cleared from a management track. This had considerably inflated the amount of debris then used to set a bench-mark for supposedly natural debris distribution throughout the forest. However, it is apparent that woody debris is not uniformly spread throughout the forest, but is naturally concentrated in heaps created by the periodic movement of floodwaters.[62]

In the face of community concerns about fire, VEAC's Chairman reportedly stated at a public forum that any resultant increased fire threat would be combatted by more controlled burning.[63] However this is impractical given that the sensitivity of river red gums to fire means that these forests are only suited to small-scale fuel reduction burns that must be very carefully controlled. Attempting larger burns with limited resources makes it difficult to control fires to the extent necessary to protect the trees. The potential for this to severely damage the forest was exemplified by a 25 hectare fuel reduction burn in the Barmah Forest (at Browns Camp) which killed 38 large red gums during October 2008.[64]

Even prior to the VEAC River Red Gum Forests Investigation, local communities lacked confidence in the capability of the Department of Sustainability and Environment to adequately deal with fire on public lands. The fragmented nature of these particular public lands, the fact that many are smallish blocks surrounded by private land, and the relative lack of government resources means that Country Fire Authority (CFA) volunteers have often been required to take the lead role in controlling bushfires in river red gum forests.

Andrew Ford, the CEO of Volunteer Fire Brigades Victoria (VFBV) which represents CFA firefighters, voiced the concern that VEAC's recommendations to increase woody fuel levels to enhance biodiversity would add another dimension to fire suppression thereby extending the time and effort required to contain and extinguish bushfires; and that the cessation of grazing would remove a valuable fuel management tool that would further exacerbate fire management problems.

Furthermore, VFBV was also concerned that VEAC's recommendations would ultimately change the demographics of the region and reduce the numbers of volunteers willing and able to fight bushfires:

> ... it is appropriate for VEAC to recognise that the people who may lose their employment due to this proposal, their grazing rights, access to firewood, and social activities, are in many cases the very same people who have in the past and will in the future be expected to defend the forest from the impact of wildfire. ... Whilst the principal beneficiaries of VEAC's recommendations will come from outside the immediate area, including metropolitan Melbourne, it is not those people who will respond in defence of the environment and other public and private assets when fire threatens.[65]

Ignoring local sentiment to ram through a political agenda

In July 2008, following a public engagement process which had generated over 8,000 submissions from key stakeholders and the wider community, VEAC released its *Final Report* containing recommended changes to public land tenure, management and use in Victoria's floodplain river red gum forests.[66] The major recommendations were, as expected, a tripling of the area of national parks, and an almost 90% reduction to the area of multiple-use State forests. Associated with this change was the virtual removal of all cattle grazing and the drastic restriction of timber harvesting to just a quarter of the forest area which had previously been available for that purpose.

In accordance with the *Victorian Environmental Assessment Council Act 2001*, the Victorian Government had a six-month period in which to decide whether, or to what extent, it would adopt VEAC's final recommendations. The Rivers and Red Gum Environment Alliance, which opposed most of VEAC's recommendations, saw this as a last opportunity to influence the government's deliberations. Accordingly they launched their own '*Conservation and Community*' plan at Victoria's Parliament House a week later.[67]

This was a glossy, professionally presented and detailed 140-page publication which outlined a suite of alternative proposals formulated on behalf of the 25 community bodies that stood to be most affected by VEAC's final recommendations. Far from being a blanket opposition to VEAC's recommendations, it was a thought-provoking attempt to find real solutions that could bridge the divide between local communities and recommendations which had been made largely on behalf of an urban demographic that had no material stake in their consequences.

This unprecedented show of well organised community opposition caused the Victorian Government to pause and ponder how it would address such local disquiet during the intervening six months before it was required to endorse or reject VEAC's final recommendations.

In early August 2008, the Brumby Government announced that a Community Engagement Panel (CEP) comprised of four appointees would be established to develop its response to VEAC's final recommendations. Their decision to take this path, instead of simply 'rubber stamping' the recommendations, was commendable but understandably outraged the environmentalist constituency which was all-too-aware of the previous Bracks government's promise, prior to the 2006 state election, to "create new red gum national and State parks if recommended by VEAC".

Soon after, a report in *The Age* newspaper voiced concerns that the formation of the Community Engagement Panel (CEP) could lead to no new national parks being declared. When asked if the CEP process might result in no new national parks, the government's Environment Minister, Gavin Jennings, replied, "That's a very unlikely outcome. Those who are most anxious about this ... would be better not to be anxious."[68]

When the Secretary of the Department of Sustainability & Environment, Peter Harris, informed his staff of the establishment of the Community Engagement Panel he noted that with six months to report to Government, "... the review group will ensure it spends this time in-part on developing community support' for VEAC's final recommendations".[69] This raised concerns amongst the Rivers and Red Gum Environment Alliance. While its members had hoped that the CEP would seriously consider adopting at least some of the alternative proposals and compromises outlined in their *Conservation and Community* plan, the DSE Secretary seemed to be signifying that the CEP's primary role was to legitimise VEAC's final recommendations under the guise of further community consultation.

This concern was soon reinforced by the CEP's Terms of Reference which specified that it must meet commitments that were "consistent with Victorian Government policy and regional commitments (including election commitments)". It was also required to "investigate how the government can address the (VEAC) recommendations while ameliorating any potential adverse impacts on regional economic activity and employment opportunities".[70] While this latter point sounded promising, the groups specifically mentioned regarding this requirement didn't include either the timber industry or cattle grazing interests. Clearly, there was never any intention to compromise with them.

The growing suspicion that the CEP process would be little more than an illusion of open and objective consultation was further entrenched when the Department of Sustainability and Environment advertised three positions to coordinate the government's response to the VEAC final recommendations. Their job descriptions included the following key responsibilities:

- "Develop, manage and lead projects that support the government's commitments to expanding the parks and reserve system in accordance with departmental project management frameworks".
- 'Promote coordinated and integrated policy development and responses ... to support improved public land estate outcomes for an expanded parks and reserve system".

As such words were being used when the government was potentially up to six months away from having to decide whether to agree to enact VEAC's recommended expansion of national parks and reserves, they strongly reinforced fears that the decisions had already been made.

Ultimately the Community Engagement Panel would go on to meet 38 'key stakeholders' and some 'individual stakeholders' in the few months before it reported back to the government in late November 2008. When the panel met with the Rivers & Red Gum Environmental Alliance in September, they explained that their role was to find community support for VEAC's final recommendations while appearing to negate the associated problems. This again suggested that no serious consideration would be given to the thought-provoking alternatives and compromises outlined in the Alliance's *Conservation and Community* plan.

Subsequently, one of the four CEP appointees would later recall, the panel's chairman, Craig Cook, was not the slightest bit interested in the RRGEA's *Conservation and Community* plan. As the only member of the panel with specific forest management expertise, this appointee tried to encourage the CEP to view its task as an opportunity to consider alternate ways of moving forward in relation to those aspects of VEAC's recommendations which were strongly contentious amongst the local community. Unfortunately, it appears that acting on this suggestion was never even countenanced.[71]

In hindsight, it is difficult to view the Community Engagement Panel as anything other than a farce which wasted the time of all those involved in order to give an appearance of serious consideration to VEAC's final recommendations. After the CEP submitted its report, the Government went on to accept almost all of VEAC's final recommendations with the exception of some concessions given to hunting as part of a last minute deal to placate opposition by hunting and shooter groups, and some limited concessions to continued grazing of some public water frontages. It appears that hunting and shooter groups which had been part of the RRGEA were 'bought-off' in the last few days before the Government accepted the VEAC recommendations, in an attempt to minimise community anger and disquiet.

The aftermath: improved forest management?

By late 2015, more than seven years had passed since the Victorian Government had locked away almost all of the state's floodplain river red gum forests in a swathe of new national parks and other conservation reserves along the Murray River. Yet, despite the passage of time, aggrieved locals in the small river town of Barmah still harbored barely concealed resentment that their lifestyles and livelihoods had been politically sacrificed to appease the urban 'green' sensibilities of inner Melbourne.

With fading 'anti-national park' signs still adorning some roadsides, the local caravan park proprietor's response to a question about the new national park – "we were all devastated by what happened" – probably exemplified the community sentiment. This was being reinforced by contemptuous tales such as of a key road bridge sitting unusable and unrepaired in the forest for three years, and of a formerly successful forest tourism venture being partly dismantled and carted away during the night because ongoing maintenance was too costly for the cash-strapped new manager, Parks Victoria – reportedly to the despair of local indigenous people who wished to continue its operation.[72]

As has become a hallmark of national park expansion campaigns throughout the country, the emotive supporting arguments had included optimistic predictions of an accompanying surge in eco-tourism which would supposedly more than make up for the socio-economic losses incurred by ending commercial forest uses. At Barmah, the caravan park proprietor was adamant that there had been little or no change in visitor stays since the national park declaration, anecdotally confirming that the region's socio-economic hit had not been counter-acted by more tourism.

The question of whether tourism can adequately replace commercial forest uses after a national park declaration was in-part examined by the 2013 NSW Parliamentary inquiry into the management of that state's public lands. The inquiry examined the Pillaga forest region where 56% of the State forests had been converted to national parks and other reserves in 2005; and the Riverina river red gum forests where 80% of the State forest had been converted to parks and reserves in 2010.

In both cases, most local participants to the inquiry rejected the notion that there had been any noticeable increase in tourism attributable to new national parks. Some noted a negative response as formerly regular visitors were discouraged by tighter restrictions on free-of-charge camping, firewood collection, and pets in areas which they'd previously frequented. In the Pillaga, it was noted that even if more people came to the region, they often by-passed the small towns where the former timber businesses had been located and camped free-of-charge, thereby providing little socio-economic benefit.[73]

An additional problem at Barmah was the increased threat of fire fostered by the removal of cattle grazing and firewood collection compounded by the plan to drastically increase the volume of coarse woody debris lying about the forest floor. By early 2015, local residents were describing the region's river red gum forests as an extreme fire threat. One who owns property adjacent to the iconic Barmah Forest noted:

> The fuel loads and fire danger at Barmah is the worst it has ever been. ….We are at great fear in the near future we will lose our forests because of their neglected management. The conditions of the forest have been deteriorating, weed burden has increased and that's just some of the issues.[74]

Another key issue was the reduced access into the forest. While a Barmah Forest tourism brochure from 1955 noted that the forest contained 200 miles (320 km) of graded roads and tracks,[75] by 2015 this had fallen to just 70 km in large-part due to lack of national park funding.[76]

Worse still, was that much of the remaining ungraded track network had reportedly deteriorated to a point where it was in parts impassable even to 4WD vehicles, thereby raising serious concerns about the capacity to quickly attend to bushfires.[77] Indeed, the risk is now reportedly so great that the Country Fire Authority's volunteer fire-fighters have refused to enter the forest.

It is naïve to believe that VEAC's River Red Gum Forests investigation ever seriously countenanced such local concerns. It seems to have been more about salving urban eco-sensibilities fostered by years of environmental campaigning under a political climate where increased 'environmental protection' was being demanded and expected. This expectation was irrespective of whether or not additional protection was warranted, or was in the best interests of the adjacent communities or even the environment itself.

Indeed, the VEAC process arguably epitomises the extent to which the prevailing 'conservation culture' has infected political and bureaucratic agendas and exemplifies how this can pervert land management policy determinations by overriding processes that were formerly founded on truly objective study and sincere community engagement.

A notable facet of this culture is an irrational refusal to accept that environmental values may be being effectively conserved on public land tenures outside of national parks and conservation reserves. The associated determination to swallow-up these other tenures – particularly multiple-use State forests – by expanding the area of national parks and reserves, is rooted in a misguided belief that environmental protection can only be achieved by quarantining areas from commercial human use. This is strongly at odds with the known benefits attributable to the integration of active environmental management in the planning and regulation of those uses.

Sadly, this VEAC process also demonstrated the futility of participating in public land use investigations undertaken by a Government pursuing an agenda of change. The community collective which invested the most into participating in the VEAC process – the Rivers and Red Gum Environment Alliance (RRGEA) – estimated that the combined value of the volunteer time of its members, plus real expenditure on consultants, had totalled around $500,000. Despite this unprecedented effort in representing the strength of local opposition to the Government's change agenda, the RRGEA drew no significant concessions from VEAC.

Arguably, this episode has significantly hardened the collective mindset of rural communities forcibly changed for no reason other than to satisfy the whims of occasional visitors from the city. Without the full support and enthusiasm of these now disaffected locals, the effective management of these forests may well be problematic.

Perhaps one thing that the RRGEA did achieve was to shine a light on a process skewed to match the requirements of a pre-determined policy outcome. Rightly or wrongly, Victorian public land use investigations have since been dogged by the perception that merely instigating and spending millions of dollars undertaking them confers an inevitability of change; and that recommitting to the status quo will never even be countenanced as an option irrespective of the strength of argument that supports it.

Unfortunately, there is little to suggest that the city-centric political establishment has learnt anything from processes such as this. In his 2012 autobiography, former Victorian Premier, Steve Bracks, confirmed just how out-of-touch he was when he noted that:

> More national forest and marine parks were created by our government than by any government in Victoria's history, but this didn't happen without a fight ... However, as each reform was completed, the protests seemed to vanish into thin air, replaced by community support for the changes and the tourism jobs that flowed from them – new work opportunities that were three to four times greater than any related job losses.[78]

As of late 2017, the Victorian Government has never verified whether there have been any socio-economic benefits associated with creating either the Great Otway National Park in 2005, or the Barmah (red gum forest) National Park in 2008.[79] Despite this, it is reportedly seriously considering declaring another huge, so-called 'Great Forest National Park' in central Victoria that is yet again being promoted by its supporters on the premise of providing a massive boost to tourism.

Chapter 6 Endnotes

1 The official website of the RAMSAR Convention on Wetlands.

2 Ramsar listing requires areas to meet the conservation management criteria of 'wise use' enshrined in Article 3.1 of the Convention on Wetlands signed in the Iranian city of Ramsar in 1971. In 1987, the term was defined as *'the sustainable utilisation of wetland resources in such a way as to benefit the human community while maintaining their potential to meet the needs and aspirations of future generations'*. That this does not preclude human resource use is significant and concurs with the reality that these river red gum forests were managed under regimes that included timber harvesting and cattle grazing for generations before, and for the duration of their Ramsar listing. The majority of the Murray River floodplain red gum forests were Ramsar-listed in 1982 (Victoria) and 2003 (NSW).

3 D. Macleod and B.D. Dexter, 2005, *Barmah-Millewa Forum: A Short History of Community Involvement in the Barmah-Millewa Forest on the Murray River*, Barmah-Millewa Forum, ISBN 1921038527.

4 Myer Foundation 2001, *Annual Report 2000-01: Environment and Water Grants*. Myer Foundation website.

5 Victorian Environmental Assessment Council 2007a, *River Red Gum Investigation Final Report*. Appendix 9 lists the Ecological Vegetation Classes (EVCs) found in their River Red Gum Forests Investigation Study Area. It is presumed that river red gum dominates the following EVCs: Intermittent Swampy Woodland, Riverine Grassy Woodland, Riverine Swamp Forest, Riverine Swampy Woodland, Shrubby Riverine Woodland, Sedgy Riverine Forest, Sedgy Riverine Forest/Riverine Swamp Forest Woodland Complex, Grassy Riverine Forest, Floodplain Riparian Woodland, and Grassy Riverine Forest/Riverine Swamp Forest Complex.

These EVCs occupy a total area of 115,300 ha on both public and private land in the Investigation Study Area, and 86,400 ha of just the public land in the Investigation Study Area.

Since 2001, the red gum timber industry in the Investigation Study Area had been restricted to the Department of Sustainability and Environment's Mid-Murray Forest Management Area (FMA) after the Victorian Government ceased to allocate licences for harvesting in the small Mildura FMA. DSE's Mid-Murray FMA Plan (April 2002) noted that 41,160 ha of red gum forest and woodland was the net area available for timber harvesting on public land, which is less than half of the red gum dominated forest and woodland on public land in the VEAC Investigation Study Area.

Similarly, although cattle grazing was permitted more widely it was largely only a regular use in around 30,000 hectares in the Barmah Forest which is less than half of total area of red gum dominated EVCs.

6 D. Macleod and B.D. Dexter, op. cit.

7 The predecessors to the Victorian Environmental Asseesment Council were the Land Conservation Council (LCC) from 1971 to 1997, and the Environmental Conservation Council (ECC) from 1997 to 2002.

8 Rivers and Red Gum Environment Alliance 2008a, *Conservation and Community: A community plan for the multiple use management of public lands in VEAC's River Red Gum Investigation Area*. A submission to the Victorian Environmental Assessment Council's River Red Gum Forests Investigation.

9 The Otways and Goolengook reports can be viewed on the VEAC website.

10 Central Victorian Private Forestry Development Committee 2003, *A socio-economic study of the forest industries in Central Victoria*, prepared for the Committee by URS Forestry (unpublished).

11 Professor J.K. Vanclay and Dr. B.J. Turner, 2001, *Evaluation of data and methods for estimating the sustainable yield of sawlogs in Victoria*, Report of the Expert Data Reference Group, Department of Sustainability & Environment.

12 ALP 2006, *Victoria's National Parks and Biodiversity Policy*, p. 9, paragraph 6. The policy was released just prior to the 2006 Victorian election.

13 This was a personal aside from a political advisor to the then Victorian Minister for the Environment, The Hon. John Thwaites.

14 Kristin Favaloro, 2006, "Tentative nod for National Parks plan", *Shepparton News*, 20 November 2006.

15 The VEAC Councillors were: Chairman, Duncan Malcolm, described as having 'a long career in natural resource management' and having held administrative roles in coastal management, tourism and the irrigation industry; Associate Professor David Mercer described as having 'a background in natural resource management, recreation and tourism' and a Fellow of the Environment Institute of Australia and NZ; Professor Barry Hart described as having 'expertise in environmental science, particularly in water quality and ecological risk assessment' and having 'received several awards for his work in the scientific underpinning of natural resource management'; Ms. Jan Macpherson, a lawyer 'with expertise in environmental and planning law' and with experience in indigenous heritage and land management in northern Australia; and Ms Jill McFarlane, with a background in farming and social work who had been on the board of the North Central Catchment Mangement Authority and had 'experience in the complexities of natural resource management issues across public and private land'.

16 This was reportedly said to a young forester who had unknowly met a female VEAC staffer at a Melbourne party and was enquiring how she earned her living.

17 Dr James Fitzsimons, 2006, "Public Land Use Planning Using Bioregions and other Attributes: Determining the Study Area of the VEAC River Red Gum Investigation". In: *Proceedings of the Royal Society of Victoria*, Volume 118, Number 1.

18 Paul Madden and Colin Wood, 2007, personal comments reflecting on their membership of the VEAC River Red Gum Forests Investigation Community Reference Group. Mr Madden from Koondrook was representing the red gum timber industry and Mr Wood from Melbourne was representing the Sporting Shooters Association of Australia.

19 Ibid.

20 Paul Madden, 2007, personal comments as a representative of the river red gum timber industry.

21 Rivers and Red Gum Environment Alliance 2008b, *Pre-ordained and Compromised: The Victorian Environmental Assessment Council's River Red Gum Forests Investigation (2005-08)*, 31 pp (unpublished).

22 The Terms of Reference were outlined in the 'Summary brochure – overview of information presented in the Final Report'. It can be downloaded from the VEAC website.

23 The principles of 'ecologically sustainable development' as defined in Part 1, Section 4 of the *Commissioner for Environmental Sustainability Act 2003* (Act No. 15/2003) [Note: My emphasis]

- Ecologically sustainable development is *development that improves the total quality of life, both now and in the future,* in a way that maintains ecological processes on which life depends.
- The objectives of ecologically sustainable development are –
- to *enhance individual and community well-being and welfare by following a path of economic development that safeguards the welfare of future generations*;
- to provide for equity within and between generations;
- to protect biological diversity and maintain essential ecological processes and life support systems.
- The following are considered as guiding principles of ecologically sustainable development –
- that decision-making processes should effectively *integrate both long-term and short-term economic, environmental, social and equity considerations*;
- if there are threats of serious or irreversible environmental damage, lack of full scientific certainty should not be used as a reason for postponing measures to prevent environmental degradation;
- the need to consider the global dimension of environmental impacts of actions and policies;
- the need to develop a strong, growing and diversified economy which can enhance the capacity for environmental protection;
- the need to maintain and enhance international competitiveness in an environmentally sound manner;
- the need to adopt cost-effective and flexible policy instruments such as improved valuation, pricing and incentive mechanisms;
- the need to facilitate community involvement in decisions and actions on issues that affect the community.

24 Victorian Environmental Assessment Council 2006, *River Red Gum Forests Investigation Discussion Paper*, Appendix 8, p. 406. It can be downloaded from the VEAC website.

25 Since 1982, both the Barmah and Gunbower State Forests that comprise the bulk of the VEAC River Red Gum Forests Study Area had been listed under the Ramsar Convention as wetlands of international importance. The Ramsar Convention does not preclude human resource use and maintains that sustainable use is entirely com-

patible with wetland conservation. Timber harvesting and cattle grazing had occurred in these forests for the duration of the Ramsar listing until the VEAC investigation.

26 Victorian Environmental Assessment Council 2007b, *River Red Gum Forests Investigation Draft Proposals Paper*, Chapter 3: Public Land Use Recommendations – National Parks, p. 24.
27 Institute of Foresters of Australia, *Response to the VEAC Riverine Red Gum Forests Investigation Recommendations*, October 2007.
28 Victorian Environmental Assessment Council, 2007b, op. cit., Chapter 2: General Recommendations, p. 8.
29 Victorian Environmental Assessment Council 2006, op. cit., Chapter 14, p. 206.
30 N. Cameron, 2015, "River red gum forests in the Murray Valley – a look at past and current management", *The Forester*, February 2015.
31 V. Jurskis, 2015, *Firestick Ecology – Fairdinkum science in plain English*, Connor Court. Chapter 11: *Saving forests that never were*, pp. 225-238.
32 Barrie Dexter, 2017, Personal comments, January 2017.
33 N. Cameron, 2015, op. cit.
34 Ernst & Young 2007, *Economic Review of the VEAC RRG Forests Investigation Draft Proposals*, prepared for Field & Game Australia as a submission to the VEAC River Red Gum Forests Draft Proposals. Section 4.2, pp. 18-19
35 Institute of Foresters of Australia 2007, op. cit.
36 Ibid.
37 Ernst & Young, 2007, op. cit.
38 Gillespie Economics, DCA Economics and Environmental & Resource Economics 2007, *River Red Gum Forests Investigation – Socio-Economic Assessment of Draft Proposals Paper*. In: *River Red Gum Forests Investigation Draft Proposals Paper*, Victorian Environmental Assessment Council, July 2007. Appendix 1: Executive summary of social and economic assessment of proposed recommendations.
39 Damien Drum, Nationals MP for Northern Victoria, speech to the Victorian Legislative Council. In: *Hansard*, 12 March 2009.
40 VicForests 2007, *VicForests Annual Report 2006*, p.29: The statewide native sawlog harvest for the 2005-06 financial year was 673,000 m^3.
41 Victorian Environmental Assessment Council, July 2007, op. cit., Appendix 7, p. 131.
42 Professor J.K. Vanclay and Dr. B.J. Turner, 2001, op. cit.
43 L. Bren, 2001, *Estimates of loss of growth due to habitat tree retention in riparian river red gum forests*, by Associate Professor Leon Bren, Department of Forestry, University of Melbourne for the Victorian Association of Forest Industries, March 2001.
44 Victorian Environmental Assessment Council 2008, *River Red Gum Forests Investigation Final Report*, Chapter 4, pp. 102-103.
45 Rivers and Red Gum Environment Alliance 2008a, op. cit., p. 48.

46 Gannawarra Shire Council 2007, Submission to the Victorian Environmental Assessment Council's *River Red Gum Forests Investigation Draft Proposals Paper*.

47 Gillespie Economics, DCA Economics and Environmental & Resource Economics 2008, *River Red Gum Forests Investigation – Socio-Economic Assessment Final Report*, pp. 125-126. In: *River Red Gum Investigation Final Report*, Victorian Environmental Assessment Council, Appendix 1.

48 NSW Forest Products Association 2012, Submission to the NSW Parliamentary Inquiry into the Management of Public Land in NSW, 15 August 2012.

49 A. Janson and A. Robertson, 2005, "Grazing, ecological condition and biodiversity in riparian river red gum forests in south eastern Australia". In: *Proceedings of the Royal Society of Victoria*, Volume 117, No.1, pp. 85-95

50 B.D. Dexter, 1970, *Regeneration of river red gum, Eucalyptus camaldulensis*, MSc.For Thesis, School of Forestry, University of Melbourne (unpublished).

51 M. Jacobs, 1955, *Growth habits of the eucalypts*, Forestry and Timber Bureau, Department of the Interior, Canberra, Australia.

52 Extract from 'The Barmah Forest and Surrounding Area – A Tourists Guide' (1955) reproduced in the Nathalia and District *Red Gum Courier*, 6 November 2015.

53 P. O'Connor and B. Wehner, 2007, *Barmah Forest Understorey Vegetation Condition Assessment to Assist in Cattle Quota Determination*, Department of Sustainability & Environment.

54 NSW Forests had already developed an *Ecologically Sustainable Forest Management Grazing Strategy for the Riverina Region* based on the capability for grazing to be carefully managed to manipulate vegetation structure towards a more desirable condition.

55 Barmah Forest Cattlemen's Association 2006, *Cattle grazing in the Barmah Forest*, a submission to the Victorian Environmental Assessment Council's *River Red Gum Forests Investigation Discussion Paper*.

56 B.D. Dexter, 1970, op. cit.

57 Victorian Environmental Assessment Council 2006, op. cit., Chapter 13, p. 202.

58 A. Janson and A. Robertson, 2005, op. cit.

59 Paraphrased comments by Amy Jansen in presenting her jointly authored paper (with A. Robertson), *Grazing, ecological condition and biodiversity in riparian river red gum forests in south eastern Australia*, to the Royal Society of Victoria meeting on the Barmah-Millewa Forest, Melbourne, 18-19 June 2005.

60 A. Dickins, 2007, Submission by Audrey Dickins, Chairperson, Gunbower Island State Forest User Group, to the Victorian Environmental Assessment Council.

61 Rueben Robinson, 1997, *Dynamics of Coarse Woody Debris in Floodplain Forests: Impact of Forest Management and Flood Frequency*. B.Sc. Honours Thesis, School of Science and Technology, Charles Sturt University, NSW.

62 V. Jurskis, 2015, op. cit.

63 A. Dickins, 2007, op. cit.
64 Prescribed burn N12Shep – Browns Camp, Barmah State Park (16 October 2008).
65 A. Ford, 2007, Submission by Andrew Ford, Chief Executive Officer, Volunteer Fire Brigades Victoria, to the Victorian Environmental Assessment's Council.
66 Victorian Environmental Assessment Council 2008, op. cit.
67 Rivers & Red Gum Environment Alliance 2008, *Conservation and Community – A Community Plan for the Multiple Use Management of Public Lands in VEAC's River Red Gum Forests Investigation Area.*
68 Adam Morton, 2008, "Government up a gum tree over protection plan", *The Age*, 9 August 2008.
69 Internal Departmental memo to DSE staff, 12 August 2008.
70 *River Red Gum Community Engagement Panel Report*, 28 November 2008.
71 Private notes of Barrie Dexter, including records of discussion and correspondence with Bob Smith (forestry consultant and veteran forestry bureaucrat), who was one of the four members of the Community Engagement Panel. The other panel members were Craig Cook, Chairman (Deputy Chair of Goulburn-Murray Water), Joan Burns (Chair of the Mallee Catchment Management Authority and dryland farmer) and John McQuilten (retired Member for Ballarat Province in the Victorian Legislative Assembly).
72 The author visited the area in December 2015.
73 *Management of public land in New South Wales*, Legislative Council General Purpose Standing Committee No. 5, Report 36 (May 2013) Chapter 5: Case Study – River Red Gum Forests (p.61); Chapter 9: Case Study – Pilliga Forest (p. 119); Chapter 13: Economic Impacts (p. 235).
74 David Lee, 2015, "Grazing ban battle reignites – Barmah residents concerned over fire danger", *Shepparton Adviser*, 25 March 2015.
75 Extract from 'The Barmah Forest and Surrounding Area – A Tourists Guide' (1955) reproduced in the Nathalia and District *Red Gum Courier*, 6 November 2015.
76 Barmah Forest Preservation League Press Release on a meeting with Parks Victoria and DELWP fire management personnel, reported in the Nathalia and District *Red Gum Courier*, 18 December 2015.
77 Ibid.
78 Steve Bracks with Ellen Whinnett, 2012, *A Premier's State*, Melbourne University Press p. 278.
79 In April 2016, the Victorian Division of the Institute of Foresters of Australia wrote to Victorian Minister for the Environment, Lisa Neville, to enquire whether any work had been undertaken to verify claimed socio-economic benefits derived from creating the Great Otway National Park in 2005 and the Barmah and Gunbower (red gum forest) National Parks in 2008. The Government response confirmed that the only work showing these supposed benefits was that done by VEAC before the National Parks had been declared.

7

Politics and bureaucracy 2:
Reserving Tasmanian forests (2010-14)

"No science is immune to the infection of politics and the corruption of power"

Jacob Bronowski

If a picture says a thousand words, the June 2014 photo of buoyant Labor Party luminaries celebrating an extension to the Tasmanian Wilderness World Heritage Area (TWWHA) speaks volumes about the common cause of left-of-centre politics and environmental extremism.

Downloaded from the Facebook page of Federal Labor MP, Mark Butler, the photo shows a beaming Butler squatting alongside former Federal Labor Environment Minister, Tony Burke, and ALP National President, Jenny McAllister. Standing behind them are a group of ecstatic supporters, including Wilderness Society leaders Vica Bayley and Lyndon Schneiders, all crowded around a crudely routed wooden sign (erected alongside a major forestry road), welcoming visitors to the 'Upper Florentine World Heritage Area'.

Timber had been harvested from the forests of southern Tasmania's Florentine Valley since the 1940s.[1] However, substantial areas of the valley's forests had been progressively placed into reserves since then. By 2010, timber production on a long-term cycle of harvest and regeneration was still permitted in a 10% portion of forests within the valley's upper catchment. The rest of the upper Florentine's forests were either already contained in various conservation or management reserves, or were otherwise unsuitable for timber supply.[2]

In spite of this pre-existing high level of reservation, the 'Upper Florentine' had been a key battleground in the campaigns waged by environmental activists against Tasmania's native hardwood timber industry over several decades. Accordingly, along with other iconic forests, it had been prioritised for reservation by environmental campaigners negotiating the so-called 'forests peace deal' instigated in 2010 to supposedly resolve the conflict over Tasmania's native forests.

The 'forests peace deal' (which ultimately led to a legislated *Tasmanian Forests Agreement (TFA) Act 2013*) resulted from two-and-a-half years of often fraught negotiations between three environmental groups (The Wilderness Society, Australian Conservation Foundation, and Environment Tasmania) on one side, and an array of timber industry representatives on the other.[3] Significantly, the state's forest management agency, Forestry Tasmania, played only an advisory role and was not party to the formal negotiations conducted from 2010 to 2013.

It should be noted at this point that the *TFA Act 2013* was repealed and replaced soon after a new Tasmanian Liberal Government took office in March 2014.[4] Prior to being repealed, over 100,000 hectares of new forest reserves had been created under the Act, and largely added to the TWWHA during 2013. This addition of long-standing wood production State forests to the TWWHA still stands despite an attempt by the Abbott Federal Government to have it partly revoked in 2014.

An examination of the 'forests peace deal' process and the associated TWWHA extension is now largely an academic exercise. Nevertheless, it is instructive in highlighting the extent of political and bureaucratic interference and subterfuge that has arguably become commonplace in relation to public native forests – seemingly to ensure that new or reformed forest policies accord with the prevailing 'conservation culture' which covets large-scale land reservation, largely at the expense of active land management.

The following chronology of key events during the Tasmanian 'forests peace deal' process is presented to improve understanding of the subsequent discussion:

9 Sept 2010	Tasmania's largest timber company, Gunns Ltd, announces it will exit native forests sector to concentrate on a plantations-only future
Late Sept 2010	The Tasmanian Government establishes a Forestry Roundtable of timber industry and environmental group delegates to negotiate the future of the native forest timber industry
October 2010	The Forestry Roundtable delegates release an agreed **Statement of Principles** which authorises environmental groups to select forests which they would like reserved in new national parks, but commits them to support the building of 'a pulpmill'.
December 2010	Bill Kelty appointed by the Government to act as an independent facilitator to assist the signatories to the Statement of Principles to progress it into agreed actions (aka the 'forest peace talks')
July 2011	Mr Kelty provides final report to the Tasmanian Government outlining the agreed actions determined from the 'forest peace talks'
7 August 2011	**Tasmanian Forests Intergovernmental Agreement** (the IGA) signed by Australian (Labor) Prime Minister Julia Gillard and Tasmanian (Labor) Premier Lara Giddings Under the terms of the Intergovernmental Agreement, the Tasmanian and Australian Governments establish an **Independent Verification Group** (the IVG) to examine and report on the veracity and practicality of the proposed actions arising from the 'forest peace talks'
Oct – Nov 2011	Greens leader, Senator Bob Brown, engages with Federal Environment Minister, Tony Burke, to develop a proposal to extend the Tasmanian Wilderness World Heritage Area[5]
23 March 2012	The Independent Verification Group presents its Final Report comprised of five categories: forest conservation, wood supply, mineral productivity, socio-economics, and social values
November 2012	The **Tasmanian Forest Agreement 2012** is signed by the Roundtable signatories. It agrees to over 500,000 ha of new forest reserves, industry-exit and other state compensation, and a guaranteed wood supply to the remaining industry. Tasmanian Government introduces the *Tasmanian Forests Agreement Bill 2012* which passes through the Lower House of Parliament.

December 2012	Tasmanian Upper House initiates a parliamentary inquiry into the 'forest peace deal' process to better inform its members prior to voting on the *Tasmanian Forests Agreement Bill 2012*.
February 2013	The Federal Government lodges a nomination for a 172,000 ha 'minor' addition to the Tasmanian Wilderness World Heritage Area with the UNESCO World Heritage Committee. Approximately 123,000 ha is comprised of agreed new reserves proposed under the **Tasmanian Forest Agreement 2012**. The nomination is lodged even though the *Tasmanian Forests Agreement Bill 2012* has not yet been passed by the Tasmanian Upper House.
30 April 2013	After months of inquiry, debate and amendments in the Upper House, the *Tasmanian Forests Agreement Act 2013* passes into law. It provided for new national parks to be progressively declared, subject to durability provisions designed to prevent anti-logging protests and guarantee wood supply to the remaining smaller industry.
June 2013	The UNESCO World Heritage Committee approves the addition to the TWWHA
September 2013	A new Australian (Coalition) Government is elected and later declares its intent to seek a partial revoke of the recent addition to the TWWHA on the grounds that it included areas of long-disturbed (non-wilderness) forests that were inappropriately nominated.
March 2014	A new Tasmanian (Liberal) State Government is elected with a mandate to revoke the *Tasmanian Forests Agreement Act 2013*, cease the progressive reservation of new forested national parks, and revive the state's hardwood timber industry.
June 2014	The World Heritage Committee rejects the Australian Government's nomination of a minor amendment to the TWWHA to reclaim about 40% of the recent TWWHA addition.

The preliminaries ...

Tasmania's recent 'forests peace deal' process needs to be considered in the context of decades of environmental activism and substantial forests reservations. This stems back to 1982 when 769,000 hectares of public lands (including 83% of the state's designated 'wilderness') was World Heritage-listed by the United Nations agency, UNESCO.

In 1987, the Federal Government's so-called 'Helsham Inquiry' examined the case for including additional forests in the Tasmanian Wilderness World Heritage Area (TWWHA). Although it found that only about 10% of the 275,000 hectare study area warranted World Heritage-listing, the Hawke Labor Government ignored this majority finding to accept the dissenting report of one of the three Commissioners, Peter Hitchcock. Accordingly, most of the study area, plus other forest reserves stemming from the earlier Salamanca Agreement, led to the addition of a further 604,000 hectares to the TWWHA in 1989.

In 1997, after several years of work, the Tasmanian Regional Forest Agreement between the Tasmanian and Federal Governments added a further 396,000 hectares to the conservation reserve system. Around 95% of the state's designated wilderness was now included in formal conservation reserves. A further 170,000 hectares was reserved under the Tasmanian Community Forest Agreement in 2005. On top of these formal reservations, evolving forest practices regulations since the mid-1980s have increasingly mandated operational management reserves to protect particular landscape and conservation features during timber harvesting.

Accordingly, and contrary to almost all media coverage, the 'forests peace deal' process which began in 2010 was never an exercise in saving Tasmania's native forests. By that time, more than three-quarters of the state's forests were either already reserved or privately-owned, and so sat outside the 'peace deal' process. As this was rarely, if ever, publicly acknowledged beyond Tasmania, the great majority of interested Australians were unaware that the 'forest peace' negotiations were concerned with only the approximately 20% portion of Tasmanian forest which was multiple-use State forest still designated for long-term wood supply.

Despite the limited focus of the process, its ramifications were much wider given that forest policy changes which adversely affect timber industries drawing logs from public lands, also impact the state's quite sizable private forests sector which also supplies wood to the industry.

The concept of a negotiated 'forests peace deal' was reportedly initiated by veteran forest activist, Sean Cadman, in conjunction with Greg L'Estrange, the Managing Director of Tasmania's biggest timber company, Gunns Ltd.[6] It arose from the heat of a hostile business climate caused by a potent combination of:

- falling export woodchip prices due to a high $Aus dollar during the global financial crisis;
- significant reputational damage to the domestic and international markets for Tasmanian wood products caused by incessant environmental campaigning against the state's forest industry;
- significant undermining of the economic viability of the timber sector by environmental activists targeting company shareholders and financiers;
- the loss of the traditional Japanese woodchip market due to various factors, that would subsequently be compounded by the March 2011 tsanami which damaged some of Japan's major pulp mills; and
- an increasing pulp and paper market preference for plantation-grown rather than native forest hardwood chips.

Compounding these adverse business conditions was a growing concern amongst hardwood sawmillers about the capability of the native forests to supply sufficient high quality sawlogs into the future.[7] It had been recognised several decades earlier that, in the face of increasing concessions to conservation, maintaining the legislated level of sawlog supply to the industry would require native forest log supplies to be increasingly augmented by plantation-grown logs. Tasmania's Forestry Commission (later Forestry Tasmania) had begun establishing hardwood sawlog plantations (of mostly Shining Gum (*Eucalyptus nitens*) with some Tasmanian Blue Gum (*Eucalyptus globulus*)) in State forests in the early 1990s.

It was presumed that these would begin supplying high-grade sawlogs to the industry within 25 years, and would comprise a significant component of the sawlog harvest from about the mid-2020s.[8,9] However by 2010, more detailed volume inventory, as well as wood quality concerns from within the industry, were suggesting that these plantations would not supply the anticipated volumes of high-grade sawlogs in such a short time frame.[10] Accordingly there was a need to reduce Tasmania's annual public native forest harvest to sustain the sawmilling industry at a lower level of production for the longer than previously anticipated period before the plantation sawlogs would come on-stream.

Gunns Ltd was most affected by these business and sawlog supply challenges because: 1) it had suffered most of the reputational, corporate, and market damage from incessant anti-logging campaigns over the previous five years; 2) it was by far the state's biggest producer of hardwood sawn timber; and 3) it produced the overwhelming majority of native forest woodchips for the then deflated Japanese export market.

At a ForestWorks conference in Melbourne in September 2010, L'Estrange used his keynote address to set the 'forests peace' process in motion by announcing that Gunns Ltd would deal with these challenges by extricating itself from the native forests sector of the industry. This would entail closing its hardwood sawmills in Tasmania and on the mainland, as well as its Tasmanian woodchip processing and ship-loading facilities.

L'Estrange's assertion that the various business and supply challenges had made the industry financially unviable would subsequently be contested by an array of industry figures, and it remains unclear whether his announcement – which reportedly even stunned some within his own company – was rooted in a personal or Board-level ideology, or was a reaction to other unspoken challenges being faced by Gunns Ltd. For example, it has been suggested that the critically important loss of its Japanese woodchip supply contracts was due to Gunns' supposed arrogance when negotiating woodchip supply contracts; and/or that its widely-announced intention to build a Tasmanian pulp mill had suddenly transformed it from a raw materials supplier to a competitor in the eyes of its Japanese pulp and paper-making customers.[11]

Indeed, Gunns' focus on building its contentious but already approved plantations-based pulp mill in northern Tasmania, together with its burgeoning estate of hardwood plantations being grown for pulp, made their decision to eschew native forests eminently understandable.

However, extending this to a blanket exit from all native forests, including their operations in WA and Victoria, where they also owned major sawmilling assets, is harder to explain. This was particularly so given that L'Estrange had reportedly acknowledged that the company's large Victorian sawmill at Heyfield (that state's largest hardwood mill) was one of its most profitable ventures.

Despite the obvious angst created amongst its employees and contractors, Gunns' sudden exit from Tasmania's native forests represented a fortuitous opportunity to resolve the sawlog supply challenge. Gunns was responsible for about half of the annual State forest sawlog harvest and their departure could immediately allow the annual harvest to be wound back to a lower level to ensure the sustainability of the remnant sawmilling industry in the longer than expected absence of plantation-grown logs. An additional advantage was that a substantially reduced harvest would have somewhat diluted its impacts by allowing coupes to be spread more broadly across the landscape, thereby affording greater flexibility to forest management.

Furthermore, Gunns' departure from native forests theoretically provided an opportunity to improve outside perceptions of the industry from being dominated by a (largely hated) corporate giant to being comprised of a far less threatening bunch of smaller players, including family-owned businesses. To the environmental movement and its supporter base, Gunns *was* Tasmanian forestry and the company had for years been routinely vilified as though it was solely responsible for all of forestry's (and even society's) actual or imagined ills. Gunns sudden disappearance off-the-scene and the consequent halving of the annual harvesting rate could potentially have cooled much of the anti-logging fervour associated with Tasmania's forests.

On the other hand, Gunns' sudden closure of its export woodchipping facilities created substantial problems. As woodchips had been a valuable sawmilling by-product, the sudden inability to sell them represented a substantial hit to the economic viability of the remaining hardwood sawmilling industry. This impact on sawmill income was magnified by the sudden logistical problem of having to physically deal with mounting volumes of off-cut waste which had formerly been processed for a profitable use. In the forests too, there was the management problem of how to physically deal with huge volumes of sub-sawlog grade logs which suddenly had no market.

But the greatest impact of Gunns' departure from native forests was the significant loss of employment amongst harvesting and haulage contractors and mill workers. This was arguably the strongest catalyst for the 'forest peace talks' as industry leaders presumably reasoned that it could be a vehicle for securing compensation pay-outs for businesses and workers suddenly forced out. Without doing a 'peace deal', these contractors and their workers would walk away with nothing.

Arguably, the prospect of compensation pay-outs and a more unionised future workforce prompted the Construction, Forestry, Mining and Energy Union (CFMEU) to flip from being a traditionally staunch supporter of the native forest timber industry to advocating its substantial closure in return for an undertaking from the environmental lobby to support the building of Gunns' contentious Tamar Valley pulp mill. The pulp mill would require a substantial and concentrated workforce which could feasibly be all paid-up union members. In contrast to this, the native forest sector's workforce had been widely scattered amongst many small businesses with a reportedly low rate of union membership.

The CFMEU's position gave considerable ammunition to environmental groups and Greens politicians eager to manufacture a public perception that the industry was united in its commitment to largely remove itself from Tasmania's native forests. But this was far from the case. In fact, the determination of some industry negotiators to secure compensation for leaving the industry was matched by the determination of others to continue to operate and to maintain the area of State forest wood production zones which the industry could continue to access into the future.

At the negotiating table, the somewhat divided industry delegates faced the unpalatable realisation that fashioning a 'peace deal' would require bowing to the demands of opposing environmental group delegates intent on securing the transfer of huge areas of State forests into new national parks. This would create problems for an ongoing smaller remnant sawmilling industry because access to the most productive forests would probably be lost. It would also create problems for the large plywood producer, Ta Ann, because its contracted long term supply of regrowth peeler logs was based on continued access to the full area of State forest wood production zones.

So, although the departure of Gunns created potential for less and more widespread future timber harvesting, the environmentalists' view of it as an opportunity to substantially reduce the area of forest designated for wood supply meant that the envisaged smaller future timber industry would be squeezed into a smaller area of less productive State forest. This would effectively re-create issues such as harvesting intensity, lack of management flexibility, and questionable resource sustainability which had largely fuelled past opposition to the industry's existence.

There were also vastly differing pressures facing the industry and environmental group delegates. While the industry delegates were negotiating against a backdrop of dire financial implications, the environmental group delegates had nothing of material value at stake and were free to make disproportionate demands for more conservation reserves. However, while the environmental group delegates may have been united on that score, they would increasingly be representing a constituency divided over the question of whether a deal should be done at all, or whether it would be better to let the (hated) native forest timber industry simply whither and die at the behest of social and market forces.

In time, Tasmania's 'environmentalist community' would be fractured once it became more widely recognised that the 'peace deal' had required the state's Greens and environmental group negotiators to agree to the continuation of a smaller native forest industry and to support the building of 'a pulp mill' in return for substantial areas of State forests being given over to new conservation reserves.[12,13] Given the uncompromising nature of environmental activism, most other Greens politicians, environmental campaigners and their supporters wanted to both completely kill-off the native forest timber industry *and* remove any prospect of a pulp mill. Some environmental extremists had even (perhaps flippantly) raised the prospect of bulldozing the eucalypt plantations that were being grown to supply it!

Gunns' role in initiating and cajoling the 'forest peace talks' (despite never formally participating in the negotiations) was presumably predicated upon the hope that leading an industry shift from native forests to plantations would sufficiently improve their community and social standing thereby softening opposition to its proposed pulp mill. This perception created some animosity amongst the broader native forests sector because it was suggestive of the company's willingness to sacrifice its industry brethren to serve its own self-interest.

This animosity peaked in July 2011 when Gunns sold its Tasmanian east coast woodchip mill and ship loading facility at Triabunna to two mega-wealthy 'green' philanthropists (Graeme Wood and Jan Cameron), despite their purchase bid being 40% lower than that of a local timber industry consortium which intended to keep the facility operating.[14] Despite a condition of sale that the mill be kept operational for potential future use,[15] it was never re-opened and within three years had been deliberately wrecked by the 'mill manager' specifically to remove that possibility.[16]

According to an article in *The Monthly*, the 'mill manager' who planned and supervised this destruction was none other than former long-serving leader of The Wilderness Society, Alec Marr.

He had reportedly acted for the new owners during the mill sale process, and with the support of co-owner, Graeme Wood, had longed to wreck the mill's infrastructure. Under a pledge of secrecy, Wood invited a photographer and a journalist to be present to document the destruction.[17] On the eve of the destructive rampage, Marr reflected on what he was about to oversee:

> I've been waiting 27 fucking years for this ... We were buying the port more so than the mill itself. It was a bullseye: we totally fucked them.[18]

In a subsequent article in the Hobart *Mercury* (on 15 July 2014), Marr would decry *The Monthly*'s account of this episode as 'a complete beat-up'.

As intended, the sale and subsequent destruction of the Triabunna mill was a disaster for the viability of the remaining southern and central Tasmanian timber industry, including the owners and investors in substantial areas of eucalypt plantations being grown under contract for the export woodchip market. Without access to the Triabunna facility, the Tasmanian Government was forced to subsidise the substantially greater cost of pulp log cartage from southern Tasmania to the only remaining woodchip facility at Bell Bay, on Tasmania's north coast.

The events surrounding the sale and subsequent de-commissioning of the Triabunna woodchip export facility would eventually be examined by a Parliamentary inquiry given the state significance of losing such an important piece of industrial infrastructure. At the time of writing, its new owners are working towards turning it into a tourism venture.

In relation to the company's motives in selling the Triabunna facility to a non-industry entity for a significantly lower price, the inquiry found evidence that there was considerable tension between Gunns (MD Greg L'Estrange) and Forestry Tasmania (CEO Bob Gordon) and that this "... may have influenced the decision made by Gunns to sell the Mill to Triabunna Investments [*i.e. 'green' philanthropists Wood and Cameron*] rather than Aprin [*i.e. the industry consortium*] which had secured a deal with Forestry Tasmania to supply logs to the mill". Furthermore it concluded that "... Triabunna Investments, despite a contractual obligation, had no intention to reopen the Mill as a woodchip export facility".[19]

The saga that surrounds the actions of Gunns and the initiation of the 'forest peace deal' process suggests that aspects of forest management and elements within the industry were in-part responsible for the challenges that the broader native forest timber industry had to confront, and that this had opened a door for those forces with a long-standing intent to cripple or destroy Tasmanian forestry.

Democracy at work?

Tasmania's 'forest peace deal' process has been variously described as an 'outstanding example of democracy' by its supporters, or a 'blight on democracy' by its detractors. While the former view may be an idealistic commentary on what can be achieved when two warring parties sit down to negotiate, it ignores the grossly uneven playing field which dictated each side's bargaining power.

The latter assessment seems more accurate given that the process effectively amounted to the government outsourcing the determination of critical public land use policy to two unelected, polarised, and self-interested groups engaged in a 'horse-trading' exercise. As such, it ensured that the views of a whole suite of other affected community stakeholders were ignored despite having legitimate claims to participation in determining public land use policy.

By effectively abrogating its democratic responsibility for making public policy, the Tasmanian Government side-stepped the political minefield associated with any tinkering of the state's significant timber industry. By facilitating and supporting an arms-length process whereby the industry could be shown to be determining its own future, the government could avoid being the 'bad guys' when businesses inevitably closed and jobs were lost. They could then legitimately throw up their hands and declare "hey, it wasn't us – this is what the industry and unions wanted!"

The problem with such a narrative is that, as the industry negotiators were representing a sector hovering on the edge of financial ruin, any 'agreement' would be only made under considerable duress and would be effectively dictated by ideologically-driven environmentalists representing only a small minority of the population. This reality needs to be remembered in relation to almost any part of the 'peace deal' process where 'the signatories' were said to have agreed to actions, such as for example, the appointment of clearly biased 'experts' during the later so-called Independent Verification Process.

The desperation of the timber industry negotiators was exemplified by their early agreement to an initial 'Statement of Principles' under which environmental activists were authorised to select 'high conservation value' forests to be proposed for new national parks.[20] This would previously have been unthinkable given the near certainty of it leading to the industry being forced out of the most productive available forests.

Conversely, the domination of the environmental negotiators was exemplified in late 2012 when the talks temporarily broke-down due to the impossibility of reconciling the industry's minimum contractual need for wood against the loss of future resource from placing hundreds of thousands of hectares of productive State forests into new national parks. Almost immediately the Wilderness Society's Lyndon Schneiders was blaming the industry for the breakdown of the 'peace talks' in an opinion article published by the Fairfax media[21] and was arrogantly asserting on ABC News that:

> The big sawmillers have refused to move so we have to go where its going to hurt... and where we're going is into the domestic market place ... we are going to the buyers of their products.

Within days, the online protest group *GetUp!* had launched a '*Save Tasmania's Forests*' campaign against wood products retailers, Harvey Norman and Bunnings. The group's campaign director and former Wilderness Society 'corporate and forests campaigner', Paul Oosting, reportedly justified this on "information that large sawmillers blocked a deal on the basis that reduced (timber) volumes would prevent them meeting contracts"[22] – as though this was somehow an unreasonable position for businesses with legally-binding contractual obligations.

A day later in Hobart, a protest against timber company, Ta Ann Tasmania, was manned by just 10 placard-bearers amidst insider reports that the 'forest protest community' were fatigued. But by the week's end, timber harvesting had been halted at a coupe blockaded by environmental activist splinter group, Still Wild Still Threatened (SWST). Indeed, despite a supposed moratorium on anti-logging protests during the 'forest peace talks', environmental groups had been campaigning against Ta Ann Tasmania since October 2011. The market-based boycott campaign would ultimately see a reduction in demand from customers of Ta Ann Tasmania, a reduction in their workforce, and a taxpayer-funded compensation package for loss of contracted supply.

SWST were one of four environmental groups which had pledged to continue protesting irrespective of whether or not a 'peace deal' was ultimately agreed and signed. The convener of one of these groups – Jenny Weber of the Huon Valley Environment Centre – later admitted to opposing the 'peace talks' once it became apparent that it would involve compromise to allow some native forest harvesting to continue. To her, only a complete end to native forest timber production was acceptable.[23] This echoed the positions of other higher profile opponents of the 'forest peace deal' concept, including Greens political luminaries Bob Brown and Christine Milne, and celebrity eco-activist Richard Flanagan. [24,25,26]

Schneiders' intemperate 'call-to-arms' and the subsequent protests served to highlight the true nature of the 'peace talks'. In particular, that they were being conducted against a thinly-veiled backdrop of threat whereby the industry was expected to either agree to an unbalanced outcome or be sabotaged by a resumption of spurious campaigns of misinformation targeting its financiers, shareholders and markets which had in large-part forced them to the negotiating table in the first place. It was common knowledge that several industry negotiators had been explicitly told by environmentalists during the talks that 'we will do to you what we did to Gunns unless you agree to our demands' – a clear reference to the reputation-eroding campaigns that had played such a role in crippling Tasmania's largest timber company.[27]

The timber industry was caught in an impossible position – they could reject the 'peace deal' and face ongoing protests, or they could agree to the deal and lose most of their best future resource in new national parks while still enduring ongoing protests from non-signatory environmental groups which had vowed to never comply with the deal. The mooted deal offered supposed protections against ongoing protest – most notably that the three signatory environmental groups would 'help' the remaining sawmilling industry to secure Forest Stewardship Council (FSC) certification.

While FSC certification is promoted by environmentalists as an unimpeachable guarantee of sustainable forestry, the mere fact that the timber industry needs the 'help' of environmental activists to attain it strongly suggests that it is not just about forestry practices. Most of Australia's (and Tasmania's) native hardwood industry had already been certified to the Australian Forestry Standard and was acknowledged to be employing practices equivalent to world's best standard.[28] Despite this, no public land native forest timber production in Australia has yet been FSC-certified despite several agencies having attempted to attain it.

The Forests Stewardship Council was established by the environmental movement in the early 1990s. The reality that even acknowledged world's best forestry practices haven't been good enough for FSC certification in Australia paints the environmental movement as its unofficial gatekeepers with the power to veto certification irrespective of how good the forestry practices are. This is hardly a surprise given that Australia's largest and most influential environmental activist groups are ideologically opposed to most native forest wood production. However, this differs from how FSC certification is being applied overseas where even native forest clearfelling is acceptable if it is acknowledged to be the most appropriate silvicultural technique.[29]

Even if FSC certification could be attained for Tasmania's remaining public forest wood production zones, the collective universe of environmental activism is so diverse that there will always be rogue groups unwilling to accept the umpire's decision. This has been evident for several years in Victoria's South Gippsland where FSC certification of the state's biggest plantation grower has been no impediment to continued protest by groups disaffected by routine management operations, such as the use of herbicides to control competing weeds.[30] In Tasmania, the four 'anti-peace deal' environmental groups declared that FSC certification wouldn't stop their ongoing anti-logging campaigns and that even if certification was attained by Tasmania's native forest timber industry, they would simply dismiss it as an inappropriate, watered-down version of the real thing.

As well as threats by environmental groups to resume protesting, the timber industry was being virtually blackmailed into agreeing to the 'peace deal' by the Gillard Labor Government's offer of hundreds of millions of dollars in industry compensation tied to state regional development grants that would only materialise when the deal was done. If the industry couldn't agree to do the deal for itself, it was to be railroaded into it for the sake of Tasmania.

At one point, Federal Environment Minister Tony Burke described his latest offer – a one-off $100 million payment spread over a 14-year period – as a "wonderful opportunity for Tasmania"[31] as though this could ever make-up for the permanent loss of more than half of an industry that, just a few years earlier, had been generating $700 million per annum and employing thousands.

The Federal Government's increasing involvement in what was a state responsibility was troubling, but appears to have been linked to an agreement brokered in 2010 by Prime Minister Julia Gillard and then Greens Leader, Bob Brown, to garner the support needed to form a minority Government. Efforts to explore the exact nature of that agreement have involved Freedom of Information (FOI) requests to view letters between Brown and the PM, as well as then Environment Minister, Tony Burke.

What is known is that in late November 2010 – when the 'forest peace deal' negotiations were in their preliminary stages – Brown wrote a letter to PM Gillard listing "the Australian Greens recommendation for a good outcome". These included: 1) funding the exit of contractors and workers from native forests and to help them find alternative employment; 2) protecting high conservation value forests in national parks and World Heritage listing if appropriate; and 3) supporting the building of Gunns' proposed plantations-fed pulpmill, presumably as a trade-off for native forest reservations and sawmilling industry closures.[32] Subsequent letters between Brown and Environment Minister Burke in late 2011 show Brown to be the architect of the proposal to extend the Tasmanian Wilderness World Heritage Area to include vital wood production forests.

At the state level as well, the Tasmanian Labor Government was dependent on an alliance with the Greens and this manifested itself in overt cajoling of the industry to force a favourable (to the government) 'peace deal' outcome. They repetitively portrayed the 'peace deal' as an agreement that would 'secure the future for Tasmania's forest industry' without disclosing that it would be a significantly attenuated future based on a substantially reduced resource base.

The Tasmanian 'forest peace deal' process is a fascinating case study of political and bureaucratic interference in unduly shaping public policy decisions that the broader populace has every right to expect to be based on objective and evidence-based analysis. This is best exemplified by the Independent Verification Group process, and the later nomination to extend the Tasmanian Wilderness World Heritage Area which flowed from it.

The myth of impartial analysis – the Independent Verification Group (IVG) process

Following-on from the initial 'Statement of Principles' agreed to by industry and environmental group delegates in October 2010, union and ALP heavyweight, Bill Kelty, had been appointed to assist the signatories to progress it into an agreed course of actions. He provided a final report to the Tasmanian Government in July 2011, and this led to the Tasmanian Forests Intergovernmental Agreement (the IGA) being signed by the Federal and Tasmanian Governments in August 2011.

The IGA delineated an area of 572,000 hectares of 'high conservation value' forest selected by the environmental group signatories for new national parks and reserves. This included a core area of 430,000 hectares which was afforded immediate 'interim protection' pending an assessment of its suitability.[33]

Clause 20 of the IGA required the establishment of an Independent Verification Group (the IVG) to be charged with the primary task of verifying the claimed 'high conservation values' within the nominated 572,000 hectares, whilst also assessing the compatibility of such a large new reservation of forests with the wood supply commitments of an ongoing, smaller timber industry.[34] In its IVG Fact Sheet, the Federal Department of Environment gushed that "for the first time, the process will provide an independent and robust assessment of conservation values and timber supply requirements …" – conveniently forgetting the many previous inquiries and assessments into Tasmanian forestry, including the comprehensive Regional Forest Agreement process of the late 1990s.

Subsequently, the Tasmanian and Australian Labor Governments appointed Professor Jonathon West to chair the IVG. A further five IVG members were selected from a list nominated by Professor West, the Tasmanian and Australian Governments and other stakeholders; with the nominees authorised by the signatories to the IGA.[35] On 20 September 2011, the appointment of the IVG's six members was approved under executive power by the then Federal Minister for the Environment.[36]

According to the IVG's subsequent Final Report, its chairman and the other five IVG members were appointed for their independence and extensive expertise in forestry, forest ecology, conservation reserves, forest modelling and geology.[37] Only one of the six IVG members had practical experience in forest management and forest agency administration.[38] Three of the other five members were academics with current or past linkages to the environmental movement and/or had publicly supported its views on forests.

The IVG's Terms of Reference required these appointees to be "independent of government and all other stakeholders" and their Federal Department of Environment engagement contracts viewed under a Freedom of Information (FOI) request, show that none declared any conflict of interests.[39] However, a subsequent investigation of the IVG process by the Australian National Audit Office (ANAO) in 2014 found no evidence that the Department of Environment had pursued the question of conflicts of interest. The ANAO investigation found that this was irregular behaviour at odds with the usual requirement for members of such non-statutory advisory committees to sign a declaration of independence and/or a conflict of interest disclosure document.[40]

Aside from the six IVG members, a further 32 third party contractors were engaged by the Department of Environment to source data and author the various IVG reports. Their engagement was overseen by IVG Chairman Jonathon West under a contract template that included a clause requiring the contractor to warrant that no conflict of interest existed or was likely at the time of signing, and to notify the IVG Chair immediately if a conflict of interest subsequently arose. During the later ANAO investigation, Professor West advised that no conflicts of interest had been declared by these third party contractors.[41] A subsequent viewing of their engagement contracts under FOI found that George Harris – a specialty timbers sawmiller and craftsman – had been the only IVG third party contractor to declare a conflict of interest.[42]

The management of conflicts of interest would seem to be critically important for a body such as the IVG which, as its title suggests, was charged with providing 'independent' advice. However, neither the Department of Environment, nor IVG Chairman Jonathon West, agreed. When the ANAO investigated the process, it was informed by IVG Chairman West that he did not believe that conflicts of interest were relevant in the context of the IVG's work, primarily because it was not making recommendations to government or participating in the decision-making process. Similarly, the Department of Environment advised the ANAO that:

> Given the history and context, the Department took a holistic rather than a case-by-case approach to conflicts of interest. Rather than managing conflicts of interest on an individual basis (arguably all players were conflicted), the Department sought to ensure that all respective interests were represented. It is also important to note the role of the stakeholders (in particular the reference group of signatories) in selecting and accepting the advice of the IVG. In this context, the Department did not rely on the contractual clauses relating to conflicts of interest.[43]

As this explanation was in response to a formal investigation into the integrity of the IVG process, it arguably represents an attempt to justify systemic flouting of due process which occurred to avoid highlighting obvious conflicts of interest that would otherwise have clearly undermined the IVG's claim to independence.

IVG Chairman West's view that 'conflicts of interest' were irrelevant because the IVG was not making recommendations or participating in decision-making rings rather hollow when it is appreciated that the IGA signatories (the 'peace deal' negotiators) and both the Australian and Tasmanian Governments had agreed to be bound by the results of the verification (IVG) process.[44] In addition, Clause 20 of the IGA had required the IVG to "design and implement an independent and transparent verification process to assess and verify stakeholders claims" and to "make recommendations on appropriate forms for land tenure for reserving native forests …"[45] This surely makes it clear that the IVG required independence and transparency, and was informing the decision-making process, including making recommendations.

Similarly, the claim by the Department of Environment that the IVG process was effectively balanced because everyone was conflicted could only be valid if there were equal numbers of contractors representing all opposing interests. However, a subjective assessment of the 38 IVG members and third party contractors suggests that those with overt links to or clear support for environmental activist groups and their 'save-the-forests' agenda, outnumbered by approximately two-to-one those who could be presumed to support the status quo on forest management and timber industry access; with the remaining 17 being either unknown or neutral.[46] Further to this, the views of other public land stakeholders, such as for example, apiarists, were not represented at all.

Despite the Department of Environment's awareness of the many undeclared conflicts of interest amongst those engaged in the IVG's work, its Deputy Secretary, Dr Kimberley Dripps, in evidence to a 2014 parliamentary inquiry described the IVG process as "extremely detailed and yes, it was thorough".[47] At the same inquiry, IVG member Professor Brendan Mackey, who had overseen the group's work plan, initially described the IVG process as having been "conducted to the highest level of professional integrity", but under strident questioning was later forced to admit that some of the IVG's 'independent assessors' were members of the Wilderness Society and the Australian Conservation Foundation – the very organisations whose claims they were meant to be independently assessing.[48]

Despite the reality that the most influential elements of the IVG process were largely conducted by members and third party contractors biased by linkages to environmental activism, it has been consistently lauded for its independence by environmentalists, politicians, and bureaucrats who agreed with or benefitted from the huge increase in

Tasmanian forested reserves. Accordingly, the undeclared 'conflicts of interest' amongst key participants in the IVG process, makes very interesting reading.

Most notably, the IVG Chairman, Professor West, had formerly been the Director of the Wilderness Society in 1986 – 87 during the tumultuous era of early Tasmanian forest protests. During this time he became a "personal friend" of forest activist, Sean Cadman, who had played a leading role in initiating the 'forest peace deal' negotiations, and would later be employed to assist the IVG.[49]

After his environmental activist days, Professor West had spent 18 years at Harvard University working in the fields of business innovation and strategy.[50] He enjoyed close links with the Labor Party through having worked as an advisor to Labor Environment Minister, Barry Cohen, in the 1980s, as well as being a long-standing friend of Craig Emerson, the then Gillard Labor Government's Minister for Trade.[51]

Professor West's appointment raised the initial concerns about the IVG's supposed independence. These were magnified by the subsequent appointment of Australian National University Professor Brendan Mackey as the lead scientist to oversee the IVG's work plan. Mackey had been the inaugural Director of the ANU's Wild Country Research and Policy Hub – a body established through a formal partnership between the university and the Wilderness Society – and he had long advised the environmental group as a member of its WildCountry Science Council.[52]

Professor Mackey's relationship with the Wilderness Society has been financial as well as advisory and stretched at least as far back as 2003 when they awarded him a $217,000 research grant to assist in establishing the WildCountry Science initiative.[53] In 2004, he received a further $220,000 grant from the Wilderness Society to develop new techniques for biodiversity conservation,[54] and in 2005 he was part of a partnership with the Wilderness Society and several other institutions and government agencies, which received an $870,000 ARC Linkage Grant to develop new techniques for biodiversity conservation, inclusive of large-scale connectivity processes.[55] Professor Mackey may well have received further grants from the Wilderness Society, but enquiries to establish this were met with advice from the ANU that such information is not publicly available.[56] However, some acknowledgement of the Wilderness Society's financial support for Mackey's research was contained in his 2008 (Mackey *et al*) *Green Carbon* report, which effectively advocated the closure of south-eastern Australia's native forest timber industry to 'save forest carbon emissions'.[57]

Mackey's IVG duties, as detailed in his Department of Environment engagement contract, included "to assess and verify stakeholder claims relating to sustainable timber supply requirements, available native forest and plantation volumes … and areas, conservation values, and boundaries of reserves from within the ENGO [*environmental groups*]-nominated 572,000 hectares as required by the Terms of Reference."[58]

In addition, Clause 15 of Mackey's IVG engagement contract clearly stated that if he was in a position where a conflict of interest existed he needed to declare it or risk his employment being terminated by the Department of Environment.[59] Yet, as has been established earlier, he provided no declaration of a conflict of interest to the Department despite his long-standing relationship with a key IGA stakeholder, The Wilderness Society.[60]

The perceived lack of independence within the IVG was further magnified when Professor Mackey engaged environmental consultant, Peter Hitchcock, as a contractor to assess the appropriateness of the proposed new forest reserves for World Heritage listing. Hitchcock had been formerly engaged by the Wilderness Society in 2008 to work on an ultimately unsuccessful proposal to add much the same areas of State forest to the Tasmanian Wilderness World Heritage Area. As his work for the IVG effectively entailed verifying or updating his earlier work undertaken for the Wilderness Society.[61]

Further to this, the three scientists selected to peer review the IVG's forest conservation reports – Dr Reed Noss, Professor William Laurance, and Dr Oscar Venter – also brought highly questionable independence to their role. All were philosophically devoted to improving conservation outcomes. One of them (Dr Noss) was closely associated with environmental activism, and the other two were demonstrably supportive of it, albeit in relation to tropical forests in developing countries. At the very least this was suggestive of a bias towards supporting more conservation reserves over maintaining a commercial forest use.

Dr Noss, is a known associate of the Wilderness Society through his role as co-founder of the US Wildlands Project (now Network),[62] an environmental initiative which advocates huge areas being reserved for conservation, reportedly aiming for 50% of the land mass of the USA to be reserved.[63] The Wildlands Project was the template used by Australia's Wilderness Society to develop its WildCountry Vision. From this, a WildCountry Science Council was established (largely through the efforts of Professor Mackey), and the ANU WildCountry Research and Policy Hub was established.[64]

Clearly Dr Noss would have been well known to Professor Mackey through their mutual involvement in the WildCountry Science initiative, yet neither publicly acknowledged their association. It was clearly inappropriate for Dr Noss to be asked to peer review the work of an associate and one with such close ideological links to the very organisation he had founded. It was obvious, in my opinion, that Noss's belief in the Wildlands and WildCountry concepts advocating the reservation of huge areas of the landscape, and his past association with the Wilderness Society, would strongly bias his peer review work.

Professor Laurance is a distinguished scientist and advocate for environmental sustainability, particularly in relation to tropical forests. As the founder and Director of ALERT (the Alliance of Leading Environmental Researchers and Thinkers), he leads a group that "is all about the environment"; is intent on "helping world class scientists to influence key environmental decisions"; and is "designed to complement .... other scientific and environmental groups."[65] While this is a worthy initiative, in my opinion it is strongly suggestive of a predisposition to support conservation initiatives over and above all else based on the well documented concerns over illegal logging, endemic corruption, and weak regulatory control that are often endemic to developing countries which are the primary focus of Professor Laurance and ALERT. The potential implications of his experience for objectively reviewing work undertaken in relation to Tasmanian forestry is evident from several subsequently published articles which suggest that Professor Laurance doesn't fully appreciate that forestry problems which may be prevalent in developing countries are virtually absent in Australia (and Tasmania).[66]

Dr Venter has been a colleague of Professor Laurance in relation to conservation in tropical forests, and had also reportedly previously undertaken consultancy work for international environmental group, Greenpeace, as well as the Green Institute.[67] The attitudes of Greenpeace to native forest wood production are well known, and the Green Institute is an affiliate of the Australian Greens whose anti-native forest timber industry stance is also well known. While the nature of his work with those groups is unclear, his association with them at least raises questions about his capacity to remain independent when assessing work pertaining to the future of Tasmania's native forest timber industry.[68]

The list of third party contractors employed to assist the IVG process also included some well known environmental activists and their academic associates, such as former veteran Wilderness Society campaigners Virginia Young and Sean Cadman, and ANU environmental law academic, Associate Professor Andrew Macintosh.[69] Further to this, both Ms Young and another IVG author, Professor Jann Williams, were both members of the WildCountry Science Council.

Associate Professor Macintosh is well known from his former role as Deputy Director of the 'progressive' think-tank, The Australia Institute, which has often produced reports critical of Australia's mining and forestry sectors. He was employed by ANU Enterprises to help author their IVG Report 8A on forest carbon value. In a 2013 article on *The Conversation*, in which Macintosh was critical of the Tasmanian forestry sector, it was disclosed that he receives funding from a forest carbon offset company, Forests Alive, managed by former Wilderness Society forests campaigner, Virginia Young.[70] As has been stated in an earlier chapter, Forests Alive facilitates the payment of carbon credits to landowners for "avoided emissions by stopping native forest logging." Reportedly, Ms Young also supervised Macintosh's IVG forest carbon work.

Veteran forests and climate change campaigner Sean Cadman's long history of environmental activism (especially for the Wilderness Society) overrides any claim to being independent.[71] Cadman was employed as a third party contractor in the IVG process. He acted as one of three principal 'verification advisors' to IVG Chairman, Jonathon West, and as a co-author of its Social Values report which included a consideration of effective reserve establishment and management.[72] His consultancy firm, Cadman and Norwood Environmental Consulting, had received over $67,000 from the Wilderness Society in the year before the commencement of the 'forest peace deal' process; while his partner, Rosemary Norwood, was a member of the Wilderness Society's executive committee and had formerly worked for the ACF and the Department of Environment.[73]

To some extent Cadman's appointment as a 'verification advisor' could be viewed as having been balanced by the appointment of Alan Hansard – then head of the National Association of Forest Industries – as one of the other two 'verification advisors'.

Following the IVG process in 2013, Cadman was engaged by the Department of Environment to assist Peter Hitchcock in conducting a 'Review of Dossier on World Heritage Boundary Extension'. As with his earlier IVG work, when Cadman undertook this project he was conflicted by owning and operating an eco-tourism business near Quamby Bluff in northern Tasmania featuring an eco-lodge and specialising in guided walking tours of nearby World Heritage forests.[74,75]

Others to assist with the supposedly objective and independent IVG process despite displaying a blatant anti-timber industry bias, included Rod Knight and Peter McQuillan. Knight's environmental consultancy firm was awarded two IVG contracts despite him being a formerly prominent Wilderness Society operative who co-authored its 1991 'manifesto', *Saving Native Forests in the Atomic Age*. This document outlined the group's goals and strategies for undermining and ultimately destroying Australia's native hardwood industry by forcing "a speedy industry transition out of native forests."[76]

Dr Peter McQuillan of the University of Tasmania's School of Geography and Environmental Studies was paid over $16,000 by the IVG to "provide expert advice",[77] despite being a well-known advocate for closing Tasmania's timber industry who had been recruited by various environmental groups over many years to add academic credibility to their campaigns. In 2007, McQuillan had travelled to Japan with then leader of the Tasmanian Greens, Peg Putt, in an attempt to orchestrate a marketplace boycott of Tasmanian wood products.[78] More recently, he had appeared in several short You Tube clips at around the time when the 'forest peace deal' negotiations were starting, in which he urged Tasmania to stop cutting down its forests. These clips were variously sponsored by the Wilderness Society, *GetUp!*, Environment Tasmania, and the Australian Conservation Foundation.[79] One of these clips had been used as a political advertisement during the 2010 Tasmanian election campaign.

Questions about the independence of the IVG were swirling around while it was undertaking its work. The Tasmanian and Australian Governments sought to block these concerns by continually emphasising the independence of those working on its various reports, to such an extent that it could be construed as a case of "he who protesteth too much". Typical of this was the joint media release by the Federal Environment Minister and the Tasmanian Premier announcing the formation of the IVG in 2011:

> This process to be overseen by this group of experts will ensure that all stakeholders and the wider community can have confidence in the independence and veracity of the process.[80]

Flowing from concerns about its independence and objectivity, the IVG's work contained some considerable flaws. The Institute of Foresters of Australia (IFA) – the professional association for the nation's forest scientists since 1935 – perhaps generously observed that this was "... largely due to constraints imposed by timing and narrow Terms of Reference that limited its ability to widely consider the proposal for new forest reserves in the context of existing Tasmanian forest management, including the substantial extent of already existing parks and conservation reserves".[81]

The IFA went on to question why the IVG's assessment of the nominated new reserves' (supposedly) high conservation values, did not involve comparing their values against those in other (unreserved) forests, as would have been expected in order to confirm that they were indeed 'high'. Instead, the IVG's assessment merely set out to show that the nominated areas had conservation values – hardly a groundbreaking revelation since any collection of trees, including the blandest plantation monoculture, has some conservation value.[82]

Furthermore, the IFA strongly criticised the method used by the IVG to assess conservation values for using criteria developed by environmental groups which "... makes judgements about them on the basis of criteria which have not been scientifically-defined or agreed upon. This ... highlights the shortcomings of a process that appears to predetermine an outcome rather than evaluating conservation claims on their merits using an objective and scientifically-accepted process".[83]

The IFA's detailed critique of the Independent Verification Group's work also included the following observations:

> Despite IVG Chairman, Professor West acknowledging that authors contributing to the IVG process held, or were perceived to hold bias, these perceptions were not addressed appropriately as no independent, transparent or credible peer review has been undertaken of their reports.
>
> Without such a review these reports remain discussion papers and do not inform the process. This conclusion is supported within the reports by authors who state that their work is presented as 'preliminary' or for 'discussion' thereby acknowledging the need for further work to verify claims and validate conclusions.
>
> The IVG reports provide only superficial consideration to impacts associated with a significant reduction in domestic timber production. Issues such as 'trade leakage', increased imports and global impacts have been ignored despite Australia having a global responsibility to not only conserve forests, but to also use them and in doing so protect global biodiversity and reduce carbon emissions.
>
> The IFA notes that the IVG reports do not examine or assess whether setting aside more areas of Tasmanian forest as conser-

vation reserves will improve the conservation value of forest ecosystems in the State, nor does it assess the effectiveness of present management of national parks and other reserves in achieving protection of natural values; predict the effectiveness of a different system (i.e., an expanded reserve system) for achieving protection of natural values; or provide any comparison of the success of the present system with likely outcomes predicted under the proposed new system.

The IFA would go on to lament that many of the IVG's report authors were clearly unfamiliar with the complex forest management system that operates in Tasmania. Accordingly, the Institute was only satisfied with the IVG's wood supply report, largely because the responsible IVG member had worked closely with Forestry Tasmania's timber resource specialists to access and understand their data, and because the draft report had then been peer reviewed by other forestry experts in wood supply inventory and analysis.

Unfortunately, the IFA could not recommend the IVG's forest conservation work:

> It is of considerable concern that the IVG reports give only limited recognition to the fact that Tasmanian forests are already well managed for biodiversity and other natural values (landscape, soil and water, geodiversity) both through one of the most proportionally extensive reserve systems in the world, and in State forests, through one of the world's most stringent regulatory frameworks governed by the *Forest Practices Act 1985* and the Forest Practices Code 2000. This is further supported through contemporary and dynamic prescriptive and guideline documents e.g. the Threatened Fauna Advisor, and the extensive lists of publications of Forestry Tasmania and the Forest Practices Authority on natural values protection – all of which are readily available on websites and as web tools.
>
> It is almost inconceivable that an assessment of claims for substantial new reservations would take no account of the reality that Tasmania's forest practices system is recognised internationally as a benchmark for best practice in forest regulation and is regularly used as a model for improvement in many countries.

In 2008, researchers from Yale University and the Australian National

University had independently compared forest practice policies in Tasmania against the policies of 38 other jurisdictions from 20 countries. Assessments were based on five criteria: riparian zone management, clear-cut size, road culverts and decommissioning, reforestation requirements, and annual allowable cut. They had found Tasmania's Forest Practices Code "to be comprehensive and amongst the most prescriptive in the world".[84]

The Independent Verification Group completed its work in March 2012. Despite the many serious concerns surrounding it, the public release of its various component reports was accompanied by a media release from Federal Environment Minister, Tony Burke, which used the word 'independent' on six occasions.[85]

The Tasmanian Forest Agreement 2012 and the flow-on legislation

Based on the findings of the IVG process, the Tasmanian Forests Agreement (known as TFA 2012) was eventually signed by the industry and environmental group 'peace deal' negotiators in November 2012. Clause 37 of the Agreement called for the Australian Government to nominate an area of 123,650 hectares of State forest as a proposed extension to the Tasmanian Wilderness Area (TWWHA). This would be submitted for consideration by the Committee at its next meeting in June 2013.

The *Tasmanian Forests Agreement Bill 2012* was prepared and introduced to the Parliament so as to enshrine this and its other agreed commitments in State legislation. It was promptly passed by the Tasmanian Lower House where Labor and the Greens held the numbers. However, the Upper House was dominated by independents sufficiently concerned by the proposed legislation to instigate a parliamentary inquiry replete with public hearings, to provide sitting members with a better basis for voting on it.

During this Upper House inquiry in December 2012, long-standing concerns about the impact of the 'forests peace deal' on Tasmania's so-called 'special timbers' sector came to prominence. This sector had not been invited to participate in the 'forest peace deal' process despite it being proportionally the most valuable component of the state's timber industry, as well as a significant contributor to the state's tourism and arts sectors. In 2010, the special timbers sector was reportedly generating $70 million per annum and employing around 2,000 Tasmanians, while a further 8,500 were participating in woodcraft activity as a hobby or on a limited commercial basis.[86,87]

The sector's concerns were based on the impending loss of the majority of its future resource – blackwood, myrtle beech, sassafras, and celery top pine – in new forest reserves under the TFA 2012. Most of this lost resource was contained in the approximately 123,000 ha of State forests earmarked for addition to the TWWHA. Although the TFA 2012 also guaranteed that an agreed volume of special timber species logs would be made available to the industry each year, representatives from the special timbers sector had inspected the remaining unreserved State forests and concluded that this committment could not be met.[88]

After a complaint to the Tasmanian Government, the environmental group signatories to the TFA 2012 were directed to nominate 24 contingency coupes within the new forest reserves from which special timbers could be harvested in the event of insufficient resource being available within the remaining State forest wood production zones. However, led by Andrew Denman, the Tasmanian Speciality Timbers Alliance examined these nominated contingency areas and were horrified to find that many contained little or no usable logs. This was also verified by several of Forestry Tasmania's foresters who were regarded as experts in the location and distribution of the harvestable special timbers resource.[89]

Accordingly, the special timbers sector made it known to the Upper House inquiry that these supposed contingency areas were inadequate and, in some cases, were virtually treeless. In response to this, an amendment to the Bill was prepared and successfully moved by Adriana Taylor MLC, to allow special timber species to be harvested if necessary from any lands, including the former State forests mooted for addition to the Tasmanian Wilderness World Heritage Area (TWWHA). Eventually the amended Bill was passed as the *Tasmanian Forests Agreement Act 2013* on 30 April 2013.

Special timbers sawlogs have traditionally been a minor (by volume) by-product of integrated harvesting of wet mixed forests primarily producing eucalypt sawlogs and pulpwood (ie. woodchips). However, special timbers rainforest species can also be specifically harvested in small-scale, low impact selective tree removal operations which differ markedly from the public perception of 'industrial logging'. This was what was being proposed if it ever became necessary to access newly reserved areas to maintain the supply of speciality timbers to the industry.

Internationally, the World Heritage Committee has allowed such low impact timber harvesting in World Heritage-listed forests subject to stringent conditions. However in Australia, the mere prospect of even such low impact harvesting in a World Heritage Area was an outrage to environmental campaigners and the Greens, as well as Federal Environment Minister, Tony Burke.

Behind the scenes and unbeknownst to Tasmania's Liberal and independent parliamentarians, both the Federal and Tasmanian Labor Governments gave a secret undertaking to the Tasmanian Greens and the signatories to the TFA 2012, that they would use Federal environmental protection legislation to block any attempt to harvest special timbers within the expanded World Heritage Area, despite the *Tasmanian Forests Agreement Act 2013* allowing it.[90] During the final Lower House debate on the legislation on 30 April 2013, the existence of this hitherto hidden undertaking was revealed and was referred to as 'a secret side deal' by the Liberal opposition.[91]

While this revelation did not prevent the Bill from passing into legislation, it was a further indication of the determination of Labor and its Federal Environment Minister, Tony Burke, to reserve more of Tasmania's forests irrespective of the social and economic consequences. In this case, the impending loss of much of the state's highest value wood manufacturing sector even despite its significant contribution to the tourism and arts sectors, which are traditionally strongly supported by Labor and the Greens. This exposed the lie of a public assurance which Burke had previously given to Tasmanian communities: "Where you get a clash between the minimum requirements for wood supply and a conservation aspiration, wood supply will win".[92]

World Heritage Area deceit

Letters obtained under FOI show that as far back as late 2011, Federal Environment Minister Tony Burke, in conjunction with Greens Leader, Bob Brown, had begun to develop a proposal to extend the Tasmanian Wilderness World Heritage Area (TWWHA).[93] In December 2011, Brown had even provided Burke with a 'shape file' outlining the proposed new TWWHA boundaries and an 'attributes table' of values that could be used to inform a nomination to the United Nations' World Heritage Committee. This was almost 18 months before the Tasmanian Parliament would pass the *Tasmanian Forests Agreement Act 2013*, which legally enshrined a committment to extend the TWWHA.

On 31 January 2013, Environment Minister Burke announced that a nomination to extend the TWWHA had been prepared and would be submitted to the World Heritage Committee in accordance with the requirements of Clause 37 of the Tasmanian Forests Agreement 2012 (TFA 2012). This was still almost three months before the Tasmanian Parliament would pass the *Tasmanian Forests Agreement Act 2013*.

The submitted nomination was actually for a substantially larger area of 172,000 hectares, almost 50,000 hectares greater than what had been agreed to in the TFA 2012. This additional area was mostly comprised of an already existing national park and various conservation reserves, but also included some private properties covenanted as conservation reserves.[94]

No explanation for this was provided at the time, but it was later revealed that the pre-existing conservation reserves were added in the interests of creating a more rational TWWHA boundary, while the private properties – owned by Bush Heritage Australia and the Tasmanian Land Conservancy – had been added after their owners approached the government requesting their inclusion.[95] Reportedly, this had implications for some neighbouring private property owners who, without any prior consultation, were suddenly bordering a World Heritage property with uncertain restrictions on what they could do on their own land within any associated buffer zone.[96]

These concerns, as well as those of others such as local apiarists and recreationalists, who had traditionally accessed the affected forests for generations, gave rise to some community opposition. Compounding these community concerns was the lack of any detailed government mapping of the intended new TWWHA boundary, which effectively limited the capacity for potential opponents to lodge a formal protest.

A further complication was that the nomination lodged with the World Heritage Committee misrepresented this substantially larger nomination as a 'minor boundary modification' even though the proposed 172,000 hectare addition equated to an approximately 12% increase to the TWWHA's pre-existing area of 1.412 million hectares.[97] This exceeded the notional 10% increase which had previously been set as an upper limit for a 'minor boundary modification' to an existing World Heritage property.[98] The significance of this was that a 'minor boundary modification' could be approved by the World Heritage Committee without any independent assessment of its values. On the other hand, a new nomination, or a greater than 10% extension to an existing World Heritage property, generally entailed an 18-month evaluation to verify its cultural and environmental appropriateness for World Heritage-listing.[99]

In evidence presented to the Tasmanian Upper House inquiry into the *Tasmanian Forests Agreement 2012 Bill* during February 2013, a senior bureaucrat from the Federal Department of Environment admitted that the Gillard Government had requested the World Heritage Committee to treat the nomination as a 'minor boundary modification'.[100] There was a strong political incentive to make such a request because, by avoiding the normal 18-month evaluation process, the extension to the TWWHA could be listed prior to the 2013 Federal election (due for September, just 6-months away). It was already looking more than likely that the Gillard Government would be voted-out of office and replaced by a Coalition Government which, in opposition, had voiced its intention to cease reserving forests.

That the TWWHA extension was pursued with undue haste for political purposes would be subsequently admitted by a gleeful Greens Leader, Christine Milne, after the nomination was accepted by the World Heritage Committee in June 2013:

> In parallel with the IGA [i.e. the 'forest peace deal'] process, Bob Brown and I worked with Minister Tony Burke to develop this extension and get this World Heritage nomination in … so that it could be decided ahead of the [September 2013] Federal Election.[101]

Ms Milne's role as a key driver of the nomination raised some eyebrows given that she had been a former Vice President of the International Union for the Conservation of Nature (IUCN) which acts as an Advisory Body to the World Heritage Committee.[102] It was also notable that Professor Mackey (who had played such a prominent role in the associated IVG process) was an IUCN Regional Councillor for its Oceania (Asia-Pacific) Region at that time.[103] In addition, all the environmental group signatories to the TFA 2012 were members of the IUCN, with the Wilderness Society being particularly active. The extent to which these affiliations exerted an influence over the outcome of the nomination is unknown, but there is a strong perception of likelihood.

The haste with which the 2013 World Heritage Area extension was pursued is also suggestive of the Federal Labor Government appropriating the process to insulate the proposed new Tasmanian forest reserves from the strong expectation of a new Tasmanian Liberal Government being elected with a mandate to revoke the unpopular *Tasmanian Forests Agreement Act 2013*. As this duly occurred in March 2014, the Gillard Government's rush to push through the TWWHA extension could also be construed as having subverted the Tasmanian Parliament's democratic right to act in the best interests of the majority of its constituents.

It is particularly significant that the TWWHA extension nomination was successful despite the values for much the same areas of State forest having been rejected for World Heritage-listing just five years earlier. In March 2008, the World Heritage Committee had ordered a UNESCO Tasmanian Wilderness Reactive Monitoring Mission to visit the disputed areas to examine incessant claims by environmental activists that these same State forests should be added to the TWWHA. The Mission had subsequently reported back that there was no reason to extend the TWWHA because it already adequately represented forest values, and that similar values found in adjacent State forests were also being "well managed, but for both conservation and development purposes".[104]

This finding vindicated the effectiveness of Tasmania's State forest management systems (outside the TWWHA) in identifying and protecting significant environmental and cultural values through appropriate local reservations, and recognised that this could provide adequate protection without the need to reserve huge swathes of the landscape, as is routinely demanded by environmental ideologues.

It is unclear if any new information on the conservation and heritage values of the nominated State forests had emerged in the interim period between the 2008 rejection and the 2013 acceptance of these same areas for World Heritage listing. However, in July 2014, environmental consultant, Peter Hitchcock, claimed that "… the concept for extension of the Tasmanian Wilderness World Heritage Area (TWWHA) … and adoption of an appropriate eastern boundary was being progressively formulated long before the Tasmanian Forests Agreement 2012 and was informed by a wide range of proposals, heritage values, documentation and other considerations".[105]

However, this doesn't necessarily answer the question of whether new information had come to light since 2008. It is notable that as late as 2010, the then Federal Environment Minister, Peter Garrett, had said that the Government had no plans to extend the TWWHA boundary to include more wood production State forests while citing the 2008 UNESCO Mission's finding that it was not warranted.[106,107] Furthermore, the Tasmanian Greens 2010 *Forests Transitions Strategy* was advocating continued limited timber harvesting in the Styx, Weld, and Florentine Valleys specifically because they did not consider these forests to be contentious or subject to future reservations.[108] Yet these areas were all added to the TWWHA just three years later on the grounds that they had suddenly acquired suitably outstanding heritage values.

Despite this, Hitchcock also questioned whether the 2008 nomination for listing had even been rejected by noting that, despite the report of its UNESCO Tasmanian Wilderness Reactive Monitoring Mission, the World Heritage Committee had, at its later 2008 meeting in Quebec, "... resolved to advise Australia that it reiterate(s) its request to the State Party to consider, at its own discretion, extension of the property to include appropriate areas of tall eucalyptus forest, having regard to the advice of IUCN ...".[109]

This is somewhat disingenuous because Hitchcock neglected to mention that the World Heritage Committee had been heavily lobbied to disregard the report of its own Reactive Monitoring Mission by a three-person Wilderness Society delegation, including himself, which had attended the 2008 Quebec meeting.[110] In addition, Christine Milne was at that time an IUCN Vice-President and it seems highly likely that she would have also been exercising her influence to convince the Committee to dismiss the findings of its own Reactive Monitoring Mission and reiterate the request to Australia to consider extending the TWWHA.

The contention that Tasmanian environmental campaigners engineered the World Heritage Committee's repeated requests for Australia to extend the TWWHA is supported by an examination of various UNESCO reports from 2008. Their Reactive Monitoring Mission had reported back that "... the TWWHA provides a good representation of well-managed tall eucalyptus forests and there is similar forest outside the property which is also well-managed ..." This report had included a table of state-wide areas of tall wet eucalypt old growth forest as at 2007 (Annex E: Forest Reservation Status). This showed that although 36% of the old growth area was contained within the TWWHA, a further 32% was protected within a range of national parks and other formal conservation reserve categories located outside of the TWWHA.[111] In addition to this, the majority of old growth forest outside of these formal reserves was also informally reserved due to State forest management systems.

That a total of 68% of Tasmania's tall wet old growth forests (which is generally agreed to possess the highest conservation values) were already formally reserved across the state is strongly suggestive of adequate representation in conservation reserves and far exceeds the international conservation target of just 17%. However, in 2008, the IUCN which advises the World Heritage Committee, had received 'new information' in a report from the Tasmanian branch of the Wilderness Society claiming that "… the ecological diversity of the tall eucalypt ecosystem is incompletely represented in the World Heritage Area, in particular, only 29% of tall eucalypt forest is included in the property. It has also been suggested that the values outside the property are different and complementary to those of the tall eucalypt forest included in the property …".[112]

While the author of this new Wilderness Society report was not disclosed,[113] by understating the area of wet tall eucalypt forest reserved in total, including outside of the TWWHA, it created a greater imperative to increase the representation of these forests in the TWWHA. That this was seemingly accepted without question by the World Heritage Committee is suggestive of a tacit willingness to misconstrue an environmental group's aggressive pursuit of an ideological 'save-the-forest' agenda as akin to objective advice.

Irrespective of how much or how little new information supported the 2013 TWWHA extension nomination, it clearly ignored some key issues as well as avoiding due process to paint a more compelling case for it. However, for most observers, such shortcomings were masked by its glossy, photogenic presentation which looked far more like an environmental campaign pamphlet than a serious government submission. This superficially impressive presentation reportedly owed much to the contribution of a former veteran Wilderness Society forests campaigner, who had reportedly been employed for eight weeks by the Bob Brown Foundation to prepare the nomination.

Firstly, the nomination made little or no mention of wilderness values despite recommending additions to a 'wilderness world heritage property'. Presumably this was because the most recent study of wilderness values, undertaken in 1997 during the Tasmanian Regional Forests

Agreement process, had found that the particular State forests included in the extension nomination actually had 'low' wilderness value.

This had even been conceded during the IVG process in 2011, when Professor Mackey had noted that: "Forest wilderness issues warrant further consideration, especially in areas adjoining the TWWHA ... it will be important to assess the current extent of and potential to restore forested wilderness in areas which warrant formal assessment for World Heritage listing".[114] Despite this, no follow-up assessment of wilderness values had occurred prior to the Government's 2013 nomination to extend the TWWHA.

Wilderness is a landscape-scale concept which encompasses features such as being 'untrammelled by man' (the USA *Wilderness Act 1964*); being 'not developed with roads ... or other industrial infrastructure' (The WILD Foundation); and 'remote from the influences of European settlement' (Commonwealth of Australia, 1998). These factors are regarded as being equally as important to the concept of wilderness as being 'intact, and undisturbed', and 'truly wild' (The WILD Foundation).

Accordingly, it is entirely possible for healthy, high quality forests to have 'degraded' wilderness values if they are easily accessible and contain evidence of recent or current human use. This was the case with a substantial portion of the State forests encompassed within the 2013 TWWHA extension nomination.

According to Forestry Tasmania, the nominated State forests contained a substantial area of regrowth or mixed-age forest which had been previously harvested for timber. This included areas harvested several times stretching back to the pre-1900 era, as well as areas that had been clearfelled and regenerated within the previous three years.[115] As this past disturbance was widespread, it had been serviced by a network of forestry roads and tracks that were either still in-use or over-grown but still evident. In addition, a major highway and a power line with an associated cleared easement passed through other parts of the nominated area, and there were also several small timber plantations included within it.[116] When a UNESCO Reactive Monitoring Mission visited the TWWHA in November 2015 it noted:

It is important to recall that the TWWHA boasts a surprisingly extensive road network. There are more than 1,100 kilometers of roads according to State Party information provided to the mission team, much of it located within the area added to the TWWHA in 2013.[117]

Secondly, the TWWHA extension nomination departed from due process by containing no report of its social and economic impacts. This contravened the requirements of the TFA 2012 which had recommended an addition to the TWWHA but had also specified that socio-economic modelling was required to examine its state-wide and regional impacts.[118] The nomination also contravened the 1997 Tasmanian Regional Forests Agreement (RFA) which both the State and Federal Governments were still bound by. Clause 40 of the RFA committed the 'Commonwealth' to "… give full consideration to the potential social and economic consequences of any World Heritage nomination of places in Tasmania and that any nomination will only occur after the fullest consultation …".[119]

Thirdly, the TWWHA extension nomination failed to provide any information on the indigenous cultural heritage values of the forests in question, or to take account of previous UNESCO work showing that such values were already adequately represented in the existing TWWHA. When the nomination was being considered, an Advisory Body to the World Heritage Committee was poised to reject it due to a lack of any information about indigenous cultural significance.[120] However, according to former Wilderness Society forests campaigner, Geoff Law, this was countered when: "Several ENGOs [*environmental groups*] and an indigenous representative attended the Phnom Penh meeting of the Committee and collaborated with the Australian Government."[121] This reportedly prompted an eleventh-hour intervention by Federal Environment Minister, Tony Burke, who promised $500,000 to fund a study of the TWWHA extension's cultural heritage values should the nomination be accepted.

Despite such gaps in the required evidence, the World Heritage Committee was still able to be convinced that the nominated areas, which had been rejected as unsuitable in 2008, had, over the course of just five years, suddenly acquired values deserving of World Heritage-listing. Accordingly, it accepted the Gillard Government's nomination to add 172,000 hectares to the TWWHA at its Phnom Penh meeting in June 2013.

Concerns surrounding the TWWHA extension nomination since its announcement in February 2013, were routinely dismissed by the Australian Government's environmental bureaucracy, the Greens, environmental activists, and by the World Heritage Committee itself largely on the grounds that the nomination had been 'agreed to' by environmentalists and the timber industry (i.e. in the TFA 2012), and because 'new government work' (i.e. the IVG process) had verified the requisite heritage values.[122]

However, such a justification was weak. Firstly, the TFA 2012 was an agreement arrived at without any input from some industry sectors – most notably, the special timbers sector – which had been excluded from the negotiations. In addition, the forest manager, Forestry Tasmania, and its forest scientists had also been excluded from the formal negotiations. Similarly, several environmental groups were excluded and had vociferously declared their opposition to the TFA 2012 and had vowed to keep campaigning against the timber industry. Supporting them in opposing the TFA 2012 were no less than Bob Brown, the founder of the Tasmanian and Australian Greens, and Christine Milne, the then Australian Greens leader. In addition, the agreement process had excluded the rest of civil society, including important stakeholders such as private forest owners, apiarists, and a wide range of recreational forest users.

Secondly, the Independent Verification Group (IVG) process (as described earlier) was highly flawed and far from independent given the undeclared conflicts of interest existing amongst many of its third party contractors and report authors. Despite this, the Australian Government continued to mislead the community, and even the World Heritage Committee, that the IVG process had been beyond reproach.

When the World Heritage Committee received letters of complaint about the nominated extension to the TWWHA, it asked the Department of Environment to respond to the allegation that "no independent scientific assessment of the heritage values has been undertaken". In a letter to Mr Kinshore Rao, Director of the World Heritage Centre, on 28 March 2013, the Department of Environment's Deputy Secretary, Dr Kimberley Dripps reiterated that heritage values described in the nomination had been largely based on earlier work undertaken by the Independent Verification Group. She urged Mr Rao to:

> Note also that the members of the Independent Verification Group were selected for their expertise and independence, and so their advice resulted from an independent process, led by experts.

Yet, as has been outlined earlier, the Department of Environment would later admit to an ANAO investigation that it was aware that many of the IVG members and third party contractors were far from being independent experts.

Irrespective of the integrity of the IVG's work, its Final Report had actually noted that: "While the IVG Heritage Report (Technical Report 5A) used the National and World Heritage criteria set out in the EPBC Act [*1999*], it does not constitute a formal heritage assessment as provided under processes established under the EPBC Act and the World Heritage Convention, respectively."[123] Accordingly, the Department of Environment's assertion to the World Heritage Committee that the IVG and the related TFA 2012 had verified the presence of heritage values within the areas nominated for addition to the TW-WHA, was not correct.

Trying to right a wrong in the face of double standards and hypocrisy

Disquiet over the manner by which the TWWHA extension had been achieved soon manifested itself in the forest policies of Liberal Oppositions at both the Tasmanian and Federal levels. In September 2013, the Tony Abbott-led Coalition Government was elected and set about delivering on a promise to overturn the TWWHA extension achieved just three months earlier.

After some consideration, they decided to submit a nomination to the World Heritage Committee for a 'minor boundary modification' that would excise a 74,000 ha portion of the 172,000 hectares which had been added to the TWWHA only months earlier. The grounds for this nominated excision were reasonable: 1) that the area in question had been multiple-use State forest for over 90 years and much of it had been disturbed by timber harvesting and roading over that period; and 2) as it had been an important future resource for the state's timber industry its loss had considerable socio-economic ramifications that had not been assessed during the development of the nomination which had led to its listing – clearly in contravention of requirements under TFA 2012 and the Tasmanian Regional Forest Agreement.

Historical records showed that past timber harvesting had affected approximately 21,000 hectares (or 28%) of the 74,000 ha planned for excision from the TWWHA, in addition to the scattered infrastructure (ie. roads, tracks, bridges) across the rest of the area which was associated with this past harvesting. This harvesting had apparently occurred over the past 150-years, and some areas were reportedly on their second or third harvest cycle.[124]

A fierce public debate ensued as the Greens and their environmental group cohorts claimed that far less of the area had been previously harvested and that, in any case, evidence of recent human use did not necessarily undermine the appropriateness of an area for World Heritage listing. Their varied claims that as little as 4% or as much as 14% having been previously harvested were seemingly based on an arbitrarily applied cut-off date of 1960, and therefore substantially understated the reality. Nevertheless, it was somewhat bemusing to observe the hypocrisy of environmental activists desperately striving to justify the high conservation value of harvested and regenerating forests – some only treated within the previous five years – after decades of deriding them as 'unnatural plantation monocultures with no biodiversity value'.

When the Abbott Government eventually submitted a nomination to the World Heritage Committee to excise the 74,000 hectares from the TWWHA, it was mercilessly vilified in the media for playing politics with the environment. This was in stark contrast to the media's willful blindness to the blatant politicking by the former Gillard Government and her Greens allies to secure the extension only a year earlier.

Notably, whereas the initial 2013 extension had been facilitated through the misuse of the 'minor boundary modification' process, the World Heritage Committee this time around summarily rejected (reportedly in less than ten minutes) the Abbott Government's attempt to correctly use the same mechanism (but in reverse). Some Committee delegates even attacked the concept of attempting to excise some forests from the TWWHA as an embarrassment that diminishes Australia in the eyes of the global community.[125]

This highlighted a double standard whereby minor additions to World Heritage properties can be listed without any independent verification of their values, whereas a proposed minor excision attracts derision with nary a thought that such a course may be well reasoned. It also casts some doubt over the objectivity being applied to managing the World Heritage concept, and arguably reflects the heavy lobbying of Committee delegates by Australian environmental activists imparting skewed views of the reality and/or the Government's intentions.

Later it would be revealed that the Abbott Government's nomination had been poorly prepared raising questions about whether it was merely going through the motions to fulfill an election promise. It would hardly be a surprise if the government was not seriously committed to excising the nominated area from the TWWHA given the prevailing political climate. The Government was already being loudly derided as an environmental pariah for undoing climate change policy initiatives (such as the Carbon Tax) which had been introduced by the previous Labor Government, and hardly needed any additional aggravation.

Subverting science and due process for political purposes

Those who had hoped that part of the 2013 TWWHA extension would be excised would arguably have been satisfied if the World Heritage Committee had deferred any decision subject to independent scientific scrutiny of the disputed area's heritage values. This would have at least represented a return to the due process that had been by-passed for political and ideological purposes when the former Labor Government and its allies had misappropriated the 'minor boundary modification' convention.

Certainly, many forest and conservation scientists closely involved with Tasmania's forests had been incensed at how the proposed new forest reserves and the subsequent 2013 nomination to extend the TWWHA had arisen. Former Forestry Tasmania conservation biologist, Simon Groves, articulated a view that the process which had led to new Tasmanian forest reserves and the TWWHA extension, had been:

> ... a perversion of science ... it rewards bad behaviour and ... those that engage in bad behaviour. It sends the wrong signals to the protagonists and the public. It sets up the rest of Australia's forested regions for similar processes ... and the conservation benefits are, I believe, largely delusional.[126]

Another consortium of five Tasmanian forest and conservation scientists concurred that "there was a complete lack of scientific rigour in the assessment process that led to the reserve proposals" and further opined that the simplistic 'lock-up' model of huge landscape reservations being implemented in Tasmania, was out-of-step with current world conservation thinking which "regards the best model as one where timber production is part of a larger landscape in which nature conservation and timber harvesting co-occur as part of a complex mosaic."[127]

In stark contrast to such concerns, Australia's broader church of environmental scientists – who are mostly remote from any practical involvement in Tasmania's forests – almost universally celebrated the World Heritage Committee's rebuttal of the Abbott Government's attempt to excise part of the 2013 TWWHA extension. Their attitude paints a disturbing picture of hypocrisy amongst the vast bulk of Australia's community of environmental scientists and activists. Undoubtedly they would be outraged if a proposed new resource use was approved without any independent scientific study, but this case demonstrated their propensity to become cheerleaders for science-free decisions to increase forest reservations when the political winds blow their way.

The only concession that could be offered to excuse such hypocrisy is that anyone whose knowledge of this matter was primarily reliant on the mainstream media would have little or no awareness of the corrupted process and political manipulation that had led to the 2013 TWWHA extension, and would therefore lack any informed appreciation of the motivation for attempting to partly revoke it.

Since the failed attempt to partly over-turn the 2013 TWWHA extension, a further development has perhaps strengthened the perception of environmental ideology perverting supposedly independent bureaucratic functions.

In mid-2014, when Tasmania's newly-elected Liberal Government repealed the unpopular *Tasmanian Forest Agreement 2013 Act*, it immediately halted the process of transferring half a million hectares of mostly multiple-use State forests into the national parks estate. This included the 123,000 hectares of State forest which had just been added to the TWWHA in June 2013.

As a result of this, unlike most other Australian World Heritage properties, the TWWHA now contains portions of land that are not a national park. Instead, these recently added former State forests had been temporarily reclassified as 'conservation areas' and 'regional reserves' by the previous government prior to impending national park declaration, and remain so now that the park expansion program has ended.

Significantly, these two interim forms of public land tenure have less onerous resource use restrictions. In fact, the state's *National Parks and Reserve Management Act 2002* allows the extraction of speciality timbers and mineral resources from 'conservation areas' and 'regional reserves'. The upshot of this is that under Tasmanian law, there are parts of the recent additions to the TWWHA which could be timber harvested or mined subject to appropriate environmental protections. This also accords with the World Heritage Convention which allows for careful and sustainable resource use (such as 'selective timber cutting') on non-national park land tenures within World Heritage properties.[128]

However, the International Union for the Conservation of Nature (the IUCN) also has a worldwide system of protected area categories that Australia's State Governments are required to annually assign to public land tenures. While some of these IUCN categories also allow for natural resource use, it appears that Tasmania's environmental bureaucracy has applied the 'no use' 'Category 1b Wilderness' to those areas within the TWWHA in which the limited-use provisions of the *National Parks and Reserve Management Act 2002* would have otherwise applied. This is suggestive of the state's environmental bureaucracy deliberately subverting the State legislation to prevent resource use within a small part of the TWWHA where it is currently legally allowed.

Despite this, the new Tasmanian Liberal Government deferred to the state legislation to include the limited harvesting of special timbers in its draft revised Management Plan prepared for the expanded TWWHA. This provision, along with proposed limited tourism developments, outraged the state's environmentalists and prompted the Wilderness Society to yet again send a delegation to lobby the World Heritage Committee to reject the Government's draft TWWHA Management Plan at its meeting in Bonn, Germany, in June 2015.

Ultimately, the World Heritage Committee did reject the draft TWWHA Management Plan. This was portrayed in the media as a stern rebuke to the Tasmanian Government by the United Nations.[129] In reality, it was an unjustifiable rebuke for simply proposing to undertake activities that are permitted under State law and which were also in accordance with the World Heritage Convention.

Aftermath

The Tasmanian Forest Agreement 2012 and the related IVG process essentially involved new conservation reserves being selected by environmental activists and then assessed for their suitability largely by environmental activist associates, supporters and former operatives engaged by a bureaucracy directed by a Government beholden to the political representatives of environmental activism. That this could be so widely portrayed as an objective and independent process is a travesty.

Purely political decisions about future forest use have been commonplace throughout Australia over the past 15-20 years. What is different about the Tasmanian forest reservation process from 2010-14 is that politicians and the environmental bureaucracy have fraudulently portrayed it as having been scientifically-verified through an independent process. This is a dire allegation, but is entirely justifiable given that those who controlled the process have since admitted to being aware of the conflicts of interest and biases amongst those who were engaged to undertake the supposedly objective IVG process. Presumably the failure to pursue conflict of interest declarations was due to fears that it would have exposed a lack of independence and invalidated much of the IVG's work.

While willfully misleading the Australian public that a forest reservation process has been independently and objectively verified may be one thing, deliberately duping a United Nations conservation agency is quite another, and arguably damages Australia's international standing.

Sadly, the recent events in Tasmania suggest that World Heritage listing is likely to be increasingly misused as a weapon by Australia's environmental extremists, and their political representatives and allies. It is significant that there is a continuing campaign to have a further 800,000 hectares of Tasmanian land (the so-called 'Tarkine' region) added to the TWWHA. If successful, this would bring the proportion of Tasmania's land mass that is World Heritage-listed up to almost one-third – arguably an unprecedented level of landscape reservation on a global jurisdictional basis.

The advantages of World Heritage-listing as a vehicle for achieving environmental activist agendas are that: 1) it is an international concept that is popularly regarded as sacrosanct; 2) it is insulated from the shifting vagaries of domestic politics; and 3) it provides an opportunity for environmental agendas to be foisted onto countries by a supposedly unimpeachable international body – the United Nations. The veracity of this last point is exemplified by recent directives from the United Nations agency, UNESCO, over the management of the Great Barrier Reef under threat of an 'in danger' listing that is feared for its potential to severely undermine a country's reputation and attractiveness to international tourists.[130]

Australia's environmental activist community long-ago grasped that the World Heritage concept and even domestic policy processes, can be manipulated to realise long-held ideological agendas by infiltrating the bureaucracies of Federal and State Governments, as well as key international advisory bodies, like the International Union for the Conservation of Nature. The IUCN provides conservation advice to the World Heritage Committee but has employed career activists in key positions – such as former Greens leader, Christine Milne, as an IUCN Vice-President. With regard to Tasmania's forests, it has shown a predisposition to confuse clearly compromised and selective commentary and advice from environmental ideologues – such as the Wilderness Society – with balanced, objective, and factual advice on the real situation.

Criticism of the influence of the Wilderness Society on the IUCN has been countered by some who have pointed out that other groups – such as the Institute of Foresters of Australia (the IFA) – are also IUCN members with the potential to provide alternative forestry perspectives.[131] However, the Wilderness Society earns in the vicinity of $15 million per annum, largely from tax-free donations, and employs well-paid career staff. In stark contrast, the IFA is essentially a voluntary organisation that survives almost entirely on the membership subscriptions of its 1,000 or so members, and employs only a part-time CEO and administrative person. Clearly the IFA could never afford the expense of dispatching delegations of lobbyists to World Heritage Committee meetings around the globe, as the Wilderness Society regularly does.

This huge disparity between the influential capacity of the corporatised environmental lobby compared to professional associations with a more informed scientific and practical knowledge of natural resource use and land management, suggests that the World Heritage Committee is prone to basing its decisions on a skewed narrative that differs markedly from the reality. This seems more than probable when decisions, such as the Committee's recent rejection of the Tasmanian Government's draft TWWHA Management Plan, are disdainfully made without even any discussion or debate.[132] This is strongly suggestive of pre-conceived decisions shaped by backroom lobbying that dismisses alternative arguments as not even worth hearing.

While Greens politicians and environmental activists clearly see World Heritage listing as a great thing that can be used to further their aims,[133] the recent Tasmanian experience suggests that it can be misused as a political plaything that eschews science and evidence for an ideology which preaches that environmental values can only be protected by quarantining them in huge reserves where few humans go. In so doing, it effectively denies a democratic voice to the millions of Australians who also value forests for recreational or commercial uses.

Those inclined to accept deceitful behaviour as justifiable in the interests of saving biodiversity, need to be mindful of the contextual reality that existed in Tasmania even before the 'forest peace deal' and related World Heritage issues arose. That is that around 80% of Tasmania's forests were already either reserved or privately-owned and were not even considered in the 'forest peace' process, and also that Tasmania already had one of the highest proportions of its land mass under World Heritage-listing of any jurisdiction in the world.

Chapter 7 Endnotes

1 Peter MacFie, 2014, *Misleading Maps? – the WHA extension 2013*. Unpublished essay by Tasmanian historian Peter MacFie.

2 Forestry Tasmania 2009, *Upper Florentine Valley*, 23rd January 2009. Forestry Tasmania brochure.

3 Groups whose representatives negotiated the Tasmanian forests peace deal:
  Environment Tasmania
  The Wilderness Society
  Australian Conservation Foundation
  Forest Industries Association of Tasmania
  Construction Forestry Mining and Energy Union
  National Association of Forest Industries
  The Australian Forest Contractors Association
  Tasmanian Country Sawmillers Foundation
  The Tasmanian Forest Contractors Association
  Timber Communities Australia

4 Arguably the repeal of the TFA Act did go far enough as the Trache 1 reserves gazetted under the Act in December 2013 remain in force at the time of writing (February 2017), although they sit in limbo and have not progressed to national park or consevation reserve status as was the intention at their gazettal.

5 The following records of meetings and letters was obtained by Andrew Denman, Tasmania under Freedom of Information request:

18 October 2011: Greens Senator Brown meets Federal Environment Minister, Tony Burke, to discuss options for extending the Tasmanian Wilderness World Heritage Area (TWWHA).

28 October 2011: Brown follows-up by writing to Burke outlining a plan for extending the TWWHA.

3 November 2011: Burke meets with Brown and Greens Senator Christine Milne to agree to proposed TWWHA extension.

14 November 2011: Brown writes to Burke promising a 'shape file' of the proposed TWWHA extension boundary and an 'attributes table' (supplied on 30 November 2011).

Early December 2011: Burke writes to Brown thanking him for the 'shape file' and 'attributes table' and alerting him that the findings of the Independent Verirication Group would inform nomination to extend the TWWHA which will be submitted to the World Heritage Committee.

6 Greg L'Estrange, 2011, *Many voices in the Tas pulp mill discussion*, ABC News online, 31 March 2011.

7 Julian Amos, 2013, in testimony presented to a public hearing of the Tasmanian Legislative Council Select Committee on the Tasmanian Forests Agreement Bill 2012, Hobart, 5 February 2013.

8 The Intensive Forest Management Program (from 1991-96) was initiated to establish eucalypt plantations for future production of sawlog and veneer logs to off-set the loss of productive native forest resource included in the Tasmanian Wilderness World Heritage Area, following the 1987 Helsham Inquiry. In addition, the Regional Forest Agreement process of the mid-1990s saw a further 535,000 ha of native forest resource (including 48,000 ha of privately-owned forest) made unavailable due to placement in conservation reserves. Compensation for this loss of timber resource was paid to the Tasmanian Government specifically to fund further establishment of eucalypt sawlog plantations.

9 Shining Gum (*Eucalyptus nitens*) grows naturally in the montane wet forests of central and easternVictoria and has been used for sawn timber for generations. As it often grows as a minor stand component in association with the high value ash-type eucalypts, especially *Eucalyptus regnans*, it is often sawn in conjunction with this species and sold as VicAsh, which along with its Tasmanian equivalent, TasOak, are Australia's two most valuable hardwoods. The Shining Gum currently being harvested in Victoria is 75-year old regrowth from the 1939 bushfires, but this regrowth was first harvested in

the mid-1980s when it was about 45 years old. As wood quality generally improves with age, it was probably always optimistic to expect 25-30 year old plantations to produce high volumes of the best grade timber. A more realistic expectation may be 40-45 years.

10 *Review of Tasmanian Forest Estate Wood Supply Scenarios*, by Mark Burgman and Andrew Robinson (University of Melbourne), Final Report to the Independent Verification Group, Intergovernmental Agreement, Version 9.9 (7 March 2012).

11 *Inquiry into the Triabunna Woodchip Mill and Future Development Opportunities for the Triabunna Region*, House of Assembly Standing Committee on Community Development, Parliament of Tasmania (February 2015), Section 4, pp. 15-20.

12 Mel Barnes, 2013, "Conflict over forest 'peace' deal", *Green-Left Weekly*, 11 May 2013.

13 The support for 'a pulp mill' agreed to be the environmental negotiators was carefully worded to distinguish it from the already approved Gunns Tamar Valley pulp mill which was opposed by virtually all of Tasmania's 'environmentalist community'. Nevertheless as the Gunns mill was the only one on the table, this support for 'a pulp mill' was widely interpreted as meaning the Gunns' mill, even though a discussion was initiated about the prospect for establishing the mill at an inland location at Hampshire where any potential environmental and human impacts could be more easily mitigated.

14 *Inquiry into the Triabunna Woodchip Mill and Future Development Opportunities for the Triabunna Region*, op. cit., p. 29.

15 *Inquiry into the Triabunna Woodchip Mill and Future Development Opportunities for the Triabunna Region*, op. cit., pp. 40-45.

16 John van Tiggledon, 2014, "The destruction of the Triabunna mill and the fall of Tasmania's woodchip industry", *The Monthly*, July 2014.

17 Ibid.

18 Ibid.

19 *Inquiry into the Triabunna Woodchip Mill and Future Development Opportunities for the Triabunna Region*, op.cit., pp. 5-6.

20 *Tasmanian forests Statement of Principles to lead to an agreement*, signed 14 October 2010 by timber industry and ENGO negotiators. The clause in question stated: *Immediately protect, maintain and enhance High Conservation Value Forests identified by ENGO's on public land.*

21 Lyndon Schneiders, 2012, 'Barking up the wrong trees: The timber industry glibly spurned a compromise that could have saved Tasmanian forests", *The Age*, 2 November 2012.

22 *GetUp!* Save Tasmania's Forests Campaign.

23 Safi, Mike, 2014, "War and peace – and war again?", *The Guardian Australia*, 18 September 2014.

24 Richard Flanagan, 2013, "I don't agree", *Tasmanian Times*, 3 May 2013

25 Bob Brown, 2013, *Forests deal 'greenmails' the green groups*, ABC Environment, 20 May 2013.

26 Transcript: Christine Milne Press Conference on the Tasmanian Forest Peace Agreement, 17 April 2013.
27 Julian Amos, 2013, op. cit.
28 The Australian Forestry Standard Ltd (AFS Ltd) oversees the operation of the Australian Forest Certification Scheme (AFCS) which applies to forest management and chain of custody certification in Australia. The AFCS is endorsed by the Program for Endorsement of Forest Certification (PEFC).

The PEFC is an international non-profit, non-governmental organisation dedicated to promoting sustainable forest management (SFM) through independent third-party certification. It is an umbrella organisation which works by endorsing national forest certification systems developed through multi-stakeholder processes and tailored to local priorities and conditions.

With over 30 endorsed national certification systems and more than 240 million hectares of certified forests, the PEFC is the world's largest forest certification system.
29 Forest Stewardship Council Canada Working Group 2004, National Boreal Standard accredited by the FSC 6August 2004.
30 Hancock Watch website.
   FSC Watch website.
31 David Beniuk, 2012, "Tassie forest wars set to go on", *Sydney Morning Herald*, 27 October 2012.
32 Letter from Australian Greens Leader, Senator Bob Brown, to Prime Minister Julia Gillard, 22 November 2010. Obtained under FOI request by Andrew Denman, Tasmania
33 Department of Environment 2011, Tasmanian Forests InterGovernmental Agreement Fact Sheet.
34 Ibid.
35 Ian McPhee, 2014, Letter from the Auditor-General for Australia, Mr Ian McPhee, to Mr Andrew Denman, President of the Tasmanian Special Timbers Alliance, Kettering, Tasmania, 15 September 2014.
36 Ibid.
37 The Independent Verification Group's members were:
   Chair, Professor Jonathan West
   Dr Robert (Bob) Smith
   Dr Michael Lockwood
   Professor Brendan Mackey
   Professor Mark Burgman
   Professor Ross Large
38 Dr Robert (Bob) Smith had formerly headed government forest management agencies in Victoria and NSW.

39 Andrew Denman, 2016, IVG engagement contracts viewed by Andrew Denman, President of the Tasmanian Special Timbers Alliance, under FOI protocols.
40 Ian McPhee, 2014, op. cit.
41 Ibid.
42 Andrew Denman, 2016, op. cit.
43 Ibid.
44 Advice provided by IVG Chairman Jonathon West to the Intergovernmental Taskforce in September 2011, as outlined in the letter from the Auditor-General for Australia, Mr Ian McPhee, to Mr Andrew Denman, President of the Tasmanian Special Timbers Alliance, Kettering, Tasmania, 15 September 2014.
45 Clause 20 of the Tasmanian Forests Intergovernmental Agreement between the Commonwealth of Autralia and the State of Tasmania, 7 August 2011.
46 Andrew Denman, 2017, personal correspondence, February 2017.
47 Dr Kimberley Dripps, 2014, in evidence submitted during a public hearing of the Parliamentary Environment and Communications References Committee inquiry into the Tasmanian Wilderness World Heritage Area, Canberra, 6 May 2014.
48 Professor Brendan Mackey, 2014, in evidence submitted during a public hearing of the Parliamentary Environment and Communications References Committee inquiry into the Tasmanian Wilderness World Heritage Area, Canberra, 6 May 2014.
49 Helen Gee, 2001, *For the Forests – A History of the Tasmanian Forests Campaigns*, The Wilderness Society, pp. 90, 210, 213, 221, 224, 231, and 244.
50 Ibid.
51 Ibid.
52 The Wilderness Society 2003, *Renowned conservation scientist leads Adelaide WildCountry Science Council meeting*, 8 September 2003.
53 Australian National University, Annual Report 2003, p. 164.
54 Australian National University School of Resources, Environment & Society, 2004 Yearbook, p. 7
55 Australian National University School of Resources, Environment & Society, *2005 Yearbook*, p. 6.
56 Email exchange between Mr Andrew Denman, Tasmanian Special Timbers Alliance, and Sue Clarke, ANU University Records Manager, May 2016.
57 B. Mackey, H. Keith, S. Berry and D. Lindenmayer, 2008, *Green Carbon – the Role of Native Forests in Carbon Storage – Part 1: A green carbon account of Australia's south eastern eucalypt forests, and policy implications*, ANU E-Press, p. 41: "We are grateful to The Wilderness Society Australia for a research grant that supported the analyses presented in this report."
58 A copy of Professor Mackey's Department of Environment IVG engagement contract was obtained under Freedom of Information protocols by Andrew Denman, President of the Tasmanian Special Timbers Alliance, Kettering, Tasmania.

59 Ibid.

60 Ian McPhee, 2014, op. cit.

61 A copy of Peter Hitchcock's Department of Environment IVG engagement contract was obtained under Freedom of Information protocols by Andrew Denman, President of the Tasmanian Special Timbers Alliance, Kettering, Tasmania.

62 The Wildlands Network (formerly Project).

63 The Wildlands Project 2013, Agenda 21 Course: Lesson 3, April 2013. This online lecture claims that the Wildlands Project aims to put 50% of the USA's rural lands under reservation for conservation purposes.

64 *WildCountry – Victoria*. Wilderness Society website (Updated October, 2013).

65 ALERT website: About Us – Our Goals.

66 Bill Laurance has written at least the following two articles supporting the tactics of environmental activism which implies a personal belief that their use is as justified in Australia as in international forests beset with vastly greater problems:

"Boycotts are a crucial weapon to fight environmental-harming firms", *The Conversation*, 7 April 2014.

"Australia needs politically-active environmental groups", *The Conversation*, 12 June 2015.

67 Curriculum Vitae of Dr Oscar Venter.

68 Andrew Denman 2017: Despite ongoing inquiries to the Department of Environment no records of the engagement of these peer reviewers have been found beyond a re-imbursement invoice for their services from IVG Chairman Jonathon West. The Department was asked to advise whether the peer reviewers were engaged as individuals or through their respective institiutions, but they refused to provide the information.

69 Institute of Foresters of Australia 2011, *Tasmania cuts Parks and Wildlife Service staff at the same time as Wilderness Society associates benefit from government largesse*, Media Release, 11 November 2011.

70 A. McIntosh, 2013, "Tasmanian Forests Agreement: Liberal society needs an alternative". *The Conversation*, 10 May 2013. The article disclosed that its author received funding from carbon offset company Forests Alive. This company's website describes its projects as "the first within Australia to be accredited under the Verified Carbon Standard, and among the first VCS projects in the world in the field of avoided emissions by stopping native forest logging". The Managing Director of Forests Alive is Virginia Young, a well-known former Wilderness Society forests campaigner, who is described on the website as being instrumental in the *Green Carbon* research (by Mackey et al, 2008), which is described as underpinning the science of forest carbon offsets.

71 During the Hobart public hearings of the 2014 Parliamentary inquiry into the Tasmanian Wilderness World Heriatage Area, Cadman responded to a question about his former work with the Wilderness Society by stating that he had "worked for almost every large ENGO in the country as a consultant and, in the case of the Wilderness

Society, for 2½ years as their forest campaign coordinator." This somewhat understates the extent of his activist career which stretches at least as far back as 1985, when he had authored the book – *Woodchips, the real impacts* – for the Forest Action Network and the Wilderness Society. As recently as 2011, Cadman was representing the Wilderness Society on the Board of FSC Australia, and he represented The Wilderness Society at the 2009 UN Climate talks in Copenhagen. A Search of the Wilderness Society website contains articles where he is described as the organisation's National Forests Campaigns Coordinator from as early as October 2004, until as late as November 2007.

72 M. Lockwood and S. Cadman, 2012, *Social Values and Considerations for Effective Reserve Establishment and Management*, Independent Verification Group Report, February 2012. Accessed through the Department of Environment website.

73 The Wilderness Society Inc, Financial Statements for the Year ended 30 June 2010: p.1 lists Rosemary Norwood as a member of the Executive Committee from 25 October 2008 until 28 February 2010; p.22 shows that $67,282 was paid to Cadman and Norwood Environmental Consultants for undisclosed services.

74 Ian McPhee, 2014, op. cit. and records of correspondence between Andrew Denman and the Department of Environment with regard to a complaint about non-disclosure of conflict of interest re this project.

75 Forest Walks Lodge, Quamby Bluff near Deloraine in northern Tasmania.

76 The Wilderness Society 1991, *Saving Native Forests in the Atomic Age: A Peoples Guide to the 'Wilderness not Woodchips' campaign*, by R. Knight, and A.J. Brown, March 1991, p. 12.

77 University of Tasmania – Profiles – Dr Peter McQuillan.

78 Rainforest Action Network 2007, *Top paper company misinforming customers about source of wood*, RAN website, 26 June 2007.

79 *Our Native Forests and Climate Change*, 2010 You Tube clip authorised by Virginia Young of the Wilderness Society on behalf of Get-Up and Environment Tasmania.

Tasmania's Native Forests Solutions, 2011 You Tube clip authorised by Phill Pullinger of Environment Tasmania on behalf of the Australian Conservation Foundation and Environment Tasmania.

80 Giddings 2011, *Government committed to Forest Agreement*, joint media release by Federal Environment Minister Tony Burke and Tasmanian Premier, Lara Giddings, 22 September 2011.

81 Institute of Foresters of Australia 2012, *Critique of the work of the Independent Verification Group (IVG) appointed to advise the Tasmanian Intergovernmental Forests Agreement (IGA)* (unpublished), May 2012.

82 Ibid: "The authors of the IVG's forest conservation reports do not attempt to rank the ENGO polygons and without an assessment of their conservation values against those outside these polygons, and between polygons, it is questionable whether this approach is valid. While the information and approach is interesting, it is meaningless unless an analysis of the relevance and significance of such findings is undertaken.

This has occurred despite Professor West agreeing with the Institute of Foresters that an assessment of the conservation values of forests required a 'ranking' system that would differentiate between multi-use and conservation uses and management".

83 Ibid.

84 L. McDermott et al (2008), *A Global Comparison of Forest Practice Policies using Tasmania as a Constant*, viewed at <http://www.yale.edu>

85 T. Burke and B. Green, 2012, *Independent Verification Group advice released*, Joint Media Release from the Federal Environment Minister and the Tasmanian Deputy Premier, 23 March 2012.

86 Forestry Tasmania 2010, *Special Timbers Strategy*, p. 10

87 Farley et al, *A review of the Tasmanian woodcraft sector for the Woodcraft Guild of Tasmania Inc and Forestry Tasmania*, March 2009.

88 Andrew Denman 2017, President of the Tasmanian Specialty Timber Alliance, personal comments.

89 Paul Harris, the Honourable MLC for Huon 2013, *Hansard*, 16 April 2013.

90 Letter from Mr Tony Burke, Federal Minister for Sustainability, Environment, Water, Population and Communities to Ms Jane Calvert, Interim Chair, Signatories to the Tasmanian Forests Agreement, 30 April 2013; and *'Tasmanian State Government and Commonwealth Government Committments to assist with the Implementation of the Tasmanian Forests Agreement'*, 29 April 2013.

91 Gutwein, Peter, the Honourable Liberal MLA for Bass 2013, Reply to Mr McKimm (Greens) and Mr Green (Labor Deputy Premier), Lower House debate of the Tasmanian Forests Agreement 2012 Bill, *Hansard*, 30 April 2013.

92 Federal Environment Minister, Tony Burke, addressing the Kingston Community Cabinet, 3 October 2011.

93 The following records of meetings and letters was obtained by Andrew Denman, Tasmania under Freedom of Information request:

18 October 2011: Greens Senator Brown meets Federal Environment Minister, Tony Burke, to discuss options for extending the Tasmanian Wilderness World Heritage Area (TWWHA).

28 October 2011: Brown follows-up by writing to Burke outlining a plan for extending the TWWHA.

3 November 2011: Burke meets with Brown and Greens Senator Christine Milne to agree to proposed TWWHA extension.

14 November 2011: Brown writes to Burke promising a 'shape file' of the proposed TWWHA extension boundary and an 'attributes table' (supplied on 30 November 2011).

Early December 2011: Burke writes to Brown thanking him for the 'shape file' and 'attributes table' and alerting him that the findings of the Independent Verirication Group would inform nomination to extend the TWWHA which will be submitted to the World Heritage Committee.

94 Tasmanian Farmers and Graziers Association 2013, *Extension of the Tasmanian Wilderness World Heritage Area*. Submission to the World Heritage Committee (unpublished), 28 February 2013.

95 Veronica Blazely, Department of Sustainability, Environment, Water, Population and Communities, 2013. Evidence presented to the Legislative Council Select Committee on the Tasmanian Forests Agreement Bill 2012, Parliament House, Hobart, 28 February 2013.

96 Greg Hall MLC, Independent Member for Western Tiers – Submission No. 114 to the Parliamentary inquiry into the Tasmanian Wilderness World Heritage Area, 11 April 2014.

97 Australian Government 2013, Tasmanian Wilderness World Heritage Area (Australia) Property ID 181bis: *Proposal for a minor boundary modification for submission to the World Heritage Committee*, 1 February 2013.

98 The International Union for the Conservation of Nature (the IUCN) which acts as an advisory body to the World Heritage Committee (WHC) had advised the WHC in 2012 (WHC-12/36.COM/INF.8B2) that: "A notional cut-off of 10% increase has generally been considered to be the absolute upper limit for a modification to be considered via the 'minor modification' process, considering the Operational Guidelines clearly define such modification as having a minor impact on the extent of the property."

99 World Heritage Committee 2012, *Operational Guidelines for the Implementation of the World Heritage Convention* WHC 12/01, UNESCO Intergovernmental Committee for the Protection of the World Cultural and Natural Heritage, Section III.1: Modifications to the boundaries, to the criteria used to justify inscriptions, or to the name of a World Heritage property - Minor modifications to the boundaries, Clauses 163-164, p. 42.

100 Veronica Blazely, 2013, op. cit.

101 Christine Milne, Leader Australian Greens 2013, *Greens celebrate World Heritage listing*, 24 June 2013.

102 According to *Wikipedia* (accessed, May 2015), Christine Milne was a Vice President of the IUCN from 2005-08.

103 The Australian Committee of the International Union for the Conservation of Nature (ACIUCN) – Executive Committee 2012-13

104 UNESCO 2008, *Tasmanian Wilderness (Australia) Report of the Reactive Monitoring Mission, March 15-20, 2008*, by Kishore Rao, (UNESCO World Heritage Centre), Nikita Lopoukhine (IUCN-WCPA), and Kevin Jones (ICOMOS), Recommendation 7, p. 16.

105 Peter Hitchcock, 2014, *Science was rigorous for Tasmanian World Heritage listing*, ABC Environment, 21 July 2014.

106 Graham Lloyd, 2016, op. cit.

107 World Heritage Committee 2010, Thirty-fourth session, Brasilia, Brazil, (25 July- 3 August 2010), WHC 10/34.COM/7B, Stae of conservation of World Heritage properties, pp. 94-98

108 Andrew Denman, 2014, in evidence tendered during a public hearing of the Parlia-

mentary Environment and Communications References Committee inquiry into the Tasmanian Wilderness World Heritage Area, Hobart, 31 March 2014.

109 Peter Hitchcock, 2014, op. cit.

110 Wilderness Society (Tasmania) 2008, *World Heritage Committee calls for increased protection of Tasmania's World Class Forests*, Media Release, 8 July 2008.

111 UNESCO 2008, op. cit.

112 World Heritage Centre 2008, *Tasmanian Wilderness (Australia) State of Conservation (SOC) Report 2008*, p. 7

113 It may have been authored by environmental consultant, Peter Hitchcock, who had previously worked for the Wilderness Society on World Heritage Area matters.

114 Independent Verification Group 2012, *Final Report on the work of the Independent Verification Group for the Tasmanian Forests Intergovernmental Agreement*, March 2012, p. 17.

115 "Tony Abbott's Tasmanian wilderness claim doesn't check out", *ABC News* Fact Check, 24 June 2014.

This was a highly flawed 'fact check' because it relied upon three IVG members who had played leading roles in shaping the nomination to extend the Tasmanian Wilderness World Heritage Area: Peter Hitchcock, Sean Cadman, and Professor Brendan Mackey. Forestry Tasmania supplied information that 21,000 hectares (or 28%) of the 74,000 ha of the area which the Abbott Government was trying to excise from the TWWHA had been previously harvested, but this was largely ignored in the ABC's finding.

116 Peter MacFie, 2014, op. cit.

117 UNESCO 2015, Reactive Monitoring Mission to the Tasmanian Wilderness, Australia, 23-29 November 2015, Mission Report by Tilman Jaeger (IUCN) and Christophe Sand (ICOMOS).

118 Tasmanian Forest Agreement 2012 signed by environmental group and industry representatives on 22 November 2012. Clause 17 requires socio-economic modelling and Clause 37 recommends that Government nominate an extension to the TWWHA.

119 Tasmanian Regional Forest Agreement between the Commonwealth of Australia and the State of Tasmania, November 1997, Clause 40, p. 16.

120 The advisory body, ICOMOS, assessed the Australian Government's 2013 nomination to extend (modify the boundary) of the TWWHA noted that: "It does not address cultural values. No information has been provided by the State Party as to the inclusion within these areas of cultural attributes of Aboriginal importance, in relation to the Outstanding Universal Value of the existing property, nor have the boundaries been justified in relation to cultural attributes."

121 Geoff Law, 2015, *The Tasmanian Wilderness – A Case for Long-lasting Civil Society Involvement in Protecting World Heritage*, paper presented to the World Heritage Watch Conference: UNESCO World Heritage and the Role of Civil Society, Bonn, Germany, 26-27 June 2015.

122 WHC-13/37.COM/INF.8B2.Add: IUCN Evaluations of Natural and Mixed Properties to the World Heritage List: *World Heritage Minor Boundary Modification Proposal – Tasmanian Wilderness (Australia) ID No. 181 quint.*

123 B. Mackey, 2012, *Tasmanian Forest Agreement: Summary Report of Conservation Values*, prepared by ANU Enterprises for the Independent Verification Group of the Tasmanian Forest Agreement, March 2012, p. 51.

124 Parliamentary Secretary for Forestry, Senator Richard Colbeck drawing on historical data from Forestry Tasmania contained in the ABC Fact Check Report: *Tony Abbott's Tasmanian Wilderness claim doesn't check out*, updated version, 24 June 2014.

125 ABC 2014a, "UNESCO rejects Coalition's bid to delist Tasmanian World Heritage forest", *ABC News*, 24 June 2014

126 Simon Grove, 2012, *Let science prevail in conservation and forestry*, submission to the Legislative Council Select Committee on the Tasmanian Forests Agreement Bill 2012, December 2012.

127 Dr Mark Neyland, Dr Peter Volker, Dr Tim Wardlaw, Dr Dean Williams and Dr Paul Adams 2012, submission to the Legislative Council Select Committee on the Tasmanian Forests Agreement Bill 2012.

128 World Heritage Committee 2012, *World Heritage Resource Manual: Managing natural world heritage*, UNESCO, IUCN, ICCROM, ICOMOS, WHC, p. 57.

129 "United Nations World Heritage Committee calls for major changes to Tasmanian Government forests plan", *ABC News*, 2 July 2015.

'UN critical of Tasmania's plan to manage forests", *Lateline*, 2 July 2015.

130 ABC 2015a, "Great Barrier Reef: Green groups 'fear mongering' exposed by UNESCO decision, Queensland Resources Council says", *ABC News*, 2 July 2015.

131 Brendan Mackey, 2014, op. cit.

132 ABC 2015, op. cit.

133 Geoff Law, 2015, op. cit.

8

Active management or benign neglect? The burning question for forest biodiversity

"The belief was that all a wilderness needed was to be left alone, that now it would be safe within the confines of the park, and we've all just realised that this isn't the case"

Professor David Bowman[1]

In late 2014, individuals of two threatened wildlife species – the long-nosed potoroo and the southern brown bandicoot – were trapped from healthy populations within NSW State forests for relocation into nearby national parks where they had long been absent.[2,3] In the same south coast region in March 2016, the NSW Government moved to save the local koala population by adding around 12,000 ha of State forest, where they are relatively plentiful, to the conservation reserve estate where they are largely absent.[4]

Such examples don't fit the dominant cultural belief about forest conservation which, after decades of environmental campaigning, typically depicts State forests as decrepit, logged-out landscapes clothed in a regrowth monoculture devoid of biodiversity. According to this narrative, threatened species populations in State forests, if they occur at all, are declining or close to extinction. They are certainly not meant to be healthy or flourishing. That supposedly only occurs in 'protected' national parks where boundless high conservation values have been secured.

According to environmentalist dogma, saving Australia's native forests and their threatened species requires converting State forests into national parks where the ecology will magically restore itself. While the above few examples show the flaw in such thinking, the need for more national parks is commonly justified on the grounds that, as so much of our original forests have been lost to settlement and agriculture, it is only right that we protect what is left by excluding damaging human uses.

While this may be a superficially persuasive argument, its veracity rests on whether or not national parks and other conservation reserves are actually protecting forests and their ecology, and conversely, whether or not State forest management, including commercial use, is actually damaging it to any significant extent.

Multiple-use State forests are typically diverse tracts of land containing a range of species associations and stand structures in-part shaped by resource use, but mostly determined by fire history, soil productivity, and the nature of the terrain. It is typical for most parts of most State forests to be unsuitable or unavailable for commercial resource use for a variety of natural, accessibility and regulatory reasons. Accordingly, their portrayal as being dominated by post-harvesting regrowth is usually far from the truth.

For example in Victoria, the net harvestable wood production zones comprise only around 20% of the State forests (and only about 9% of the total area of the state's forested public land). Where wood production zones occupy a much higher proportion of the State forests (such as in NSW), it typically reflects a smaller remnant State forest estate following the rebadging of much of its former extent as national parks and other conservation reserves.

Timber harvesting has typically been geographically scattered over time throughout broad zones resulting in a landscape mosaic of managed disturbance interspersed with a much larger area of unharvested (or long since harvested) forest. There are certainly areas/catchments where timber harvesting has been more concentrated, and in-part this also reflects the loss of State forest to other public land tenures, thereby reducing the management flexibility to more widely spread the harvest across the landscape. Some ecologists have acknowledged that there are regional biodiversity benefits associated with creating a mosaic of different forest age classes with its wider range of associated habitats housing a broader array of species, compared to large swathes of forest with uniform age and structure.[5]

Environmental activists have typically dismissed post-timber harvesting regrowth as a 'biological desert'. However, regrowth is a natural phase of forest development with its own characteristic biodiversity, which can include threatened species. For example, the long-footed potoroo of Victoria's East Gippsland was first discovered in 1980 within 30-year old post-harvest regrowth virtually at the backdoor of the Bellbird Hotel near Club Terrace. Since then it has been found to be geographically widespread, but with a particular liking for regrowth to such an extent that it is not uncommonly found in regeneration as soon as two-years after timber harvesting.[6]

The healthy State forest populations of the long-nosed potoroo and the southern brown bandicoot along the NSW south coast have been attributed to active management of feral carnivores. The NSW Forestry Corporation began fox baiting in State forests around Eden in 2007. By 2013 this had substantially reduced feral predation and allowed bandicoot and other native marsupial populations to recover quite strongly. According to NSW Forestry Corporation ecologist Peter Kambouris:

> It's become clear that the most significant threat to the southern brown bandicoot and other threatened forest fauna is not timber production or fire, which they tolerate and recover from well, but predation from the fox.[7]

Indeed, the pre-eminent role of feral and invasive pests in the decline of Australian flora and fauna was recently highlighted by the Invasive Species Council which blamed foxes and feral cats for most native mammal extinctions since European settlement, and nominated them as a far greater ongoing threat than habitat loss which has formerly been assumed to be the main threatening process in Australia.[8]

The relative success of koala populations in these same regional State forests is also an interesting case study. In the early 1990s, the then Tantawangalo State Forest had been rebadged as the Tantawangalo Koala Nature Reserve after a campaign by local conservationists assisted by Bob Brown. According to the environmentalist rhetoric this would supposedly create a 'regional stronghold' for koalas.[9]

Twenty years after the creation of the Tantawangalo Koala Nature Reserve, it would be reasonable for the community to expect the koala population in this 'protected area' to have expanded and stabilised. However, recent intensive surveys have shown that the Tantawangalo koala population "is now very small and possibly extinct."[10] Subsequent surveying found that koala numbers were far higher in nearby State forests, thereby leading the NSW government to transfer an additional 12,000 ha of State forest into new koala reserves in 2016.

An overlay of recorded koala sightings with forest disturbance history shows that State forests subjected to past timber harvesting are not associated with a loss of koalas.[11] Indeed, given the recent explosion of local koala populations in blue gum (*Eucalyptus globulus*) plantations in south-western Victoria, it seems likely that a mix of old trees interspersed with vigorously regenerating younger trees – as is usually more prevalent in multiple-use State forests – may well be a preferred habitat.

Victoria's Central Highlands' mountain ash forests provide a further example of the conservation value of multiple-use State forests. Following recommendations by the Victorian Government's Leadbeater's Possum Advisory Group in 2014, a renewed intensity of surveying found 116 new possum colonies within 18-months. Of these, 71 were in multiple-use State forests and 45 in national parks.[12] This discrepancy was reportedly due to more surveying in State forests given the relative difficulty in accessing national parks which lack roads and tracks, but again demonstrates that State forests are far from devoid of biodiversity.

By January 2017, the number of new possum colony detections had soared to 537 in a little under three years, mostly in State forests.[13] This dwarfs the previous average detection rate of about 15 new colonies per year from 1998-2014 which had fuelled earlier fears for the possum's survival. Understandably, these fears have now largely dissipated except amongst those campaigning for the declaration of a new, so-called 'Great Forest National Park'.

There are many other examples showing that State forests either have equivalent or superior biodiversity values compared to adjacent national parks. It is also significant that so much State forest has been deemed to have sufficiently high conservation values as to be worthy of national park status, although, as pointed out earlier, to a large extent this reflects a recent diluting of the national parks concept through its misuse as a strategy to end commercial uses, especially timber production.

The co-opting of wood production forests into the conservation reserves estate has even extended to some privately-owned forests. For example, thousands of hectares previously owned by Tasmanian timber company, Gunns Ltd., which was – after decades of wood production – purchased for conservation purposes in 2010 and subsequently included in an extension to the Tasmanian Wilderness World Heritage Area in 2013.[14]

While it should be good news that high forest conservation values often occur in conjunction with commercial use, this finding has often been appropriated by environmental groups and Greens politicians as further evidence of the need to rebadge virtually all State forests as new national parks, ostensibly to protect such values from damaging activities, primarily on-going timber harvesting.

Clearly both multiple-use State forests and national parks can and do have significant conservation values. Accordingly, effective environmental protection is not as simple as just changing land tenure by turning State forests into national parks, but requires management strategies targeted to specific conservation needs irrespective of land tenure. As pests, weeds, and fire are equally problematic for all public lands, the propensity for environmental activists and their political associates and supporters to pretend that such threats can be mitigated by simply proclaiming a change of land tenure is highly disturbing. That this creates little more than an illusion of environmental protection is all-too-evident as pests and weeds continue to flourish while unnaturally damaging fire regimes arguably worsen in national parks and reserves.

None of this is meant to suggest that we shouldn't have national parks or other types of conservation reserves – they are critically important to landscape-level environmental protection. However, it is equally as important to recognise that forests are dynamic entities that are always changing. They are not static museum exhibits which will forever provide representative examples of particular values, as is often the justification for creating new parks or reserves. That today's old growth forests are tomorrow's regrowth forests will become all-too-apparent in the future as national parks declared specifically to 'save' old forests, ultimately degenerate into far less attractive and rarely visited stands of dead and dying trees mixed with increasingly dense scrub – at least until the next fire.

Challenges for effective environmental management

All forests and woodlands face the same environmental management challenges. Arguably none are presently being adequately addressed in any public land tenure. These challenges lie in: 1) addressing unnatural fire regimes; 2) controlling feral pests and weeds; and 3) reducing the fragmentation of forested habitat separated by past agricultural development.

Habitat fragmentation, particularly in the more developed landscapes of southern Australia, is at this stage an issue that transcends the role of land managers. Addressing it would require a huge public and private committment to acquiring farmland for massive revegetation programs to recreate strategic linkages between major forest blocks that were previously part of contiguous tracts of biodiverse habitat. Given the economic and societal importance of agricultural production, it is hard to envisage this happening to any significant extent any time soon.

On the other hand, the challenges of fire and pest control can potentially be mitigated by land managers right now. However, their capability to do so (even to an unsatisfactory degree) is typically more inhibited in national parks than in multiple-use State forests in which commercial activities, such as wood production, allow greater management capability and flexibility.

For example, in NSW in the 2011-12 financial year, the total revenue generated from its two million hectares of State forest (including native forests and plantations) was $318 million with an operating profit of $14 million.[15] Conversely, the total revenue generated from its seven million hectares of national parks was just $51 million, while the costs of park management amounted to $277 million.[16] Clearly, such a shortfall has to come from the public purse.

In the nation's parks and conservation reserves, the overall lack of funding dictates the extent to which conservation issues are actually being addressed. In Tasmania in 2014–15, the Parks and Wildlife Service (PWS) allocated about 10% of its budget to fire management, but an analysis by the Tasmanian Audit Office concluded that "only a small percentage of priorities ... are related to PWD [*pest, weed and disease*] control, with most being allocated for infrastructure work and visitor services." Indeed, the collective budget allocation to visitor management in Tasmania's South West and North West PWS regions in 2014–15, was 98%.[17] The parks and reserves in these regions include some of Australia's most iconic 'protected areas' and yet only 2% of their management funding is allocated to directly improving conservation outcomes. It is extremely ironic that areas 'protected' in-part to exclude human uses, are overwhelmingly being managed to attract human visitors and control where they can go. This alone suggests that such 'protected areas' make little contribution to improved conservation values.

As national parks management is more dependent on government budget appropriations, it is more vulnerable to changing political priorities. For example, funding for Victorian park and reserve management fell by 37% from 2011-14, with 'insiders' reportedly warning that national parks: 1) were in danger of being overrun by weeds and feral animals; 2) contained degraded tourist facilities which have had to be removed due to lack of maintenance; and 3) that there was a huge backlog of urgent road maintenance.[18]

Whatever one thinks of wood production, there is little doubt that the timber industry's scattered presence throughout a large forested landscape has always been a cornerstone of effective land management. It raises revenue; necessitates a scattered workforce with heavy machinery and operators skilled in bush work; and entails a larger Government workforce skilled in planning and managing bush operations, including the controlled use of fire as a silvicultural necessity in some forest types.

The economic imperative associated with the presence of such an industry and its regulators also necessitates the routine upkeep of roads and tracks scattered throughout the landscape for both log haulage and fire control, and creates a stronger demand for active fire management to protect the future wood resource (usually through the controlled use of cool fire).

While commercial forest use has arguably been less beneficial for pest and weed control, its immeasurable benefit for fire management has become more evident as the industry has been gradually forced out by the declaration of new national parks or other conservation reserves largely created to appease an ideological re-creation of the former 'wilderness'. As this area of reserved 'wilderness' has expanded, the inherent difficulties facing its management have become more apparent, including:

- Over-reliance on Government budgetary appropriations which are subject to intense competition from more pressing societal concerns such as law and order, education and public health.
- Lower priority for broadscale land management given limited resourcing and the inherently more pressing priority to manage visitors.
- Increased difficulty in undertaking broadscale land management due to declining accessibility caused by limited maintenance of the full road and track network, and the greater priority afforded to maintaining only the few major roads used by the great majority of visitors.
- The influence of challenging stakeholders (particularly environmental groups) with philosophical and ideological beliefs that can restrict land management strategies. For example, there is typically only lukewarm acceptance of the need to intervene in nature by culling or poisoning pest animals, using efficient herbicide control methods on weeds or other pest plants, or managing the fire threat through broadscale fuel reduction burning.

These difficulties are not entirely absent from State forests, but are generally less prevalent given that multiple-use forests are more accepted as places for human use.

While considered intervention in nature has been a cornerstone of effective State forest management for generations, it seems only in recent years that national park managers have come to terms with the ideology of 'naturalness' which, as espoused by their most dominant stakeholders, dictates that nature must be left to its own devices. Past adherence to this philosophy has at times led to disastrous outcomes, such as the reported loss of 10,000 hectares of ash-type eucalypt forest in Victoria's Alpine National Park which was allowed to revert to a tangle of wattle and scrub after successive bushfires in 2003 and 2006-07 had rendered the eucalypts incapable of adequate natural regeneration.

For several generations, similar circumstances in State forests have been routinely addressed by artificial re-seeding programs to restore the original forest, especially if it had value as a future wood resource. Indeed the iconic forest drive through a virtual cathedral of huge mountain ash (*Eucalyptus regnans*) trees on the Black Spur, north-east of Melbourne, was created by a Forests Commission replanting after the devastating 'Black Friday' 1939 bushfires. Similarly, fire-killed forests on the Toorongo Plateau were artificially regenerated in the 1940s, and re-seeding programs have been commonplace after severe bushfires in the ash forests where regeneration can otherwise be problematic, including in 1983, 2003, 2006 and 2009.

In more recent years, national park managers have apparently learned this lesson and have artificially re-seeded severely burnt forests that were failing to naturally regenerate due to too-frequent fires. In 2014, this was bemusingly presented on ABC TV's *Catalyst* program as though national park managers had hit upon some wondrous new invention.[19]

Despite this advance, there is still much opposition to an active, interventionist approach to managing conservation values. Chiefly, this is reflected in a lack of enthusiasm for fuel reduction burning amongst both environmentalists and some conservation scientists (as discussed later), but there are also other examples, such as the strong opposition of Australian National University ecologists to the use of nest boxes and artificially-created tree hollows to provide supplementary habitat for Leadbeater's possum in forests recovering from bushfires. In 2016, the ANU Long Term Ecology Group's blog was asserting that:

> There is no evidence whatsoever to suggest that nest boxes will be effective for Leadbeater's possum in these forests.
>
> There is no evidence that creating artificial hollows will be effective for Leadbeater's possum – the technique has not been appropriately trialled in a scientific way.[20]

Contrary to such negativity, a nest box installation and monitoring program initiated by Victorian Government scientists during 2014 was showing a 53% rate of possum occupancy in 493 boxes after just two-and-a-half-years. Furthermore, within eighteen months of the start of an artificial tree hollow-creation program, 37 of 72 artificial hollows were either occupied or showed signs of impending occupation, and there were strong expectations that proportional use would increase with time.[21]

Given that several of these ANU ecologists had been publicly advocating the establishment of a huge new national park as the only way to save Leadbeater's possum, their refusal to acknowledge the successes of active conservation measures smacks of fear that, if successful, they could undermine their obvious preference for a new park (and closed timber industry).[22] Allowing blinkered ideology to block real conservation gains is surely a case of 'cutting off your nose to spite your face'.

Further to this, the role of the commercial forestry agency VicForests in supporting the active conservation of Leadbeater's possum through in-kind contributions and part-funding of habitat surveys and the nest box/artificial tree hollow program, shows the importance of generating revenue to improving ecological outcomes. Indeed, it is highly ironic that: 1) the closure of the timber industry being sought by environmental activists and their political allies would be a significant (and perhaps terminal) blow to arguably the state's most high profile forest fauna conservation program; and 2) that this program is mostly being undertaken in multiple-use State forests rather than the parks and reserves which have been specifically reserved for conservation.

Dealing with fire – the case for active forest management

Arguably, unnatural fire regimes are the greatest threat to the integrity of Australia's forest and woodland biodiversity. While introduced pests and invasive weeds are also having a massive impact on indigenous flora and fauna species everywhere, unnaturally severe and/or too-frequent forest fire has the potential to destroy both the habitat and its capability to regenerate.

Completely eradicating feral pests and weeds is probably impossible without the development of biological control methods through laboratory-based research. In the interim, the best that can be hoped for is localised control using strategies that reduce their presence and impact through concerted regular treatments at a regional or sub-regional scale. If there is a drop-off in control effort, perhaps due to reduced levels of commitment or funding, pests soon rebuild to their former presence and impact.

Arguably, there is greater potential for land managers to actively manage fire across bigger areas to eventually revert to something approaching the natural low fuel state that existed prior to European settlement. This can be achieved with more broadscale fuel reduction burning which is relatively cost-effective. However, public discussion about undertaking more active forest and fire management is invariably met with contrary assertions, such as, "It seems to me that forests were doing just fine before we turned up a few hundred years ago ..."[23] While for many, leaving forests to their own devices may be a superficially attractive notion, it will not rewind the fire regime back to its natural, pre-European state.

This was obvious within decades of the first graziers establishing their runs. In 1890, the explorer, Alfred Howitt, drew on his extensive observations over many years to explain why the fire regime had changed and the impact of this on the nature and ecology of the forests of eastern Victoria:

> The influence of settlement upon the Eucalyptus forests has not been confined to the settlements upon lands devoted now to

> agriculture or pasturage ... It dates from the very day when the first hardy pioneers drove their flocks and herds down the mountains from New South Wales into the rich pastures of Gippsland [*in about 1840*].
>
> Before this time graminivorous marsupials had been so few in comparative number that they could not materially affect the annual crop of grass which covered the country, and which was more or less burnt off by the aborigines, either incidentally or intentionally, when travelling or for the purpose of hunting game.
>
> These annual bush fires tended to keep the forests open, and to prevent the open country from being overgrown, for they not only consumed much of the standing or fallen timber, but in a great measure destroyed the seedlings which had sprung up since former conflagrations. The influence of these bush fires acted, however, in another direction, namely, as a check upon insect life, destroying, among others, those insects which prey upon the Eucalypts.
>
> Granted these premises, it is easy conclude that any cause that would lessen the force of the annual bush fires, would very materially alter the balance of nature, and thus produce new and unexpected results.
>
> The increasing number of sheep and cattle in Gippsland, and the extended settlement of the district, lessened the annual crop of grass, and it was to the interest of the settlers to lessen and keep within bounds bush fires which might otherwise be very destructive to their improvements.
>
> The results were two-fold. Young seedlings had now a chance of life, and a severe check was removed from insect pests. The consequences of these and other cooperating causes may be traced throughout the district, and a few instances will illustrate my meaning.[24]

Howitt went on to give many examples where the changed fire regime had resulted in a significant expansion of dense forest into areas that were formerly grassed and only sparsely or very sparsely treed. For example:

> Within the last twenty-five years, many parts of the Tambo Valley, from Ensay up to Tongio, have likewise become overgrown by a young forest, principally of E. hemiphloia and macroryncha,

which extend up the mountains on either side of the valley. This dates especially from the time when the country was fenced into large sheep paddocks, when it became very important that bushfires should be prevented as a source of danger to the fences, and even when fire occurred the shortness of the pasturage checked its spread.

I might go on giving many more instances of this growth of the Eucalyptus forests within the last quarter of a century [from about 1865-1890], but those I have given will serve to show how widespread this re-foresting of the country has been since the time when the white man appeared in Gippsland, and dispossessed the aboriginal occupiers, to who we owe more than is generally surmised for having unintentionally prepared it, by their annual burnings, for our occupation.[25]

Debate has raged for some time about the veracity of this and many similar historical observations and the conclusions drawn from them. However, the publication of historian Bill Gammage's 2011 book on the subject seems to have at last fostered a stronger consensus that aboriginal burning and/or fires naturally-ignited by lightning were responsible for far more annual burning of the landscape during pre-European times.[26] Furthermore, this fire was generally of low intensity because its regularity prevented the build-up of heavy fuel accumulations and, in all but the wettest regions, this regular burning kept forests and woodlands far more open and less scrubby than they are today.

However, this view wasn't widely appreciated at the turn of the 20th century when the flammability of the Australian landscape was a largely foreign concept to our earliest European-trained foresters. Accordingly, and perhaps understandably, they were mired in indecision about the best approach to managing forest fire. In the early 1900s fire was feared, and there was realistically little that could be done to prevent it entering forests from uncontrolled pastoral burning on adjacent lands or to control illicit burns lit to promote grazing in the reserved forests themselves. Adding to the difficulty was the huge areas of remote forest with little or no access.

As early as 1923, Victoria's Forests Commission warned the State government that fire would continue to be a major threat and was indeed "a tragedy waiting to happen". Major efforts were made to exclude fire from forests and educate the public in its safe use on adjacent lands, but with only moderate success. All the while, small advances were being made in improving the capability to detect fires (including from the air), and developing effective fire-fighting tactics which were nevertheless primitive by today's standards.[27]

In the 1920s, two schools of thought prevailed about the management of forest fire. Field-based forestry officers and local bushmen maintained that regular use of fire as a tool to 'clean-up' the forest floor and foster persistently light fuel loads, was the key to controlling bushfires. However, this was heresy to the academic professional foresters – particularly those with European training – who believed that bushfires would largely vanish as tangled wilderness was converted to organised, tended forest.[28]

In 1927, Western Australia's forestry authorities were the first to recognise that the field foresters and bushmen were right, but there was a big difference between knowing what to do and having the skills and resources to do it.[29] The result was that, during the 1930s, some burning was being done in Australia's forests, but not generally in accord with any organised plan or coordinated approach.

In January 1939, the huge 'Black Friday' conflagrations burnt over 1.5 million hectares of Victoria's forests, razing many settlements and killing 71 people.[30] This would ultimately lead to a new era of forest fire management after the subsequent Stretton Royal Commission's recognition of the absurdity of attempting to exclude fire, and its strident advocacy of the concept of using fire against itself.[31] In spite of this breakthrough, it wasn't until almost a decade after World War Two that the concept of controlled fuel reduction burning began to be firmly adopted as the primary tool for managing forest fire.[32]

At the time, this was a strategy that no other developed nation had dared to deliberately adopt. It was rooted in a sensible recognition and acceptance of the country's indigenous, fire-dependent flora and fauna; the long tradition of Aboriginal burning; and the on-going use of fire in other rural land uses. Gradually, integrated systems of planned and controlled burning were introduced into the nation's public forests – firstly in Western Australia, and then extending throughout the country by the early 1960s.[33]

By the mid-1960's, the use of aerial incendiaries had been developed and pioneered in WA and had then spread to eastern Australia by 1967.[34] This enabled large areas to be lit quickly and inexpensively when conditions were right. In Victoria's forests between 1972 and 1982, the gross area annually treated by fuel reduction burning varied from 37,000 hectares in (1973-74) to 477,000 hectares (1980-81).[35] This large range reflects the variability of seasonal conditions but also highlights the benefit of having structures, resources, and political backing that enabled suitable conditions to be fully exploited whenever they arose.

Unfortunately since the mid-1980s, arguably starting in Victoria and NSW, the capability to undertake extensive fuel reduction burning programs has progressively declined. In WA, the optimal burning program was maintained until the late 1990s and started to decline thereafter.[36] There are a range of social and demographic reasons for this decline but a significant factor has been the substantial loss of forestry expertise associated with the dramatic expansion of national parks, and more recently, the separation of commercial forestry from broader forest management functions.

This has been widely acknowledged since the record 2002–03 bushfires in Victoria and south-eastern NSW. Whilst drought conditions created some extreme, uncontrollable fire intensities, disquiet over the actions of the responsible government agencies (both before and during the fires) spawned seven official inquiries. These included a Federal inquiry conducted by a House of Representatives Select Committee, an inquiry by the Council of Australian Governments (COAG), and an in-house Victorian government inquiry chaired by its Commissioner of Emergency Services.

In part, these addressed growing public concerns over politically-expedient government policies that had progressively transferred huge areas of State forest into national parks and reserves at the expense of former multiple-use management regimes that included wood production and more active fire mitigation strategies. These concerns have largely been fanned by rural communities and farmers directly affected by what they regard as poor management of adjacent public lands. In addition, former foresters and fire scientists within groups, such as Forest Fire Victoria and the Bushfire Front of WA, have also been critical of trends in public land fire management.

The capability of public land managers to effectively manage fire is largely a question of funding and commitment. This is integral to acquiring and maintaining appropriate levels of equipment and trained, experienced personnel for fire prevention and damage mitigation activities, such as fuel reduction burning and maintaining forest road networks, including bridges.

In the State forests' component of southern and eastern Australia's public lands, wood production has traditionally been associated with a strong culture of active land management and heightened fire readiness. This, plus the associated revenue raised from timber licence fees, royalty charges, and roading levees had always made a positive contribution. The loss of this contribution was articulated by a number of submissions to the 2003 Federal House of Representatives Select Committee inquiry. In particular there was a strong view that the fire expertise evident in the management of production forestry had not been transferred over to the national park managers that had effectively replaced them.

A submission from the National Association of Forest Industries quantified the loss of forest fire-fighting capacity within north-eastern Victoria as declining from about 150 foresters, overseers and forestry workers in the mid-1980s, to less than 40 by 2003. The Victorian Association of Forest Industries also noted that of the 85 timber industry bulldozers and their crews operating on the Gippsland sector of the Alpine fire in early 2003, half were expected to be unemployed once government-sanctioned reductions to sawlog production came into force in the coming months.[37]

Bureaucratic changes may also have played a part in reducing the capability and effectiveness of fire management. A Victorian investigation into the impact of changing forest policy has noted that from 1982 to 1995 there were four restructures associated with functional rationalisation and multi-skilling in the responsible public land management agency. These led to respective 37% and 44% reductions in head office and field-based personnel with native forest management skills. Since then, demarcation between activities in the substantially amalgamated responsible Department has blurred more recent changes that have allegedly further eroded forest and fire management capability.[38] Undoubtedly, other states have endured similar changes.

Following Victoria's 2006-07 bushfires, which burnt a further 1 million hectares of public forest, a Victorian Parliamentary Inquiry into the impact of public land management practices on bushfires found:

> … that the reduction in the extent of timber harvesting on public land and associated loss of local knowledge and expertise, machinery available for fire prevention and suppression, and a decline in the number and accessibility of vehicle access tracks, has had a negative impact on land and fire management, particularly the bushfire suppression capacity of relevant agencies.[39]

Two years later these same concerns would be reiterated by the 2009 Victorian Bushfires Royal Commission that investigated the 'Black Saturday' bushfires which killed 173 people. After drawing on a range of expert advice, the Commission would go on to recommend a tripling of Victoria's annual area of fuel reduction burning as a key to mitigating the potential for future severe bushfire damage.[40]

However, by 2013, it was becoming obvious that the capability to achieve this new fuel reduction target was lacking. As a leaked Victorian Department of Primary Industries (DEPI) internal discussion paper lamented:

> In the past, DEPI has been able to rely on the support of an active timber industry to provide the training ground, skill and equipment base for its ongoing suppression and burning requirements, however this is a rapidly diminishing convenience. There is no suitable replacement industry with similar requirements and DEPI will find it increasingly difficult to access these specialist contract services as they continue to disappear.[41]

Arguably, the capability to suppress most forest fires is now being impacted to at least some extent by a lack of preparatory fuel reduction and/or neglected road and track access. However, the extent of this impact is rarely publicly documented. An exception was the Goongerah – Deddick Trail Fire which burned in Victoria's East Gippsland region for 70-days from January to March 2014 after several uncontrolled lightning strikes coalesced into a major conflagration. Although largely confined to public lands (mostly in the Snowy River National Park), the fire eventually caused some significant stock and property damage in adjoining farmlands. This was accompanied by a storm of community outrage about the missed opportunities to stop the fire growing so large, including a claimed lack of preparatory fire season readiness.

One of the more vocal critics was Bonang cattle farmer and former long-serving national parks ranger, Dave Ingram, who had resigned from Parks Victoria a decade earlier citing concerns about the lack of forest fire management and its potential environmental consequences:

> To protect lives and property, you do your work in the bush, you do your preventative burning, you maintain your track networks and maintain the facilities. That hadn't been done for five years in our area. There were other fires [*elsewhere in the State*] … but we had this fire going for so long before there were any resources on it. There was just no comprehension of how big the fire was.

Mr Ingram was keenly aware of the damage caused by the fire, including later impacts when heavy rains lashed the exposed burnt soils.

> Three hundred year old ash trees are gone. There's a huge amount of wildlife lost, like tiger quolls and potoroos. There's something like 120 special protection zones in the Snowy River National Park and I'd say 70% of those were burnt. I doubt whether there's been anyone back in there to see what happened. There's been hardly any recognition of what we've lost.
>
> We had a storm not long ago at McKillops Bridge – something like 160 mm of rain fell in three hours. Now its 45 to 50-degree slopes (leading into the Snowy River) – there's nothing on it. There's probably 20,000 tonnes of topsoil and sediment that went straight

into the river. If the fire was put out, or it hadn't got there, that rain wouldn't have been as damaging.[42]

In the aftermath of this fire, a local Community Reference Group was formed to liaise with fire management authorities in the hope of informing better future outcomes. In-part this process demonstrated the difficulty faced by fire authorities in dealing with divergent community expectations especially given that the community includes the hamlet of Goongerah which for 30 years has been the regional hub of eco-activism focussed on closing down the East Gippsland timber industry.

Accordingly, the community's call for a return of the former culture of strong fire season readiness and aggressive fire suppression tactics – although worthy – carried considerable irony given the role that its environmental activist cohort had played in destroying that culture. By helping to force a substantial government down-sizing of the regional timber industry they bore a burden of responsibility for the associated loss of most of the forestry and industry workforce which, with its heavy equipment, had formerly been central to effective forest and fire management. Further to this, the government report stemming from this community liaison process noted that although many in the community supported fuel reduction burning, "other community members felt that fuel reduction burns are as damaging as (bush)fires, with long-term health effects, and questioned the effectiveness and evidence behind the benefits of fuel reduction burning".[43]

The influence of incessant environmental activism from Goongerah-based anti-logging activists (including half-a-dozen legal challenges against the government's commercial forestry agency) is arguably exemplified in the fact that 85% of the region's extensive public forests are now in some form of formal, informal, or incidental reservation, including substantial areas of national park.

In contrast to State forests where road and track access has traditionally been better maintained, national park management philosophy has been typically based around channelling visitors to developed tourism infrastructure at selected sites, whilst limiting broader park access to reduce human disturbance. Where this has led to road and track closures, or this has been forced by lack of funding for basic maintenance, it concurs with the philosophy of environmental activism that has traditionally viewed vehicular access and active fire management across broad forested landscapes as unnatural or undesirable.

A comparative evaluation of forest fire management in State forests and national parks undertaken in NSW over the ten-year period from 1993 to 2003, found that higher annual rates of fuel reduction burning under cool conditions translated into smaller areas burnt by hot summer bushfires. In the State forests, the proportion of the total area being annually fuel reduced (3%) was almost 8 times greater than was being achieved in NSW's national parks (0.4%) at that time. This greater proportional commitment to planned burning in State forests was reflected in an average 2.8% of their area being annually burnt by summer bushfires, whereas around 4.8% of NSW national park was being annually burnt by bushfires under the same weather patterns over the same period.[44]

This significant difference aligns with their respective land management philosophies. In the NSW national parks at that time, fuel reduction burning was primarily focused on community protection and largely restricted to boundary areas in close proximity or adjacent to urban and rural communities. Conversely, in State forests, prescribed burning was undertaken for a broader range of values and was both more extensive and more widely spread across the landscape.[45]

Based on the Western Australian experience (described below), it could be argued that the proportional extent of prescribed burning being undertaken in the NSW State forests (3%) during the decade ending in 2003, was substantially less than optimal. However despite this, more than half of the fire occurring in the State forests each year was being planned and controlled. Conversely, less than 10% of the annual fire in national parks was controlled, with over 90% being wildfire burning out of control often in hot and windy summer conditions when threats to neighbouring communities, in-park infrastructure, and environmental values were highest.

An Australia-wide analysis of the period from 2001–06 also showed that higher rates of planned burning undertaken in cooler months translated into lower rates of unplanned hot summer bushfires. The region least affected by damaging summer bushfires was south-west WA where planned fuel reduction burning was responsible for 73% of the total annual area burnt. In Victoria during the same period, only 20% of the total annually burnt area was attributable to planned fuel reduction, with unplanned bushfires responsible for the other 80%. Over the whole of Australia during that period, planned burning was responsible for just 19% of the annually burnt area with 81% burnt by unplanned and often damaging bushfires.[46]

Since the disastrous 2003 bushfires in Victoria and NSW, claims of better fire management in multiple-use State forests has sparked a greater committment amongst state government land management agencies to undertake more burning in national parks and other conservation reserves. The term 'tenure blind' was coined to describe planned annual burning programs implemented irrespective of public land tenure. However, while this may be the intention, it is harder to safely burn areas where roads and tracks have been deliberately closed or rendered unusable through neglect. So it is likely that planned fuel reduction burning continues to be more prevalent in State forests typified by better access and managed by more enthusiastic and experienced fire practitioners.

For around 50-years starting from the early 1960s, the forests of south-west Western Australia reflected the effectiveness of an extensive annual program of fuel reduction burning in averting major bushfires. Following the disastrous 1960–61 bushfires, WA's foresters implemented a burning program aimed at treating 6 to 8% of the region's public forests each year. This represented a substantial 2.5 times increase in the amount of annual burning undertaken during the previous decade. In the first 10-years after this change, there was both a 17% reduction in the number of summer forest fires and a 250% reduction in the average total area which they burnt each year.[47]

Subsequent studies have showed that this initial success persisted over the longer term. For example, in the mid-1980s it was found that substantially higher rates of fuel reduction burning in the forests of south-west WA over the previous 25-years had been responsible for much smaller bushfires in comparison to those occurring in similar climate, terrain and forest types in the eastern states. While the average WA forest wildfire was just 15 hectares in size over this period, fires were found to be 18 times larger in Tasmania,[48] and respectively 12 and 13 times larger in Victoria and NSW.[49] Further WA research of longer term patterns has reiterated that summer bushfires decline in both numbers and area burnt as the annual area of fuel reduction burning increases.[50]

Unfortunately fuel reduction burning being undertaken annually in the forests of south western WA has now substantially declined (to around 3% of the forest area per annum) and consequently the region is now experiencing a return to larger and more threatening bushfires, such as those of the 2014–15 fire season.[51] This represents a fall from grace for a region that was once the envy of the eastern states for being able to largely avoid damaging bushfires.

The part played in this by the populist 'conservation culture' was recently articulated by retired senior forester and former Deputy Director of the WA Department of Conservation and Land Management, Roger Underwood:

> For many years in Western Australia, all through the 1970s, '80s and early '90s, those in favour of reducing bushfire hazards were in the ascendancy. The [*fuel reduction burning*] program was achieved, year by year. It didn't cost much and nobody took much interest in it. The main thing was that the community and the bush were spared the horror, ugliness, waste and heartbreak of big, angry bushfires. Then the Greens, the lefties in the ALP, the doctors' wives, the grape-growers, the bush hippies and the inner-city academics got control of the game and the whole thing unravelled.[52]

Whilst much of the community has an ingrained belief that national parks are always environmentally beneficial, the often severe ecological impacts of damaging summer bushfires is the price being paid for embracing a 'consequence-free' notion of parks expansion that has ignored the critical role that human resource use and active forest management has always played in the capability to manage the fire threat.

Well before the Western Australian experience, the capability to undertake fuel reduction burning had started to similarly decline in Australia's eastern states. In addition to the factors outlined by Underwood, societal and demographic changes have also made planned burning more difficult. This includes more people living on small blocks within the forest; the growth of outlying suburbs which abut forested lands and have lengthened the vulnerable urban-public land interface; and the evolution of a more risk-averse and litigious society. While none of these factors are related to public land tenure per se, they have made the planning and conduct of burning on public lands (either in national parks or State forests) more demanding, and have effectively restricted the amount of burning that is achievable in an already narrow window of opportunity governed by weather and safety constraints.

A hugely significant factor has been a demonstrable shift away from the formerly more equitable balance between fire prevention (such as planned burning) and emergency bushfire suppression, particularly through the increased use of costly airborne water-bombers. This is arguably a response to the now far greater likelihood of uncontrollable bushfires entering into outlying suburbs with potential for disastrous loss of life and property. Under such circumstances aerial water-bombing, particularly from helicopters, can be exceptionally good in accurately targeting particular houses or hot spots far more quickly, safely and effectively than ground crews.

The downside to the increased use of aerial fire-fighting technology is that emergency bushfire suppression has arguably become disconnected from the traditional fire management package in which approximately equal budgetary weighting was previously given to both in-fire season emergency suppression and off-season fire prevention/mitigation practices. The effectiveness and sheer theatre of aerial water-bombing conducted under appropriate circumstances has – with the help of dramatic media footage – engendered a popular impression of it as a panacea for our bushfire threat. This has especially taken root amongst those who – for whatever reason – lack enthusiasm for fuel reduction burning and like to argue that aerial fire-fighting obviates its need.

Forests, fire and a flawed conservation culture

Following Victoria's disastrous 'Black Saturday' bushfires in February 2009, the Institute of Foresters of Australia expressed a concern that the increased reliance on emergency fire-fighting response effectively amounted to a trend away from the proven, time-worn concept of using fire against itself. They feared that the substantial expense incurred in aerial fire-fighting would correspondingly reduce the budget available for off-season fire prevention/mitigation activities, such as planned burning:

> It is appreciated that this situation has slowly evolved since the 1980s in response to a complex raft of political and social changes which have progressively reduced the commitment to traditional methods of forest and fire management. The resultant decline of activities such as fuel reduction burning has gradually consigned the state to increasingly severe bushfires, undoubtedly exacerbated by long term drought, which has only increased the need to bolster suppression forces. This need is only being further entrenched by changing demographic patterns which have put more people at risk in fire-prone urban fringes and the political realities of dealing with this.
>
> Unfortunately, the 'emergency response' model is typified by huge expense without any great improvement to bushfire outcomes when compared to mitigating the frequency of severe 'mega fires' by reducing fuels before the event. The rise of 'emergency response' at the expense of declining levels of preventative action can be represented by a public health analogy whereby efforts to prevent problems by vaccinating and/or encouraging better lifestyles are diluted in favour of a greater focus on surgery to treat cancers. Very few would support such an approach to public health and this should also be the case regarding forest fire management.
>
> The reality that we have firmly arrived at this place was confirmed by events just prior to the 2008-09 fire season when the Victorian Government announced that it would reject the recommendation of its own Parliamentary Inquiry to triple the amount of fuel reduction burning, but then soon after announced the deployment of two Elvis water-bombing helicopters to Victoria for the fire-season. The reported cost of these helicopters is $60,000 per day on stand-by and $15,000 per operational hour. At the height of the [*Black Saturday*] emergency on February 7th [2009], weather conditions were such that these helicopters were unable to operate.[53]

Notwithstanding the weather-related limitations on aerial fire-fighting, it is less effective on forest fires, especially in tall forests with dense canopy cover.[54] Although water-bombing can temporarily slow or hold parts of forest fires, it still takes fire-fighters on the ground to completely contain and extinguish them. The capability of ground crews to contain fires is governed by a combination of accessibility, fuel and weather parameters. If the embrace of aerial fire-fighting significantly erodes budgets for preventative activities that would otherwise maintain access and reduce fuel loads, the likelihood of more frequent large and uncontrollable bushfires increases. Accordingly, an unbalanced budgetary focus on costly fire suppression technology at the expense of preventative actions creates a self-sustaining cycle of large damaging fires that then justify the need for continued high suppression expenditures.

The greater reliance on aerial water-bombing also somewhat reflects the evolution of a risk-averse operational health and safety climate in which fighting forest fires aggressively from the ground, as was traditionally standard practice, is now pursued with less enthusiasm. As a consequence, fires are more often growing larger and ultimately becoming far more costly to control and extinguish. The Wye River fire in southwestern Victoria during the 2015–16 fire season, arguably exemplifies the current era of 'safety-first' fire fighting. It was only a moderate 2,500 hectares in size, but, in what would have once been unthinkable, took 33-days for 200 to 300 fire-fighters and substantial use of aerial resources to bring under control. Excluding the value of property losses, it is likely to have been Australia's most expensive ever fire-fight on a per hectare basis. Conversely, the last time that the Wye River area had experienced a serious bushfire was in January 1962, when the total cost of forest fire suppression over the whole of Victoria for the 1961-62 season, was less than $1 million in today's money.[55]

No-one can dispute the critical importance of fire-fighter safety. However, effective forest fire-fighting carries inherent risks and trying to avoid them becomes counter-productive if it restricts suppression operations to such a degree that small bushfires are enabled to burn longer and grow larger. Under such circumstances the safety risk to both fire-fighters and the broader community increases exponentially by making it more likely that the still-going fire will coincide with severe weather conditions which make it uncontrollable. It may be that today's far greater emphasis on risk-avoidance reflects the relative inexperience of fire-fighters compared to the past when calculated risks were taken and generally well-managed by personnel with decades of experience behind them. It has been said that the greatest risk in life lies in taking no risks at all, and this also holds true for forest fire-fighting.

Internationally acclaimed fire historian, Steven Pyne, has long lamented the emergence of a 'paramilitary' emergency response culture in the USA (and in other developed countries) that massively increases the cost of fighting forest fires.[56] In accord with his concerns, the proportion of the US Forest Service's annual budget expended on fire suppression grew from 16% in 1995 to over half in 2016, much of it eaten-up in by aerial resources protecting suburbs abutting flammable forests. As a result 'non-fire programs' have been cut back, including fuel reduction that could significantly mitigate the fire threat.[57]

In an interview on the ABC's *7:30 Report* in 2006, Pyne urged Australia not to go down the path of other developed countries which had largely ceased to undertake broadscale planned burning. According to him, those countries had since realised the adverse consequences of reduced levels of preventative burning but were having difficulty in reinstating it:

> ... look to your heritage of the fire stick and find a way to do the burning that needs to be done. Not simply for hazard reduction, that is, to make fire control easier, but for all the ecological benefits that fire brings.[58]

There has thus far been little in-depth analysis of the benefits and costs associated with Australian forest fire management. This may soon change given an expectation of greater fire threat in a drying climate combined with a realisation that employing state-of-the-art fire suppression technology at increasingly greater cost, has not reduced the incidence of large, damaging 'mega-fires'. In fact, the frequency of such fires is arguably increasing along with a disturbing propensity to routinely attribute them to climate change without even considering that their may be other more important factors at play, such as changed forest management paradigms and increasingly risk-averse fire-fighting strategies.

In the early 1990s, when Victoria's public lands were being managed in accordance with a more equitable balance between fire prevention and suppression expenditure, a University of Melbourne study estimated that a $24 benefit from averted property loss was being obtained for every $1 spent on wildfire suppression and prevention (mostly fuel reduction burning) by the then responsible department.[59]

In 2014, a Deloitte Access Economics (DAE) scoping study evaluated a hypothetical increase in annual fuel reduction burning from the then current 0.5% to 5% proportional area of the Blue Mountains National Park. It found that a consequential reduction in damaging summer wildfires would equate to a benefit:cost ratio of almost 6. This analysis included estimates of savings associated with reduced property loss, human health impacts and carbon dioxide emissions attributable to a reduction of summer wildfires. In 2017, an unpublished analysis by Dexter and Macleod, which extended the DAE model by including savings attributable to reduced wildfire impacts to water supply, found that this would increase the benefit:cost ratio of increasing the proportion of annual fuel reduction burning to more than 8.[60]

A more recent analysis of the costs and benefits associated with planned fuel reduction burning in the south-western forests of WA has also found that it delivers significant economic and social benefits. On average, the region's current burning program delivers a $31 million per annum saving in fire suppression expenditure, and a $169 million per annum saving in averted property loss/damage compared to a 'no burning' scenario. Long term modelling of various proportional annual burning options, suggests that every dollar invested in planned burning generates between $10 and $47 of benefit compared to the 'no-planned burning' option.[61]

Fire is inevitable in the vast majority of Australian forest types, and we essentially have a choice that can substantially influence how damaging it is. It is evident that more fire planned and introduced into the landscape during stable weather in the cooler months when it is easier to control and far less damaging, translates to fewer and less damaging mid-summer bushfires. The prospect of a drying climate and the gradual dilution of traditionally effective fire-fighting strategies is further reinforcing the need for more planned burning.

The divergence of views on fuel reduction burning

The question of how to deal with fire exposes a gaping flaw in the prevailing 'conservation culture' based on the emotion-charged premise of the bush being exceedingly fragile and requiring protection from human disturbance. This just doesn't match the reality of the Australian bush having been shaped by tens of thousands of years of fire, much of it deliberately ignited by humans. Over most of the continent, the bush and its flora and fauna has evolved to be resilient and ultimately dependent on periodic disturbance for its survival and renewal.

Despite unnaturally severe fire always posing a far greater threat to forests than popular villains such as timber production or mining, the environmental movement had for decades avoided articulating any firm position on the use of planned burning despite an obvious distaste for the notion of deliberately disturbing the bush. Perhaps this is unsurprising. Given the obvious ecological benefit of restoring natural fire regimes, no person or group which purports to be committed to saving the environment would want to appear to be opposing efforts to improve forest fire management.

In recent years there has been an increased emphasis on forest fire management within the 'conservation culture' narrative. Arguably, this can be traced to Victoria where a series of huge 'mega fires' burnt around half of the state-owned forested land (over 3 million hectares) in just a six-year period ending in 2009.

Following the severe 2003 and 2006 Victorian bushfires, the environmental movement was forced to confront the issue of planned fuel reduction burning in the face of a public backlash over their perceived lack of support for it, including accusations of wilfully obstructing some local burns. This created a dilemma because, while the movement wanted to maintain its inherent concerns about planned burning, it had no desire to appear insensitive to rural communities which regard it as essential for their safety.

The Wilderness Society (TWS) confronted this dilemma by proclaiming support for small-scale strategic burning – basically in a narrow strip adjacent to private land/town boundaries – whilst opposing broadscale burning which they typically characterise as being unplanned and of little strategic value. Its Tasmanian forests campaigner, Vica Bayley, articulated this distinction between different types of burning in 2007 when he said that "all conservationists are not opposed to all deliberate burning and indeed TWS supports ecologically-based prescription burning". However, he considered this to be "a very different concept to landscape-scale fuel reduction burns… with no consideration of ecological principles and biodiversity values."[62]

Also in 2007, the Victorian National Parks Association more clearly articulated the dichotomy of different levels of support for different types of burning – small-scale 'strategic' versus much larger 'broadscale' burns:

> We are strongly opposed however to broad-scale burning that is likely to alter ecological functioning … If current strategic fuel reduction burning is increased to broad-scale burning, there will be concomitant losses in biodiversity and environmental services, such as water quality. In the medium term, some of the forests will be made more flammable by repeated burning.[63]

This position would be reinforced in the Wilderness Society's '6-Point Bushfire Plan' released in 2008, which advocated only small-scale 'strategic' burns alongside other measures such as the greater use of aerial water bombing and the cessation of timber harvesting, which would supposedly remove the necessity to conduct broadscale fuel reduction burns.[64]

Although there is definitely a place for very small, meticulously planned strategic or 'ecological' burns, they obviously will not reduce fuels over broad areas to anywhere near the extent required to significantly lessen the intensity and damage of large summer conflagrations.

Accordingly, adopting this stance has effectively left the environmental movement advocating a completely unnatural 'no-fire' scenario in the vast bulk of the landscape. This would foster a massive fuel build-up that will invariably burn (often uncontrollably) with a heightened potential to cause severe ecological damage as well as human life and property loss.

Despite the rhetoric from the larger and most prominent environmental groups, most of the on-ground opposition to planned burning has emanated from much smaller grass-roots protestors such as, for example, the Strathbogie Sustainable Forests and the Euroa Environment Group which combined to oppose a 3300 ha fuel reduction burn planned by the Department of Environment, Land, Water and Planning in central Victoria during 2015.[65]

Despite their advocacy of small strategic local burns, the pervasive influence of the large environmental groups on the broader 'conservation culture' has meant that even these are difficult to achieve in practice. What this now commonly means at the community level was the topic of a recent memoir by former high-ranking NSW Rural Fire Service volunteer fire-fighter, Geoff Walker:

> ... we must backtrack some 30 years, to what now seems the golden age of local preventative bushfire control. Unlike today, it was an era when common sense and tradition prevailed. Back then I was an active bushfire brigade member, and we never wasted a good burn-off day. The patch of scrub I refer to was half the size of a football field ...
>
> Some twenty years later I actually asked the shire council to get it burnt off, since my fire-fighting days were over. They insisted on three separate environmental impact studies of the local reserves. Some 23,000 words later they recommended rotational burning with a three year cycle. When it didn't happen, I asked why. They replied saying that it was now officially classified as a 'riparian zone', which is a fancy way of saying it was too close to a watercourse. It couldn't be burnt off and, what's more, it wasn't a fire hazard. Meanwhile the undergrowth became a dense thicket which choked out the native plants.
>
> After more complaints, the Rural Fire Service inspected it and agreed with the council. Imagine our disbelief when a young cou-

ple started to build their new home across the road and were told that they had to spend some $40,000 to 'fireproof' their property against the threat of land that, officially, was not deemed a threat in the first place![66]

Similar examples of unfathomable bureaucratic intervention are apparently not uncommon across the nation, and the central role of 'save-the-forest' environmentalism in 'greening' local government functions wasn't lost on rural Victorians in the aftermath of the disastrous 2009 'Black Saturday' fires. During that period, public debate about the merits of planned burning reached new heights. Reportedly, it was by a large margin the most discussed topic in public submissions to the 2009 Victorian Bushfires Royal Commission.

In broad terms, community attitudes to planned fuel reduction burning can be grouped into three categories:

- Strong support – including support for increasing the area burnt each year as part of a professionally conducted program.
- Conditional support – subject to small, targeted burn areas with a lengthy time between burns and increased research into its environmental impacts.
- Opposition – based on a view that prescribed burning is environmentally destructive, unnecessary, and ineffective.

Quite understandably, support for fuel reduction burning is strongest amongst those who work within or who have lived close to forests or woodlands and have a practical appreciation of the link between fuel level and fire threat. This includes bushfire scientists, foresters and farmers. On the other hand, it is highly significant that the strongest opposition to fuel reduction burning is amongst those living in urban situations where there is little or no bushfire threat.

While the formal position adopted by Australia's most prominent environmental groups is one of grudging 'conditional support' for limited small-scale burning, after 2009 they have almost always advocated this position alongside qualifying statements deriding its effectiveness as a bushfire management tool.[67]

Immediately after the 2009 'Black Saturday' fires, the environmental movement endured considerable hostility over both its lack of support for most planned burning, and for its part in driving vegetation protection and forest policies which many believe to have pre-disposed the landscape to more catastrophic fire.

In the face of these attacks, and amidst fears that the resultant Royal Commission would recommend a greatly expanded planned burning program, environmental group supporters and some conservation scientists became more vocal in their condemnation of planned burning. One of the defining statements of the time was a newspaper opinion piece written by Andrew Campbell, a trained forester who had not worked in native forests for several decades by dint of his rise to prominence within the Federal environmental bureaucracy:

> In particular, the suggestion that having had more fuel reduction burning over larger areas more frequently during the drought of the last decade in Victoria would have prevented these fires – and by extension that doing even more of it is essential in the hotter, drier climate we are moving into – is not backed up by the best available science.[68]

While this was seized upon by environmental groups looking for credible validation of their position, it greatly angered foresters and fire managers who are intimately acquainted with the fire control benefits of planned fuel reduction burning and were insulted at the inference that it is not supported by evidence.

It is in fact self-evident that low intensity fires burning in light fuels are far easier and safer to control. Indeed, the concept, logic, and practical experience of prescribed burning as an aid to bushfire management is well established. However, scientifically quantifying and formally measuring the effectiveness of planned fuel reduction burning has always been problematic due to an array of inter-related variables which both determine its effectiveness yet present practical difficulties for controlled and repeatable experimental study.[69] Accordingly, there has been a relative paucity of published peer-reviewed research papers empirically quantifying the effectiveness of prescribed burning.

What this means for perceptions of the value of planned burning reflects a difference between how scientifically-trained field practitioners and academic researchers define 'acceptable evidence.' If it is limited only to peer-reviewed research papers published in prestigious scientific journals, then there may be some merit in the claim that the benefits of planned burning are 'not backed by the best available science'. Although conversely, there is a lack of published peer-reviewed research quantitatively proving that past planned burning doesn't reduce bushfire intensity or spread.

However, relying only on such a purely academic definition of 'acceptable evidence' effectively dismisses decades of unpublished applied research and documented case studies undertaken in-house by government scientists working for Australian land and fire management agencies (including the CSIRO), and unfairly denigrates its integrity. When applied research and case studies (there are at least 27 documented in WA, Victoria and NSW) are taken into account, it becomes clear that there is a considerable body of knowledge which presents a compelling case that prescribed burning plays an important role in mitigating bushfire frequency, extent and damage.[70,71]

In his award-winning book on Australian aboriginal burning, historian Bill Gammage also noted the inclination of academic researchers to dismiss observational evidence – in relation to Australia's pre-European fire regime – due to a preference for published, peer reviewed research based on empirically-tested hypotheses, over and above the presumed subjectivity and unreliability of observational records. Not withstanding the impossibility of applying empirical research methodology to land management hypotheses from several hundred years ago, Gammage articulated his concerns about the dismissal of such a large bulk of historical observations and anecdotes which seemed to be pointing to a common conclusion. In so doing, he highlighted the assumptions often used by scientists to dismiss the worth of mere observations, including the following comments from one prominent fire ecologist:

> Even if people [*Aborigines*] did plan [*pre-*] 1788 fire it is unwise to say so, because this would license ill-informed burning and extensive environmental damage.

> ... no matter how sociologically or psychologically satisfying a particular environmental historical narrative might be, it must be willing to be superseded with new stories that incorporate the latest research discoveries and that reflect changing social values of nature. It is contrary to a rational and publicly acceptable approach to land management to read a particular story as revealing the absolute truth.[72]

At the risk of over-generalising, it seems that many conservation scientists are overly cautious about the deliberate use of fire as a forest management tool. Amongst them is a healthy sprinkling of ecologists who – presumably in deference to the 'precautionary principle' – are disproportionately focussing on the question of whether or not planned fuel reduction burning damages the environment seemingly with the aim of advocating its restriction or cessation.

In recent years an upsurge in research papers of this type has coincided with a recommendation from the 2009 Victorian Bushfires Royal Commission which compelled Victoria to triple its planned fuel reduction burning program. One of the central arguments being used to discredit or downplay the benefits of broadacre fuel reduction burning is that it has little effect in limiting forest fires burning under extreme weather conditions when the most damaging human outcomes occur.

Bushfire science specialists have acknowledged for some time that past fuel reduction burning only minimally mitigates forest fires burning under extreme conditions when weather supplants fuel and topography as the driver of fire spread and behaviour. However, even under such conditions, forests that had been fuel reduced up to ten-years earlier exhibit better ecological outcomes through increased burnt area patchiness and decreased canopy loss compared to forests with heavier fuel loads.[73] Accordingly, they recover far quicker than areas with heavy fuels that are burnt more severely. Victorian research also suggests that vegetation burnt by low intensity fire can recover 3 to 5 times quicker than similar vegetation types burnt by high intensity fire.[74]

However, the most important point is that an overwhelming majority of forest fires do not burn under extreme conditions and can be more quickly controlled in fuel reduced areas. Such fires are therefore more likely to be contained and extinguished before the advent of adverse weather conditions which could otherwise have turned them into uncontrollable holocausts. Yet this considerable indirect benefit of fuel reduction burning is rarely recognised by academic researchers because these mitigated fires have been prevented from having any noticeable impact.

Many papers critical of fuel reduction burning have put forward alternative fire management strategies – usually including only small-scale burning adjacent to private land boundaries – but have typically failed to acknowledge what is then likely to happen to the vast bulk of the bush that would consequently be left to develop heavy fuel loads. Associated with increased academic criticism of planned burning has been a rise in the number of ecologists and related conservation scientists publicly denigrating, undermining, or questioning its effectiveness in the mainstream or online media. Some examples are:

> … most planned burning patches never encounter a bushfire during their effective lifetime [and] in any case, bushfires can burn even through one-year-old patches.[75]
>
> … there is no evidence that fuel reduction burning has any benefit in wildfire control.[76]
>
> For 26 out of 30 bioregions in south-east Australia, there is no evidence that prescribed burning has reduced bushfire sizes.[77]
>
> That recommendation has the potential to encourage burning large, remote areas - a strategy that will not afford increased protection to houses on days like Black Saturday.[78]
>
> A state-wide [*planned burning*] target, in contrast, encourages burning in remote locations where the benefits are negligible and fire-management resources are wasted.[79]
>
> In the forested regions of southern Australia prescribed burning is less effective in mitigating unplanned fire. To get a hectare less of wildfire you have to burn three to four hectares with prescribed fire.[80]
>
> Firefighters and farmers around Australia are calling for more hazard reduction burns to reduce the risk of future fires, but the evidence isn't in that this is the best approach.[81]
>
> Despite a focus on hazard reduction burning by many commenta-

tors after bushfires in Australia ... it was not as effective as the measures listed above.[82]

This is a fundamental conundrum of prescribed burning: though it is quite effective in theory, the extent to which we would need to implement it to affect fire behaviour across the entire state is completely unachievable.[83]

These sorts of comments emanating largely from researchers who have never been involved in operational fire management, have sparked concerns about the potential for misuse of academic credibility to support environmental campaigns opposing broadscale planned burning. Invariably when examined more closely, it is apparent that most of these comments have been drawn from research that is demonstrably lacking in practical appreciation of critically important context about planned burning exemplified by:

- Referring to or implying that planned burning is a failure if it doesn't physically stop a fire, even though its measure of success lies in mitigating fire behaviour to an extent that makes for easier control.

- Presuming that previous fuel reduction is ineffective in mitigating bushfire behaviour beyond a certain headfire intensity even though the relative fire intensity on the flanks and rear of such a bushfire may be an order of magnitude lower and thereby able to be significantly mitigated by lower fuels in previously burnt areas.[84]

- Failing to distinguish between the low ecological impact of planned burning undertaken under cool, stable conditions compared to the eminently greater impacts of unplanned summer bushfires burning under hot and unstable conditions.

- Wrongly equating cool burnt areas to wildfire burnt areas in a leverage context (e.g. x hectares of fuel reduction is needed to reduce wildfire extent by 1 hectare).

- Being unaware of the logistical capabilities and regulatory and planning functions of land management agencies and fire

services which has at times spawned wildly errant assumptions about planned burn timing and potential frequency.
- Presuming that all forests are burnt (or burnt in the same way) due to a lack of appreciation of those forest types that are unsuited to planned burning.
- Lacking an understanding about how much planned burning is done and that its effectiveness in mitigating bushfires is largely related to its proportional landscape extent.
- Neglecting to outline what would happen to the landscape if there was no planned broadscale burning.
- Failing to appreciate that broadscale planned burning, even when undertaken well away from settlement and property, helps to prevent small remote fires from developing into larger fires that may eventually emerge to threaten life and property.[85]

Much of the ecological concern surrounding planned burning stems from knowledge gaps about its potential impacts on the many thousands of species of flora and fauna which inhabit Australian bushland. There is no disagreement about the need for ongoing research to close these gaps. However, citing the 'precautionary principle' as a licence for doing little or nothing while waiting for the decades or centuries of research effort required to know everything before we can (supposedly safely) burn, is to effectively consign the bush to an eco-disaster in the form of periodic hugely damaging conflagrations in heavy fuel accumulations.[86]

In reality we are already living through this scenario, in large-part because there is not nearly enough planned burning being undertaken. This current lack of burning partly reflects an era where extensive bureaucratic layers of planning, notification, and approval must be negotiated before any burn can be lit; as well as the increased difficulty of burning safely given the greater numbers of houses integrated with bushland, and the high fuel loads which stem from of decades of insufficient burning.

To actively worsen this through strictly conforming to the 'precautionary principle' would be irresponsible in the extreme. After around 60-years of study there is much that is already known about planned burning – certainly enough to recognise that, even if based on imperfect knowledge, it is far less damaging than severe summer bushfires burning in heavy fuel loads accumulated in the prolonged absence of fire.

The propensity for so much academic effort to be devoted to questioning the wisdom and effectiveness of broadscale planned burning arguably represents a skewing of research priorities that sits uncomfortably against societal expectations of our scientific community as being free of agendas. For example, it is notable that research into the potential impacts of planned burning is not matched by an equivalent volume of study into the consequences of not burning, despite the prolonged absence of fire having been already identified as having ecological impacts.[87] Of greatest concern is that a research focus on the short-comings of planned burning elevates the importance of what are only potential and (most likely) subtle impacts, above the known and infinitely greater and more immediate ecological and physical degradation associated with severe mid-summer bushfires.

For example, after the huge bushfire which burnt through north-eastern Victoria and south-eastern NSW over a 60-day period in 2003, one former CSIRO scientist estimated that it had killed 370 million mammals, reptiles, and birds.[88] In addition, it was predicted that in the most severely affected half of the burnt area, up to 430 billion litres of water per year would be absorbed by regenerating trees in the Murray River headwaters until 2050.[89] The consequent reduction of stream flows would mean that Commonwealth and State government commitments under the National Water Initiative, including the Living Murray Initiative, Victoria's new water initiatives, and the return of water to the Snowy River, would be unable to be met.[90] The extent of this damage could have been significantly averted if widespread fuel reduction burning had not substantially declined over the preceding 20-years.

Foresters learnt the lessons of fire 60 to 70 years ago, whereas environmental activists and their political associates and supporters remain stuck in denial. It had been hoped that the tragic loss of life on 'Black Saturday' 2009 may have at least mitigated their lack of enthusiasm for planned burning. Unfortunately, it has only further strengthened their reticence, albeit through a concession that advocates its concentration in a narrow band adjacent to private land boundaries, seemingly as a trade-off specifically designed to remove the majority of burning from across the vast bulk of the landscape.

It seems reasonable to presume that this distaste for the deliberate use of fire to manage the bushfire threat is linked to the flawed cultural belief that has grown from decades of exposure to environmental causes in which forests have been portrayed as exceedingly fragile and needing to be saved from human disturbance. It is hard to find any other explanation for why so many with so little knowledge or practical experience can be so determined to overturn, or at best substantially dilute, a fire management strategy that has evolved from so much practical experience, applied research and observation over such a lengthy period.

The inevitable upshot of making effective forest fire management more difficult is ultimately that forest ecology is being increasingly sacrificed to the catastrophically worse environmental impacts wrought by high intensity summer bushfires burning in heavy fuels. This is especially ironic as it is being largely driven by those who regard themselves as conservationists, but who seem to be in denial that periodic extensive low intensity fire is a natural component within most of the Australian landscape. Worse still is that it threatens to overturn sensible forest fire management on the basis of misplaced ecological priorities peddled and supported by those with little or no practical experience of what they are opposing.

Chapter 8 Endnotes

1 Nicole Gill, 2016, "Rupture in Tasmania", *The Monthly*, April 2016. Professor Bowman was addressing a public meeting of Tasmanians concerned about the burning of ancient rainforest remnants by the 2016 Tasmanian bushfires.

2 "Potoroos bounce back", *Town and Country magazine*, 3 November 2014.

3 "Bandicoots thrive in Eden's forests", *Town and Country magazine*, 15 December 2014.

4 "New flora reserves formed to save remaining south-east NSW koalas" *ABC News*, 1 March 2016.

5 Graham O'Neill, 2004, "A bigger picture of disturbance," *Ecos* 18: 118 (January-March 2004).

6 DELWP 2015, *30-years of Long-footed potoroo monitoring at Bellbird*, Department of Environment Land Water and Planning News, Government of Victoria, 20 March 2015.

7 Sarah Chenhall, 2013, "Rise in bandicoot numbers delights ecologists", *The Magnet*, 21 February 2013.

8 Tim Low, 2017, *Invasive Species: A leading threat to Australia's wildlife*, published by the Invasive Species Council, February 2017.

9 V. Jurskis, 2016, "Too many koalas, too little science", *Quadrant Online*, 25 June 2016.

10 TSSC 2011, *Advice to the Minister for Sustainability, Environment, Water, Population and Communities from the Threatened Species Scientific Committee (the Committee) on Amendment to the list of Threatened Species under the Environment Protection and Biodiversity Conservation Act 1999 (EPBC Act)*, Threatened Species Scientific Committee, 25 November 2011.

11 SETA 2016, *Koalas in Mumbulla and Murrah State Forests*, South East Timber Association fact sheet.

12 DELWP 2015a, *Supporting the recovery of the Leadbeater's possum: Progress Report October 2015*, Department of Environment Land Water and Planning, Victorian Government.

13 DELWP 2017, *A Review of the Effectiveness and Impact of Establishing Timber Harvesting Exclusion Zones around Leadbeater's possum colonies*, Department of Environment, Land, Water and Planning, July 2017.

14 *Skullbone Plains, Tasmania*, Department of Environment Fact Sheet.

15 Forests NSW, *Annual Report 2011-12*, p. 16

16 NSW Government 2012, Submission No 332 to the NSW Parliamentary Inquiry into the Management of Public Land in NSW, Legislative Council, General Purpose Standing Committee No. 5.

17 Tasmanian Audit Office 2016, *Report of the Auditor-General No. 5 of 2016-17: Park management* (November 2016).

18 Josh Gordon, 2016, "On a path to ruin – Victoria's national parks are being run on the smell of an oily rag" *The Age*, 5 January 2016.

19 "Earth on Fire", *Catalyst*, ABC TV, 3 June 2014.

20 ANU Fenner School's Long Term Ecology Group blog: "A Scientific response to the Institute of Foresters of Australia Leadbeater's possum Pamphlet" (2016).

21 DELWP 2016, op. cit.

22 ANU Long Term Ecology Group website.

23 Blog comment in response to the article "Native forests are worth more unlogged, so why are we still cutting them down" by Caitlin Fitzsimmons, published in *The Age* and the *Sydney Morning Herald*, 4 October 2016.

24 A.W. Howitt, 1890, *The Eucalypts of Victoria*, by A.W. Howitt, F.G.S., F.L.S., Reprinted from the transactions of the Royal Society of Victoria for 1890.

25 Ibid.

26 B. Gammage, 2011, *The biggest estate on earth – how aborigines made Australia*. Allen and Unwin.

27 F.R. Moulds, op. cit., pp. 81-82.

28 S. Pyne, 2006 *The Still-Burning Bush*. Scribe Publications, pp. 54-56.

29 WA Forests Department 1927, excerpt from: *The Foresters' Manual*, Bulletin 39 Part III *Fire control*, pp. 23-69. "Controlled burning, together with popular education, should go far towards solving the fire problem. All areas which do not require complete fire protection will be burned systematically by light, controlled fires" In *Steven Kessell and fire in the jarrah forest*, by R. Underwood. In *The Forester* newsletter, Institute of Foresters of Australia, February 2015.

30 W.S. Noble, 1977 *Ordeal by Fire – the week the State burned up*. Jenkin Buxton Melbourne, 85 pp.

31 S. Pyne, 2006, op. cit., p. 57.

32 Ibid., p. 58.

33 R.H. Luke and A.G. McArthur, 1978, *Bushfires in Australia*. CSIRO Division of Forests Research, Australian Government Publishing Service, pp. 133-146.

34 R. Underwood, 2016, "The first aerial burns", *The Forester*, February 2016.

35 Government of Victoria 1983 *Report of the Task Force Appointed to Examine Fire Protection and Fuel Reduction Burning by the Forests Commission to the Hon. R.A. McKenzie, Minister of Forests* (April 1983).

36 R. Underwood, 2016, op. cit.

37 Commonwealth of Australia House of Representatives Select Committee 2003, *A Nation Charred: Inquiry into the Recent Australian Bushfires*. Chapter 6 – *Fire fighting resources and technology*.

38 B.D. Dexter amd A. Hodgson, 2005, *The Facts Behind the Fire – A Scientific and Technical Review of the Circumstances Surrounding the 2003 Victorian Bushfire Crisis*. Forest Fire Victoria.

39 Environment and Natural Resources Committee, Parliament of Victoria 2008 *Inquiry into the Impact of Public Land Management Practices on Bushfires in Victoria*. Chapter 5, Finding 5.2, p. 176.

40 2009 Victorian Bushfires Royal Commission 2010 *Final Report*. Recommendation 56, p. 295.

41 Victorian Department of Primary Industries 2013 *Discussion Paper: Current and future challenges in meeting DEPI's increased planned burning access, preparation, support and rehabilitation works requirements*, (unpublished).

42 Kath Sullivan, "East Gippsland residents fuming over Snowy River National Park fire", *The Weekly Times*, 18 June 2015.

43 *Goongerah – Deddick Trail fire January – March 2014 Community Report*, compiled by Emergency Management Commissioner, Craig Lapsley, Emergency Management Victoria, July 2014.

44 V. Jurskis, B. Bridges and P. de Mar, 2003, *Fire management in Australia: the lessons of 200 years*. In: Proceedings of the Joint Australia and New Zealand Institute of Forestry Conference, 27 April–1 May 2003, Ministry of Agriculture and Forestry, Wellington / Queenstown, New Zealand, pp. 353-368.

45 Ibid.

46 R. Thackaway, M. Mutendeudzi and G. Kelley, 2008, *Assessing the extent of Australia's forests burnt by planned and unplanned fire*, Bureau of Rural Sciences, Australian Government.

47 R.H. Luke and A.G. McArthur, op. cit., pp. 244-245.

48 A.B. Mount, 1985, *The case for fuel management in dry forests*. Paper prepared for Research Working Group No. 6 on Fire Research, Hobart.

49 R.J. Underwood, R.J. Sneeuwjagt and H.G. Styles, 1985, *The contribution of prescribed burning to forest fire control in WA: Case Studies*. In: Fire Ecology and Management of Western Australian Ecosystems, WA Institute of Technology, Environmental Studies Group Report No. 14 of Symposium Proceedings, Perth, WA.

50 WA forest research papers which have quantified the benefit of fuel reduction burning in reducing summer bushfires, includes:

R.J. Sneeuwjagt, 2008, *Prescribed burning: How effective is it in the control of large bushfire?* In: Fire, Environment and Society: From Research to Practice, Bushfire CRC; Australasian Fire and Emergency Service Authorities Council, Adelaide, SA, pp. 419-435. This paper cited the work of Lang who analysed fire patterns in the jarrah forests of the Collie District in south west WA from 1937 to 1987 and found a rapid decline in summer bushfires once the prescribed burning program began to treat more than 10,000 hectares per year (or 6% of the district's forest) in the early 1960s.

I. Abbot, 1993, "Ecology of the pest insect jarrah leaf miner (depidoptera) in relation to fire and timber harvesting in jarrah forest in WA", In *Australian Forestry* 56(3). In studying the history of prescribed burning and wildfire in an area near Manjimup in south west WA from 1940 to 1990, this paper reported on the dramatic decline in the size and number of serious bushfires after the introduction of prescribed burning by the Forests Department in 1958.

M.M. Boer, R.J. Sadler, R.S. Wittkuhn, L. McCaw and P.F. Grierson, 2009, "Long term impacts of prescribed burning on regional extent and incidence of wildfires – Evidence from 50 years of active fire management in SW Australian forests", In *Forest Ecology and Management* 259, pp. 132-142. This paper examined wildfire and prescribed burning records dating back to the early 1950s in the ~1 million hectares of forest in the Warren Region of south west WA. It found that the area treated annually by prescribed fire had had a significant effect on the annual number and extent of unplanned bushfires over a 52-year period. During this period, an average of more than 80% of the annual burnt area was attributable to prescribed fuel reduction burning. They concluded that a six-year cycle of prescribed burning significantly reduced summer bushfire hazard.

51 WA Senator Chris Back, *Bushfires Speech*, Parliamentary Debate in the Senate, 12 February 2015.

52 Roger Underwood, "How they made a pyre of the bush". *Quadrant Online*, February 2015.

53 Institute of Foresters of Australia 2009, Submission to the 2009 Victorian Bushfires Royal Commission, May 2009.

54 Craig Lapsley, 2015, "Bushfire message this summer is 'leave and live'" *The Age*, 30 December 2015.

55 FCV 1962, *Forests Commission Victoria 43rd Annual Report Financial Year 1961-62: Protection*: A total of 784 fires were attended by Forests Commission personnel during the season. The total season cost of fire suppression was £71,959, which equates to $952,051 in today's money (as at 2014).

56 S. Pyne, 1991, *The Burning Bush: A fire history of Australia*, Henry Holt and Company, pp. 356-360.

57 Professor J.R. Short, 2015, "The west is on fire – and the US taxpayer is subsidising it" *The Conversation*, 23 September 2015

58 *Fire expert on the threat facing Australia*, transcript of interview of Steven Pyne, Arizona State University, ABC TV, 7:30 Report, 15 December 2006.

59 J. Bennetton, P. Cashin, D. Jones and J. Soligo, "An Economic Evaluation of Bushfire Prevention and Suppression", *The Australian Journal of Agricultural and Resource Economics*, Vol 42:2, pp. 149-175 (June 1998).

60 B. Dexter and D. Macleod 2017, *What is the true cost of Forest Fire Management on Public Land in Victoria? What actions are required to increase transparency and accountability in reporting these costs in the public interest?* (unpublished report, November 2017).

61 V. Florec, D. Pannell, M. Burton, J. Kelso and G. Milne, 2016, *Think long term: The costs and benefits of prescribed burning in the south west of Western Australia*, Non-peer reviewed research proceedings from the Bushfire and Natural Hazards CRC & AFAC conference, Brisbane, 30 August-1 September 2016.

62 *Wilderness Society supports prescribed burn in Douglas Apsley National Park*, The Wilderness Society (Tasmania) Inc., Media Release, 16 May 2007.

63 Victorian National Parks Association 2007, Submission to the Victorian Parliamentary Inquiry into the impact of public management practices on bushfire in Victoria, prepared by Jenny Barnett, May 2007.

64 The Wilderness Society's 6-point Bushfire Plan, *Wilderness News*, Issue No. 173, Winter 2008 (p. 13).

65 Will Murray, "Logging postponed in Strathbogie forest, for now", *Euroa Gazette*, 2 November 2016.

66 Geoff Walker, "An old fire-fighters sorry saga", *Quadrant Online*, 28 December 2015. Walker has written a book about his bushfire experiences entitled, *White Overall Days*.

67 *Planned burns and clearing will not stop catastrophic fire events: report*, 10 September 2009.

Media release by the Victorian National Parks Association to launch a report jointly commissioned by them, The Wilderness Society, and the Australian Conservation Foundation entitled '*Victorian February 2009 Fires – A Report on Driving Influences and Land Tenure Affected*', by Chris Taylor.

68 Andrew Campbell, quoted in the Australian Conservation Foundation's submission to the Victorian Bushfires Royal Commission (May 2009).

69 W.L. McCaw, J.S. Gould and N.P. Cheney, 2008, *Quantifying the effectiveness of fuel management in modifying wildfire behaviour*, presented at the International Bushfire Research Conference, Adelaide Convention Centre, September 2008.

70 Victorian Parliamentary Environment and Natural Resources Committee, 2008, Chapter 2, p. 79.

71 Australasian Fire and Emergency Service Authorities Council and the Forest Fire Management Group 2015, *Overview of Prescribed Burning in Australasia*, Report for the National Burning Project: Sub-Project 1, March 2015, pp. 26-40.

72 Bill Gammage, 2011, *The Biggest Estate on Earth – How Aborigines made Australia*, Allen and Unwin, Appendix 1, p. 327.

73 K.G. Tolhurst and G. McCarthy, 2016, "Effect of prescribed burning on wildfire severity: a landscape-scale case study from the 2003 fires in Victoria", *Australian Forestry*, Vol 79: 1 (1-14).

74 K.G. Tolhurst, 2012, "Fire severity and ecosystem resilience – lessons from the Wombat Fire Effects Study (1984-2003)" Proceedings of the Royal Society of Victoria, Symposium on Fire and Biodiversity in Victoria, Volume 24:1 (2012).

75 Owen Price, University of Wollongong, In: "Reducing bushfire risk – don't forget the science", *The Conversation*, 11 October 2013.

76 Neal Enright and Joseph Fontaine, Murdoch University, In: "Climate change and the management of fire-prone vegetation in south western and south eastern Australia", *Geographical Research*, 52:1 33-34 (February 2014).

77 Phillip Zyllstra, University of Wollongong, In: "New modelling of bushfires shows how they really burn through an area", *The Conversation*, 22 August 2016.

78 Phillip Gibbons, Australian National University, In: "Fuelling rational debate – evidence counts for more than opinion in reducing the risk from bushfire", *The Age*, 19 January 2012.

79 Luke Kelly, Katherine Giljohann, and Micheal McCarthy (University of Melbourne), In: "Percentage targets for planned burning are blunt instruments that don't work", *The Conversation*, 30 March 2015

80 Matthias Boer (University of Western Sydney) and Ross Bradstock (University of Wollongong), In: "Burn bush, reduce emissions: evaluating costs and benefits of prescribed burning", *The Conversation*, 31 May 2011.

81 David Bowman (University of Tasmania), In: "Bad bushfire planning burns money", *The Conversation*, 18 June 2013.

82 Phillip Gibbons (Australian National University), In: "Which homes will survive this bushfire season?" *The Conversation*, 6 January 2014.

83 James Furlaud (University of Tasmania) and David Bowman (University of Tasmania), In: "To fight the catastrophic fires of the future we need to look beyond prescribed burning", *The Conversation*, 15 December 2017.

84 K.G. Tolhurst (University of Melbourne), personal comments, December 2017.

85 Australasian Fire and Emergency Service Authorities Council and the Forest Fire Management Group 2015, op. cit., pp. 40-43.

86 Ibid, p. 42

87 Ibid, pp. 57-58

88 Noeline Franklin, (former CSIRO scientist) 2003, Comments made following the Alpine Fires which burnt two million hectares in Victoria and NSW in early 2003. Reported in *SOS News*, 29 January 2007.

89 Cooperative Research Centre for Catchment Hydrology 2003, in: the joint National Association of Forest Industries and Timber Communities Australia submission to the National Water Initiative, April 2004.

90 B.D. Dexter and A. Hodgson, 2005, op. cit.

9

Going 'Green': Conservation or preservation?

"Do not confuse motion and progress. A rocking horse keeps moving but does not make any progress"
Alfred A. Montapert

Decades of largely unwarranted alarmism emanating from environmental campaigns against Australian forestry has created a self-perpetuating 'conservation culture', especially amongst our most environmentally-concerned citizens. This demographic includes arguably the majority of those employed in white collar professions that are most capable of promoting and advancing the culture and/or setting and implementing political and bureaucratic agendas.

This culture is exemplified by a conventional wisdom that forests are being saved by increasing national park declarations, while all other forested lands remain vulnerable to being trashed unless they too become national park. However, this entrenched belief is underpinned by outdated presumptions about the extent to which non-park forests are being used and damaged, and is rooted in ignorance or denial of the evolution that has occurred in the planning and regulation of forest use (particularly wood production) over the past 40-years. Albeir it must be acknowledged that forestry has never been perfect, and less than acceptable forestry practices have periodically given critics the opportunity to misportray exceptions to good management as though they exemplified standard practice.

Unfortunately, the widely accepted misconception that conservation values need saving in non-park public forests has encouraged State governments to ignore common sense rural realities when making forest policy decisions that have been clearly aimed at currying favour with the politically influential inner-urban, so-called 'green' demographic.

Indeed, the past 15-20 years of substantial national park expansion has been founded on the political realisation that there are vastly fewer disaffected rural voters compared with urban voters who strongly approve of new national parks despite mostly living remote from nature, being largely misinformed about forestry disputes, and having no direct material stake in their outcome.

This supposedly 'progressive' form of environmental politics which covets popularity amongst urban voters above the socio-economic consequences for rural communities, is a product of the combined influence of environmental activism (latterly supported by a like-minded academic cohort) and a mostly accommodating media. It has effectively promoted 'save-the-forest' campaigns by uncritically disseminating their messages whilst minimising or ignoring alternate and often better informed views.

Over time, this has fostered the disrespect that much of Australia's rural demographic now harbours for environmental activism, the mainstream media, and arguably many national parks themselves – particularly those inflicted upon them despite being at odds with the best interests of themselves and the forests, especially in relation to fire.

Critics of such a summation would undoubtedly claim that national parks have expanded in response to changed community values and expectations. While no-one could dispute the superficial reality of such an assertion, the question largely ignored by those who cite it is how this has come about. Should concerns of impending eco-catastrophe manufactured by exposure to endlessly repeated myths, half-truths and context-free exaggerations ever be acceptable as a legitimate basis for changed community expectations? One would hope not, but this has largely been the reality in relation to forests.

As most foresters have expected, it is becoming clearer with time that expanding the national parks estate has not improved the conservation of forest biodiversity. While there may now be far less commercial use of

Australian forests, it was never the massive problem that it was portrayed to be. The real problems have always been feral and invasive pests and unnatural fire regimes which can impact every hectare of every forest, and are just as bad or worse (arguably in the case of fire) than they ever were. This reality makes a mockery of the populist selling point that declaring a national park 'saves' (or restores) a forest.

There is now an urgent need to discern the future role and management of Australia's forests based on hard, truly objective evidence rather than political imperatives derived from misinformation. Critical to this is an acknowledgement that forest conservation and use can be complementary rather than competing paradigms, and that humans are a part of the natural world rather than a virus that needs to be largely excluded. Indeed, it needs to be recognised that it is human use of forests which necessitates the workforces, infrastructure, and economic imperatives that underpin the capability to effectively manage the threats to their ecological integrity.

Rightly or wrongly, some conservation scientists have become the (at times dubiously) credible champions of those who are personally or politically affronted by forestry. The importance of these scientists to any re-casting of the national parks-only conservation culture lies in the hope that they can see reason and are capable of change – although some are now so closely aligned with uncompromising environmental activism as to be surely a lost cause. On the other hand, Australia's most powerful and influential environmental groups and their political associates have a long-entrenched incapacity to compromise on their simplistic 'land-grab equals conservation' agenda. They are simply too reliant upon it and the donations that it attracts, for their continued relevance and existence.

Already, some international conservation scientists are seriously questioning formerly accepted conservation paradigms such as 'no-go' national parks. In 2012, Peter Kareiva, the former chief scientist for one of the world's biggest environment groups – the Nature Conservancy – actively challenged the past mythology of environmental protection as something only achievable by separation from human influence, and continues to advocate a new direction for the movement he has long been part of. His vision for the future:

> ... requires conservation to embrace marginalised and demonised groups and to embrace a priority that has been anathema to us for more than a hundred years: economic development for all. The conservation we will get by embracing development and advancing human well-being will almost certainly not be the conservation that was imagined in its early days. But it will be more effective and far more broadly supported, in boardrooms and political chambers, as well as at kitchen tables.
>
> None of this is to argue for eliminating nature reserves or no longer investing in their stewardship. But we need to acknowledge that a conservation that is only about fences, limits, and far away places only a few can actually experience, is a losing proposition. Protecting biodiversity for its own sake has not worked. Protecting nature that is dynamic and resilient, that is in our midst rather than far away, and that sustains human communities -- these are the ways forward now. Otherwise, conservation will fail, clinging to its old myths.[1]

Two years earlier, in 2010, prominent tropical forest conservation scientists, Douglas Sheil and Erik Meijaard, also argued that environmentalists needed to undergo a paradigm shift if biodiversity is to be conserved, especially in developing countries.[2] Furthermore they opined that some of the environmentalists' most deeply held beliefs were actually hurting the cause:

> Conservation needs to change. We need to recognize that pragmatic conservation solutions aren't about black and white, good and evil, or nature versus non-nature. Long-term conservation solutions have to involve compromises otherwise we will just be wasting our time.
>
> I don't think people are against conservation, but many are against the type of conservation which we are trying to impose on them...
>
> Compromise will be the key to making conservation acceptable and resilient in the context of the world's emerging democracies. We have to be pragmatic and realise that we can often win more lasting conservation gains by reconciling conservation needs with other human demands ... In the long-term we shall need to be more engaged with local motivations and ensure we can find com-

mon-ground. Otherwise conservation is simply a new colonialism and is unlikely to be sustainable.[3]

These thoughts are undoubtedly more relevant to conservation in the developing world where subsistence communities are still living within and relying on forests for much in their lives. However, the same need to avoid disenfranchising and dislocating rural communities by integrating their material and economic needs with the requirements of conservation, also applies to Australia.

Some Australian environmentalists recognised this quite early. In 1971, Judith Frankenberg, writing a book on nature conservation for the Victorian National Parks Association, noted that:

> There are many areas in Victoria rich in wildlife which have already been partly settled, or for which there are valid objections to their conversion to national parks of the American type. The British have solved this problem by creating multiple-use national parks. In these, scenic values and the needs of nature conservation are met by a system of control which does not preclude farming, forestry, village development of tourism and recreation.[4]

In so doing she was essentially arguing for multiple-use forest management which was already the philosophy underpinning the management of Australia's State forests at the time, but in practice up till then had been largely achieved only incidentally and was just starting to seriously evolve.

Not long after, such thoughtful musings emanating from Australia's environmental movement would be overridden by its adoption of an uncompromising determination to exclude all commercial uses from all public forests by creating ever more national parks and conservation reserves. This ideology has been in-vogue for decades now and the increasing area of what has been reserved, particularly in national parks, is largely how the nation's most powerful activist groups – such as the Wilderness Society and the Australian Conservation Foundation – measure their success. Latterly, it has also become how some politicians and political parties advertise their environmental credentials.

At the time of writing, there is little to suggest that Australia's most prominent environmental groups are contemplating any profound shift in their thinking and direction in relation to forests. They are largely continuing with their modus operandi of the past 30-years, which, with the help of a mostly supportive media, aims to convince ever more Australians and their politicians, that ending commercial resource use is necessary to guarantee the protection of forests and their ecology via an expansion of national parks and other reserves.

This differs somewhat from international environmental groups which have largely developed a pragmatic acceptance of commercial forest use as being part of an integrated approach to meeting global social and economic aims. For example, central to the World Wildlife Fund's advocacy of forest certification as a mechanism for improved environmental outcomes is that "WWF understands the threats facing forests today, but trying to prohibit the use of forest resources isn't a viable solution."[5] In contrast to this, while Australia's major environmental groups also ostensibly support forest certification (albeit only their own brand), they have largely endeavoured to use it as a means of wedging timber production out of native forests.

In Australia, the greatest hope for a more pragmatic form of environmental advocacy may well rest with the emergence of new groups, such as Planet Ark. In stark contrast to traditional environmental activism, it professes to be non-political and non-confrontational, proudly defining itself by what it supports rather than what it opposes. It espouses positive environmental actions through working with governments, businesses, and people to foster genuine environmental change. Its attitude to forests is far more nuanced than that of the most powerful and influential activist groups. Although it articulates the same need to protect forests with genuinely high conservation values, it also accepts wood production from well managed forests, and actively promotes the use of sustainably produced wood as having superior environmental credentials.[6] This represents a more evidence-based approach that is focussed on attaining real benefits rather than engaging in political lobbying for ideological outcomes with often only illusory benefits and seriously adverse socio-economic consequences.

That the continuing focus of Australian environmentalists on creating national parks is out-of-step with international thinking on forest conservation is also exemplified by the lengthy Ramsar-listing of Victorian and NSW river red gum forests and woodlands.[7]

Ramsar-listing is contingent on areas meeting the international conservation management criteria of 'wise use' enshrined in Article 3.1 of the Convention on Wetlands signed in the Iranian city of Ramsar in 1971. In 1987, 'wise use' was defined as "the sustainable utilisation of wetland resources in such a way as to benefit the human community while maintaining their potential to meet the needs and aspirations of future generations". That this does not preclude human resource use is significant and reflects the reality that these forests were able to maintain their Ramsar-listing for decades while being managed under regimes which included selective timber harvesting and cattle grazing which had also occurred for generations before the Ramsar- listing.[8]

The Ramsar *'Wise Use Guidelines and Additional Guidance'* resource continues to emphasise that human use on a sustainable basis is entirely compatible with international wetland conservation conventions. Yet respectively in 2008 and 2010, the Victorian and NSW river red gum forests were declared as national parks at the behest of concerted environmental campaigning based on the premise that timber harvesting and cattle grazing were incompatible with biodiversity conservation. On the basis of these decisions it is not unreasonable to conclude that what constitutes 'conservation' in Australia is very different to what it means internationally.

As Frankenberg had noted back in the early 1970s, Australia was adopting the American (rather than European) national parks philosophy whereby the land is sacrosanct and must be quarantined from human use or disturbance. This approach polarises land management into either conservation or use categories, rather than integrating both within the same landscape.

Understandably then, the land-use conflict that has long been evident in relation to Australia's forests mirrors that which has occurred in much of the United States, particularly in the West. Given the greater pre-occupation with citizen's rights in the US, it is perhaps unsurprising that government decisions meant to improve biodiversity conservation by restricting public land access and use have at times met strong, organised resistance.

During the late 1980s, a so-called 'Wise use movement' was formed in the American West as a response to the premise that environmental ideology is excessive, radical, and unfairly disadvantages rural communities to appease urban elites.[9] Inflammatory rhetoric reportedly emanating from 'wise use' advocates has led to critics (invariably from the environmental movement) variously accusing the 'Wise use movement' of being anti-environment, anti-government, and comprised of many so-called 'astroturf' front groups orchestrated and bank-rolled by extractive industries seeking to expand their access to resources.[10]

The environmental movement's associate, Source Watch, has also accused the 'Wise use movement' of being linked to militia groups and even religious sects, and claims that: "In every state of the US, relentless Wise Use disinformation campaigns about the purpose and meaning of environmental laws are building a grassroots constituency. To Wise Users, environmentalists are pagans, eco-nazis, and communists who must be fought with shouts and threats."[11]

The 'Wise use movement' may be none, all, or some of these things, but such vociferous protestations from environmental activists smacks of hypocrisy borne of outrage that their own campaign tactics are being used against them to effectively prosecute an attractive alternative view that is widely supported and politically influential. In a somewhat more scholarly examination of the 'Wise use movement' in 2002, Penn State University academic, James McCarthy noted that:

> The Wise Use movement is a broad coalition of over a thousand national, state, and local groups. Its existence by this name dates from a 1988 'Multiple-Use Strategy Conference' attended by nearly 200 organizations, mainly Western-based, including natural resource industry corporations and trade associations, law firms specializing in combating environmental regulations, and recreational groups. The conference produced a legislative agenda intended to

'destroy environmentalism' and promote the 'wise use' of natural resources – an intentionally ambiguous phrase strategically appropriated from the early conservation movement.[12]

McCarthy's reference to 'wise use' as 'an intentionally ambiguous phrase' derived from the early conservation movement is interesting. In fact, the phrase 'wise use' was originally coined by renowned pioneering forester, Gifford Pinchot, during his term as the first Chief of the US Forests Service (1905–10). He used it to convey the importance of sustainably managing natural resources to protect the productivity of the land and its capability to serve future generations. Accordingly, it has been likened to the concept of 'multiple-use forestry' devised a decade later in Europe.

Pinchot's personal objections to the appalling exploitation of American forests prior to 1900 have turned him into an iconic champion of conservation, which he variously described as:

> Conservation means the wise use of the earth and its resources for the lasting good of men
>
> Conservation is the application of common sense to the common problems for the common good.
>
> World-wide practice of Conservation and the fair and continued access by all nations to the resources they need are the two indispensable foundations of continuous plenty and of permanent peace.
>
> The object of our forest policy is not to preserve the forests because they are beautiful-or because they are refuges for the wild creatures of the wilderness – but for the making of prosperous homes – every other consideration becomes secondary.[13, 14]

That such descriptions of conservation, which clearly do not exclude human resource use, are being attributed to 'the early conservation movement' gives pause to reflect on the massive change in the meaning of 'conservation' over the past century. Whereas it was originally largely about controlling and regulating human use to minimise environmental impacts, today's most extreme and influential environmental campaigns, certainly in relation to Australia's forests, don't countenance any human resource use and aim to entirely eliminate it.

Going 'Green'

It seems that, at least in regard to Australia's forests, going 'green' by adopting the uncompromising agenda of environmental activism is not about conservation at all. Instead, it represents a transformation from a former regime of more actively managed multiple-uses towards a supposed idyll of full environmental 'preservation' where management is minimised. This is both ill-advised and unsustainable given the changeable nature of our forests, especially due to the prevalence of periodic fire, which ultimately most Australian forests and woodlands are predisposed to and naturally reliant upon for their renewal.

While national parks and other conservation reserves are an essential component of an effective landscape-scale conservation strategy, hard experience shows that they are rarely sufficiently resourced to be well managed. That is not to say that we shouldn't have them, but it is clear that the capacity to manage them is greatly improved when they exist amongst a mixed landscape of different tenures and uses that generate wealth, require workforces, and entail the maintenance of a road access network. If all public lands were national parks, most road and track access would eventually disappear and with it would go most of the capability to manage and protect environmental values, especially from unnaturally severe fire.

Amongst all the predictable derision it directs at those who support 'wise use' in the USA and multiple-use forestry in Australia, the environmental movement has seemingly never countenanced the possibility that, just maybe, those who advocate these concepts are not anti-environment at all. Those who live and/or work in or near forests are generally far more aware of the real threats and can see far more clearly what is and isn't going to protect the environment. What they think should carry far more weight than the arms-length protestations of environmental activists and their largely urban-based supporters.

The creation of new, ever-more undeserving national parks simply to satisfy a political ideology ultimately disenfranchises those rural communities which live cheek-by-jowl with forests by ending livelihoods and/or limiting their lifestyles on the pretence of 'saving' the environment. While these communities have found this hard to accept, the realisation that their pain has often been inflicted for little or no discernible environmental benefit – or even worse, has resulted (albeit unintentionally) in exacerbated environmental threats – must have made it even harder to accept.

Going 'green' over the last 15 to 20-years – during which considered public land use determinations have largely been supplanted by politically-expedient land grabs – represents the manifestation of a 'conservation culture' which has evolved from decades of unholy war waged against contemporary forest management. The resultant cultural belief that all public forests should be 'preserved' in undisturbed national parks has now infected the polity, the bureaucracy, and parts of academia to such an extent that the loss or marginalising of forestry expertise is crippling the capacity to sensibly manage forests.

This is far from just an Australian problem. Recently, Thomas Maness, Dean of the College of Forestry at the University of Oregon in western USA expressed similar concerns:

> When it comes to proper management of our public forests, some would like to take a page from the ancient Chinese philosopher Lao-Tzu. He posed the concept of non-action as an approach to life. In our forests, if we do nothing and let nature take its course, this line of reasoning goes, these landscapes will return to a more "natural state" on their own.
>
> The trouble is, the natural state of forests includes fire – a lot of fire. They will never return to a state that existed in the past, because the conditions that created them no longer exist. ... We need to recognize that fire has a role to play and that, at the same time, we can reduce the risk of catastrophic loss.
>
> I and foresters around the country grow increasingly concerned with the health of our federally managed forest lands. We also worry about the health of rural communities. Due to many factors – a changing climate, political inaction, the financial burden of managing a huge land base that produces very little – our approach to these forests has created a landscape ripe for large fires.[15]

Perhaps with some exceptions, going 'green' over the past two decades has achieved little more than create an illusion of 'protected forests' for millions of urban Australians who rarely, if ever, visit them. On the flip side, it has created rural anger and ongoing bitterness, and an otherwise avoidable social welfare burden. This is a downward spiral that could only be arrested if politicians collectively consider forest issues on the basis of all the evidence rather than just unquestioningly accepting the disingenuous exaggerations of environmental activism and its supporters; resists the temptation to access the political capital that it promises; and stands-up to the abuse and electoral backlash they will inevitably receive from disaffected green-left-voters.

Politicians need to recognise that care-free national park declarations are more than just a benign response to the unpopularity of logging, but can seriously weaken the capability to effectively manage forests. Typically in the past, only the socio-economic consequences of such decisions have been seriously considered while their adverse environmental consequences have been ignored. This needs to change in order to avoid further overturning the capability to manage Australian forests in the proven, time-worn manner needed to mitigate their critical ecological threats, especially that posed by unnaturally severe fire, notwithstanding the damage that has already been done to this management capability.

The world is undoubtedly confronted by a myriad of serious environmental problems, but well planned and highly regulated renewable use of a portion of forests in developed countries, such as Australia, is surely not one of them. Indeed, as is now being accepted internationally, maintaining an enduring workable balance between forest reservation and use is a far more sustainable conservation approach than effectively locking-up the whole landscape and throwing away the key.

Chapter 9 Endnotes

1 *Failed metaphors and a new environmentalism for the 21st Century*, a lecture by Peter Kareiva, In: *Peter Kareiva: An Inconvenient Environmentalist*, by Andrew Revkin, Dot Earth, *New York Times*, 3 April 2012.

2 Douglas Sheil and Erik Meijaard, 2010, "Purity and Prejudice: Deluding ourselves about biodiversity conservation", *Biotropica*, 42 (5) 566–568, September 2010.

3 Douglas Sheil and Erik Meijaard, interview on Mongabay website, 18 October 2010.

4 *Nature Conservation in Victoria*, by Judith Frankenberg for the Victorian National Parks Association (1971)
5 World Wildlife Fund 2017, Forest Certification.
6 Planet Ark website.
7 The Victorian Barmah and Gunbower forests were listed as Ramsar Wetlands of International Importance in December 1982, and the NSW Central Murray Forests were listed in May 2003.
8 The Ramsar Convention on Wetlands website.
9 The 'Wise use movement', Wikipedia, accessed November 2016.
10 W.J. Burke, 1993, "The Wise Use Movement: Right Wing Anti-Environmentalism", *The Public Eye Magazine*, Volume 7, No. 2, July 1993.
11 The 'Wise Use Movement', Source Watch, accessed November 2016.
12 James McCarthy, 2002, "First World Political Ecology: Lessons from the Wise Use Movement", *Environment and Planning A*, Volume 34:7 1281-1302, July 2002.
13 Gifford Pinchot (1865-1946), US Forest Service History.
14 Gifford Pinchot quotes – AZ Quotes.
15 Thomas Maness, 2017, *The Tao of Forest Management – Doing nothing in our forests could lead to catastrophic fire*, Terra – inspired stories from the edge of science, Oregon State University, 10 February 2017.

Index

Abbott Coalition Government, 48, 119, 211, 304-7

Abetz, Senator Eric, 73, 103-4

academic institutions and 'left-wing' ideology, 164-7

Albrechtsen, Janet, 98

Alpine fire of 2003, 209

and its ecological impacts, 359

Alpine National Park, 160, 330

Amazonia, Brazil,

deforestation rates, 35-37, 58

Arthur Rylah Institute of Ecological Research, 169

Ashbarry, Alan, 102

Attenborough, David, 130

Australian Broadcasting Commission (ABC), 133, 136, 275; *Australian Story* double-episode, 'Something in the Water' (2010), 108-11; *Catalyst*, 112-3, 330; Fact Check Unit, 119-20; *Four Corners* episode, 'Lords of the Forest' (2004), 106-8, 111-12; *Four Corners* episode, 'The Wood for the Trees' (1990), 107-8; *Lateline*, 59, 113; *Media Watch*, 116; perceptions of 'green-left' bias, 116-20; *Q and A*, 59, 113; *The Drum*, 57, 87; *7:30 Report* and *Stateline*, 112, 136, 347

Australian Bureau of Statistics, 91

Australian Conservation Foundation, 15, 18-19, 45, 46, 48, 55, 59, 129, 165, 185, 221, 263, 282, 288, 371

Australian Greens, 45, 72-73, 79, 91, 129, 132, 136, 162-4, 185-6, 188-9, 194, 203, 211-12, 264, 278, 286, 288, 302-5, 310-11; and influence in local government, 204-5

Australian Labor Party, 191, 194; and forestry issues, 188-9, 202-3; relationship with environmental movement, 44-5, 186; Labor Environment Action Network (LEAN), 186

Australian Museum Eureka Awards, 111-12

Australian National Audit Office (ANAO), 280-1, 303

Australian National University (ANU), 130-8, 165, 184, 291, 330; Academic Expertise and Public Debate – Policy 000359, 154; Code of Research Conduct – Policy 007403, 153-4; divestment of shares in seven resource use companies, 161; Fenner School of Environment and Society, 131, 133, 136, 138, 139, 145, 148, 150-1, 153-4, 164, 166; Long Term Ecology Group, 135, 137, 141, 331; Media and Outreach Awards, 157; Strategic Communications and Public Affairs Unit, 157; Wild Country Research and Policy; Hub, 140, 144, 283, 285

Australia's State of the Forests Report 2013, 11

Bali Climate Conference 2007, 142-3

Banyule City Council and the Ethical Paper pledge, 204-5

Barmah Forest, 21, 220, 236, 243-4, 246, 248, 252-3; Barmah Grazing Advisory Committee, 244; Barmah Forest Cattlemen's Association, 244

Barmah-Millewa Forum, 222

Bayley, Vica, 262, 350

Beattie, Peter (former Queensland Premier), 9

Berry, Sandra, 143

Beyond Zero Emissions, 165-6

'Black Friday' 1939 bushfires, 330; and the Stretton Royal Commission, 335

'Black Saturday' 2009 bushfires, 166, 193, 338, 345, 352-3, 359; and the 2009 Victorian Bushfires Royal Commission, 154, 194, 338, 352-3, 355

Bleaney, Dr Alison, 43, 108-9

Bligh, Anna (former Queensland Premier), 9

blue gum plantations, 325

Blue Mountains, 20, 24, 348

Bowman, Professsor David, 322

Bracks, Steve, 186-7, 255; and the decision to close the Otways timber industry, 214; and the red gum forests, 226

Brazil, 35-7, 65

Brown, Bob, 13, 43, 49, 71-3, 77, 113, 123, 203, 264, 276, 278, 294, 296, 302, 325; Bob Brown Foundation, 300

Brumby, John (former Victorian Premier), 187, 249

Bunnings, 275

Burke, Tony, (Federal Labor MP), 79, 186, 199-200, 262, 264, 278, 291, 293-4, 296, 302

Burns et al 2015, 134

Burton, Bob, 43

Bushfire Front of WA, 337

Bush Heritage Australia, 295

Butler, Mark, (Federal Labor MP), 262

Buttrose, Ita, 124

Cadman, Sean, 119, 267, 283, 286-7

Cadman, Tim, 65

Cain, John (former Victorian Premier), 183, 208

Cameron, Jan, 101-2; and Triabunna Investments, 101-2, 272-3

Campbell, Andrew, 353

Canada, 26, 29, 65; British Columbia, 23; Canadian Boreal Forest Agreement, 69-70, 77-80; Forest Products Association of Canada, 69

Cann River, xi, xii

Canopy, 79

CAR conservation reserve system, 22

Carr, Bob (former NSW Premier), 8, 187

Carron, Dr Leslie, 1

Carson, Rachel, 39

Carstensen, Kim, 70

Caswell, Tricia, 46

cattle grazing, alpine grazing, 160; red gum forest grazing, 220-1, 239, 243-6, 250

Construction, Forestry, Mining and Energy Union (CFMEU), 203, 270

Central Highlands forests, 10-11, 131-8, 151, 154; Black Spur, 330; mountain ash forests, 128, 129, 130, 134, 330-1; timber industry value, 136, 185

Chase, Alston, 39

ChipStop, 50

Choice modelling, 230, 237-9

climate change, 131, 138

Club Terrace, 324

Coetzee et al (2014), 28

Cohen, Barry (former Labor Environment Minister), 283

Colvin, Mark, 107, 112

Comrie, Neil, (Vic Bushfires Royal Commission Implementation Monitor), 209

Concerned Residents of East Gippsland, 55

'conservation culture', 13-15, 36, 63, 80, 89, 349, 360, 367, 377

Courser, Dr. Geoff, 43

Cousins, Geoffrey, 59, 113

Creighton, John, death of, 51

Creswick, ix, x, 3

Crikey, 57

CSIRO, 22, 26, 354

Curr, Edward, 236

Deloitte Access Economics, 136, 185, 348

Denholm, Matthew, 105-6

Denman, Andrew, 292

Department of Environment, Federal, 279-82, 284, 295, 303; Dr Kimberley Dripps, 282, 303

Dexter, Barrie, 236, 348

Dimassi, Maryann, 112

Doctors for Forests (Tasmania), 43

Doctors for Native Forests, 55

Drum, Damien (Victorian Nationals MP), 239

Ecological Vegetation Classes, 234-5

eco-tourism, 25, 47, 99, 252-5

Eden, 324

Emerson, Craig (Federal Labor MP), 283

Endangered Species Act (USA), 74

environmental campaigning, corporate bullying (or 'brand-mailing'), 58-61; cost of dealing with anti-logging field protests, 52 ; dispute resolution by direct negotiation between environmental and timber industry interests, 77-80; field protest methods, 49-52; litigation, 58, 71-7; social media, 57-8, 80-8; the internet and e-activism, 53-8

environmental movement, donations, 48; ideology, 47

Environmental Justice Australia (formerly the Environment Defenders Office (VIC)), 76

environmental protection commitment across the population, 90-1

Environment East Gippsland, 75-7

Environment Liaison Office (NSW and Victoria), 206-7

Environment Protection and Biodiversity Conservation (EPBC) Act 1999, 72-3, 304

Environment Tasmania, 263, 288

ESA Journals, 150

Eucalyptus globulus plantations, 267-8

Eucalyptus nitens plantations, 109-10, 267-8

Eucalyptus regnans forests, 130, 330

Euroa Environment Group (EEG), 196, 351

Evelyn, John, 2

Fairfax media outlets, 104, 133, 275

Finland, 23

Fire, Aboriginal burning, 17, 333-4, 354; comparative forest fire management in State forests and national parks, 29-30, 327, 328, 329-30, 340-1; disregarding professional forestry expertise, 209; ecological threat of unnatural fire, 15, 17-19, 327, 349, 360, 378; fuel reduction burning, the development and refining of, 335-6; fuel reduction burning, cost-benefit analysis, 348; fuel reduction burning, divergent views on, 166, 335, 340, 349-60; fuel reduction burning and ecological research, 355-9; pre-European fire frequencies, 17-19, 332-4, 354; reducing capacity to manage forest fire, 192, 336-40, 345-6; the shift from preventative management to emergency bushfire response, 345-7; the Western Australia success story, 342-4

Fitzsimmons, Caitlin, 114-5

Flanagan, Martin, 105

Flanagan, Richard, 47, 103-5, 112-3, 276

Flora and Fauna Guarantee Act 1988, 76

Foley, Luke (NSW Labor politician), 201

Ford, Andrew (CEO of Volunteer Fire Brigades Victoria), 248

Forests Alive, 163, 287

forest certification, 64-71, 372; Australian Forestry Standard, 64-7, 277; Forest Stewardship Council (FSC), 64-71, 276-7; Programme for the Endorsement of Forest Certification (PEFC), 64; different application between developed and developing countries, 68-69, 71

Forest Fire Victoria, 337

forest policy, planning and management, and bureaucratic instability, 192-3; and socio-economic consequences, 201-3; direct Ministerial interference in field management, 195-6; dominant influence of political apparatchiks, 194-5; politicisation of, 191-201; preference for administrative experience over technical expertise, 193-4; reduced influence or marginalising of professional forestry expertise, 206-10; rural opposition, 212-13

forest research, decline of, 170; shift from land management agencies to academic institutions, 168-171

ForestWorks, Melbourne conference, 2010, 190, 268

Forest Rescue, 50

Forest and Wood Products Support Group (Vic), 2015 launch of, 183-4, 191

Forestry Tasmania (now Sustainable Timbers Tasmania), 67-8, 72-4, 84-5, 119-20, 168, 267, 290, 292, 300, 302, 306; Bob Gordon (former CEO), 73, 273

Frankenberg, Judith, 371, 373

Franklin River campaign, 13, 42-4, 186

Friends of Leadbeater's Possum, 75

Friends of the Earth, 55

Fullerton, Ticky, 107, 112, 118

Gammage, Bill, 19, 354-5

Gandevia, Simon, 152

Gannawarra Shire Council, 242

Garnaut Climate Change Review 2008, 143-4

Garrett, Peter (former Federal Environment Minister), 297

George River catchment (Tasmania), 109

George River Water Quality Panel, 109-10

Germany, 2, 4, 64, 70

GetUp!, 54-5, 275, 288

Giddings, Lara, Tasmanian Premier, 264

Gillard, Julia (former Australian PM and her Government), 200, 264, 277, 283, 296, 302, 305; and agreement with the Bob Brown to form a minority government 2010, 203, 278

Goodall, Jane, 130

Goolengook forest, 225

Goongerah – Deddick Trail Fire 2014, 339-40

Grampians, 24

Grant, Will, 149

Grayling, Professor A.C., 122-3

'Great Forest National Park' proposal, 10-11, 129-136, 138, 139, 142, 152, 166, 183-5, 204, 212, 325

Green Carbon report 2008, 143-5, 149, 153, 156, 284

Greenpeace, 45, 46, 48, 69-70, 79, 142, 286; anti-coal industry campaign, 162

Greens Institute, 286

Greens, politicians, 47, 311, 326

Groves, Simon, 306

Guardian Australia, 57

Gunbower Forest, 220

Gunns Ltd, 15, 59, 103, 106, 190, 264, 267-71, 276, 326; Greg L'Estrange, 59, 190, 267-9, 273; 'Gunns 20' case, 71, 113; John Gay, 59, 113 ; proposed Tasmanian pulp mill, 15, 59-60, 62-3, 116, 270, 278

Guy, Matthew (Victorian Liberal politician), 205

Hansard, Alan, 287

Harris, George, 281

Harris, Peter (Secretary of the Victorian Department of Sustainability & Environment), 250

Harvey Norman, 60, 275

Hawke Labor Government, 266

Hay, Peter, 39

helmeted honeyeater, 134

Henningham, John, 115

Henry, Don, 163, 185

Hitchcock, Peter, 119, 266, 284, 287, 297-8

Horton, Richard, 128

Howard, John (former PM) 142, 202

Howitt, Alfred, 332-4

Huffington Post, 57

Huon Valley Environment Centre, 276

Iemma, Morris (former NSW Premier), 8

Independent Australia, 57

Ingram, Dave, 339-40

Innamincka Regional Reserve (South Australia), 27

Institute of Foresters of Australia (the IFA), 135, 148, 197, 288-91, 310, 345

Intergovernmental Panel on Climate Change, 146

International Union for the Conservation of Nature (IUCN), 23-4, 296-9, 308, 310; Red List of Endangered Ecosystems, 134

Invasive Species Council, 324

JANIS criteria, 22, 234-5

Jansen, Amy, 243, 245-6

Jennings, Gavin (former Victorian Environment Minister), 250

Journalism Code of Ethics formulated by the Media, Entertainment and Arts Alliance, 114

journalism, 'green-left' bias, 115, 190; teaching of, 115

Jolly, N.W., 3

Jurskis, Vic, 236, 247

Kambouris, Peter, 324

Kareiva, Peter, 28, 369-70

Keck, Margaret, 58

Keith, Heather, 143, 155; Keith et al, 2014, 147-51, 153

Kelty, Bill, 264, 279

Kessell, S.L., 3

Knight, Rod, 287

koala, 201, 325

Krien, Anna, 50

Kyoto Protocol, 142

Lake Pedder, 41, 43

Lalasz, Robert, 28

Lamberts, Rod, 149

land clearing for agriculture in NSW and Queensland, 37

Lane-Poole, C., 3

Latham, Mark, 202

Laurance, Professor William, 285; and ALERT (the Alliance of Leading Environmental Researchers and Thinkers), 285-6

Law, Geoff, 43, 301
Lawyers for Forests, 55, 75
Leadbeater's possum, 128-132, 134-7, 155-6, 183-5, 213; habitat range, 134-5; intensive surveying and growth in colony detections, 130, 136-7, 325; Leadbeaters Possum Advisory Group, 189, 325; nest box installation and artificial tree hollow program, 135-6, 331; upgrading to critically endangered status, 134-5, 136-7
Lee, David, 60
Lennon, Paul (former Tasmanian Premier), 203
Lessard, Laurent, 70
Liberal/Coalition Governments and forestry issues, 188-9
Lindenmayer, Professor David, 132-3, 135-7, 143, 146, 148, 151, 154, 184
Llewellyn, David, former Tasmanian Labor MP, 110
long-footed potoroo, 324
long-nosed potoroo, 322
Loyn, Richard, 151

Macintosh, Associate Professor Andrew, 286-7
McCarthy, James, 374
Mackey, Professor Brendan, 119, 143-4, 282-5, 296, 300
Magna Carta, 1
management challenges for forests and woodlands, 327-31
Maness, Thomas, 377
Markets for Change, 60
Marr, Alec, 102, 272
Marvier, Michelle, 28
Mathoura, decline of, 25
McAllister, Jenny, (former ALP National President), 186, 262
McHugh, Peter, 192
McQuillan, Dr Peter, 287-8
media, current affairs weblogs, 56-7; community trust of mainstream media, 120-1; Independent Complaints Review Panel, 106; steorotypical angles in forestry coverage, 101
Melbourne's domestic water supply catchments, timber harvesting in, 38
Melbourne University Early Learning Centre, 128
Millewa forest, 21, 220, 236, 247
Milne, Christine, 276, 296, 298, 302, 310
Moore, Patrick, 46
Multiple use forestry, 4-19, 220, 323, 326, 367-78; concept of, 4-5, 20; declining importance of, 11-12, 20, 27, 211; definition, 4-5; in NSW, 8, 12, 323; in Tasmania, 12; in Western Australia, 8-9, 12; in Victoria, 5-7, 12, 22, 27-28, 323, 325, 331; management intervention to restore fire-killed forests, 330; role of wood production in effective forest management, 327-8
Murray Valley National Park, 25

My Environment, 118, 129, 132

National Association of Forest Industries, 90, 287, 337
National Forest Summit 1996, 65
National parks and other conservation reserves, 159, 322-6, 367-78; along the Murray River in Victoria and NSW, 21-22, 27-28, 253, 255; and ecological research, 170-1; and fire, 330, 368-9; and political expediency, 200-2; American philosophy, 24, 373; Australian forested parks, 10; Australian Labor Party and national parks, 186, 187-8; definition, 20; growth of NSW parks and reserves, 8, 29, 187, 200-1, 253; growth of Victorian parks and reserves, 7, 29, 184, 187, 200; management constraints, 328-9; management expenditure per hectare, deficiency of, 25-6; questioning the concept of national parks, 20-30, 322-3, 367-9; revenue versus management costs, 328; south-east Queensland parks and reserves, 9-10, 29, 187; Tasmanian parks, 10, 20, 22, 29, 187; value of, 28-9; Western Australian parks and reserves, 8-9, 29, 187, 200
National Reserve System and the JANIS criteria, 25-6
Neumann, Fred, 151
Neville, Lisa (Victorian Government Minister), 184
Newman, Campbell (former Queensland Premier), 9, 210-11

News Ltd media outlets, 104
New Matilda, 57
New South Wales, 169; cypress pine/ironbark forests and woodlands, 24; Forestry Corporation, 205, 324; Forests Products Association, 215, 243; National Parks and Wildlife Advisory Council, 202; Natural Resources Commission, 201, 243; Natural Resources Advisory Council, 207; Parliamentary Inquiry into the management of public lands 2013, 27, 202, 253
Nicholson, Hugh and Nan, 43
Norwood, Rosemary, 287
Norman Wettenhall Memorial Lecture 2006, 197
Northcote by-election, 2017, 136
North East Forest Alliance, 50, 165
Noss, Dr Reed, 17, 285

Obama, Barack (US President), 98-9
OMICS International Group, 150-1
Online Opinion, 87
Oosting, Paul, 275
orange-bellied parrot, 134
Orbost, xi, xii
Otway Ranges (Victoria), 22, 196, 200, 201; and timber production, 225-6; Great Otway National Park declaration, 214, 255; VEAC Investigation, 225-6

Palaszczuk, Anna (Queensland Premier), 10

Pillaga Scrub, 24
Pinchot, Gifford, 375
Planet Ark, 372
plantations, 4, 39, 289; hardwood, 9, 45, 214, 267-8, 325; softwood, 9, 38-9, 41-2, 45, 214
'post-truth', Oxford Dictionary Word of the Year 2016, 122-4
Pressey, Professor Bob, 29
Putt, Peg, former leader of the Tasmanian Greens, 288
Pyne, Stephen, 347

Quadrant Online, 57
Queensland, 169; native forest wood production in south east, 9-10

Rainforest Action Network, 60
Rainforest Alliance, 64, 70
  Neville, Anita, 67
Ramsar Convention on Wetlands, 220, 373; and the definition of 'wise use', 373
Rao, Mr Kinshore, Director of the World Heritage Centre, 303
Redwood, Jill, 43
Rees, Nathan (former NSW Premier), 8
Regional Forest Agreements (RFA), 25-6, 66, 72, 75, 122, 189, 200-1, 202, 209-10
Regional Economic Modelling and Planning System (REMPLAN 2.0), 242-3
Register of Environmental Organisations, 48, 76

Resolute Forest Products, 69-70
Ricketts, Aiden, 165
Ritchie, Euan, 29
river red gum forests and woodlands, floodplain, 220; and timber production, 220-1, 239-43, 250; and cattle grazing, 220-1, 239, 243-6, 250; and fire management, 246-8; Mid-Murray Forest Management Plan (Victoria), 229; overturned flooding regime due to river regulation, 220-2; pre-European distribution, 236; Ramsar-listings, 220
Rivers and Red Gum Environment Alliance, 222-3; '*Conservation and Community*' plan, 249-51
Robinson, Rueben, 247
Rolley, Evan, 118
Routley, Richard and Val, 39-40, 139,
Rudd, Kevin (former PM), 141, 202-3

salvage logging, 137-8
Scammell, Dr Marcus, 108-9
Schein, Edgar, 35, 89
Schlich, Sir William, 4
Schnieders, Lyndon, 186, 262, 275-6
Schultz, Dr Beth, 43
Schwartz, Morrie, 105
Sheehan, Paul, 54
Sheil, Douglas and Meijaard, Eric, 91-92, 370-1
Smartwood, 64
social licence, 58, 61-64

Soule, Emeritus Professor Michael E., 16
Source Watch, 374
southern brown bandicoot, 322
Sparkes, Dr Gillian, (Victorian Commissioner for Environmental Sustainability), 209
St Helens (Tasmania), 109-10
Still Wild Still Threatened, 275
Strathbogie Sustainable Forests, 351
Sturt, Charles, 236
superb parrot, 231
Sustainable Forestry Iniatiative, 71
Swain, E.H.F., 3
Sweden, 23, 65
swift parrot, 72

Ta Ann Tasmania, 271, 275
Tantawangalo Koala Nature Reserve, 324-5
Tasmania, 43, 47; Bartlett, David (former Tasmanian Premier), 110; bushfires of 2016, 99; business threats to the timber industry prior to 2010, 267; Florentine Valley, 49, 262-3, 298; forest communities, pre-European extent, 23; 'forest peace deal' process, 47, 68, 78, 105, 199, 203, 209, 263, 265, 273-9, 311; giant freshwater crayfish, 99; Helsham Inquiry (1987), 266; Hodgman Liberal Government, 47, 211, 297, 307; Hodgman, Will (Tasmanian Premier), 99; Labor-Greens minority government 2010, 110; Lapoinya Forest dispute, 49, 77; *National Parks and Reserve Management Act 2002*, 308; plantations expansion, 40; proportion of forests in parks and other conservation reserves, 10, 23; Salamanca Agreement, 266; Southern Forests, 22, 49, 52; special timbers sector, 292-3; Styx and Weld valleys, 298; Tarkine region, 200, 309; Tasmanian Community Forest Agreement (2005), 266; Tasmanian Regional Forest Agreement (1997), 266, 300-2

Tasmanian Forests Agreement, 12, 79; and 'Statement of Principles', 264, 274, 279

Tasmanian Forests Agreement 2012, 264-5, 291-4, 297-8, 300-4, 309; and Upper House Inquiry, 292-3

Tasmanian Forests Intergovernmental Agreement (the IGA), 264, 279-80, 282, 284

Tasmanian Forests Independent Verification Group (the IVG), 264, 274, 279-94, 302-3

Tasmanian Forests Agreement Act 2013, 79-80, 263, 265, 293-6, 307

Tasmanian Forests Agreement Bill 2012, 264-5, 291-4, 296

Tasmanian Greens, 270-1, 293, 298
Tasmanian Land Conservancy, 295
Tasmanian Special Timbers Alliance, 292

Tasmanian Times weblog, 57, 80-88
Tasmanian Wilderness World Heritage Area, 119-20, 262-6, 284, 307-11, 326; attempt to partially revoke

the 'minor boundary amendment' addition, 211, 265, 304-7; 'minor boundary amendment' addition, 199, 278, 291-306

Taylor, Adriana, Tasmanian politician, 293

Taylor, Dr Chris, 166; Taylor et al (2014), 153

Terania Creek protest, 14, 42-3

The Australia Institute, 286

The Age/ Sydney Morning Herald, 105, 113, 114, 118-9, 155, 165

The Australian, 98, 104-5, 163

The Conversation, 132, 137, 152, 158-63

The Hobart Mercury, 154, 155, 163

The Monthly, 101, 103, 105

The Weekend Australian, 99

Threatened Species Scientific Committee (TSSC) Conservation Advice re Leadbeater's Possum, 134-5

Thwaites, John, (Victorian Environment Minister), 226

Timber Communities Australia, 156

timber harvesting, 323-4; decline of in native forests, 11, 29; faster than sustainable harvest rates, 188; separation from broader land management functions, 205

timber imports, tropical hardwood, 11, 30

Timber Industry Strategy, Victoria, 7

Toolangi forest, 195

Toorongo Plateau, 330

Twain, Mark, 54

Underwood, Roger, 107, 343-4

UNESCO, 27, 310; Tasmanian Wilderness Reactive Monitoring Mission (2008), 297-98; Tasmanian Wilderness Reactive Monitoring Mission (2015), 301; World Heritage Committee, 199, 265, 293-311

United Kingdom, 65; Brexit referendum, 122-3

United Nations Food and Agriculture Organisation (FAO), 36

University of Melbourne, Melbourne Sustainable Society Institute, 163-5; School of Forest and Ecosystem Science, 169

USA, climate change and forests in Oregon, 88-9; Forest Service fire budget, 347; park management expenditure, 26; Trump, Donald (US Presidential candidate), 98; Trump campaign and election victory 122-3, 213; Wildlands Network (formerly Project), 15-16, 140, 285

van Tiggelen, John, 101

Venter, Dr Oscar, 285-6

VicForests, 12, 67, 75-6, 148, 196-7, 205-6, 331

Victoria, 47; Andrews Labor Government, 26, 129, 185, 194; Bracks Labor Government, 206, 213, 222-3, 226-7; Department of Conservation, Forests and Lands, 168, 192; Department of Conservation and Natural Resources, 208-9; Department of Environment, Land, Water and

Planning, 76, 193, 196, 206, 351; Department of Environment and Primary Industries, 168, 338; Department of Sustainability and Environment, 194, 196, 205-6, 226, 244-5; East Gippsland, Prologue, 52, 197, 200, 201, 214, 339-40; Forests Commission Victoria, ix, xi, 6, 167, 192, 208, 243, 330, 334; Forest Industry Taskforce, 152, 184-5, 209; Heyfield, 213, 269; Kennett Liberal Government, 213; Latrobe Valley region, 212; mega fires, 2003-9, 138, 336, 338, 349; Napthine Liberal Government, 185, 189; Parks Victoria, 26, 252, 339; pre-1970 National Parks, 6; river red gum forests, 21; School of Forestry, Creswick, ix, 3; South Gippsland, 277; Victorian Government, 168, 183-6

Victorian Association of Forest Industries, 52, 337

Victorian Environmental Assessment Council (VEAC), 201, 212, 222-55; and the Environmental Conservation Council, 223; and the Land Conservation Council, 222, 234; and national park expansion, 223, 227-8; and the *Victorian Environmental Assessment Council Act 2001*, 227, 231-2, 249; and the *Commissioner for Environmental Sustainability Act 2003*, 232; Duncan Malcolm (former Chairman), 228; River Red Gum Forests Investigation (2005-8), 222-55

Victorian National Parks Association, 221, 229, 371
  and fuel reduction burning, 350
von Carlowitz, Hans Carl, 2

Wade, Felicity, 186
Walker, Geoff, 351
Warrumbungles National Park, 24
water quality and yield, affect of timber harvesting on, 38
Weber, Jenny, 276
Weilangta State Forest court case, 72-4
Western Australia, 169; and fire management, 335-6, 341-4, 348; Conservation Commission, 207-8; Department of Conservation and Land Management, 107; WA Forest Alliance, 43; WA Forest Products Commission, 12, 67, 205
West, Professor Jonathon, 280-1, 283, 287, 289
wilderness, definitions of, 300-1
Wilderness Society, 15-16, 18-19, 43, 46-50, 55-6, 59, 66, 129, 142-5, 154, 165, 186, 221, 262-3, 272, 275-6, 282-5, 287-8, 298-301, 308, 310-11, 371; and fire management, 18, 350-1; Forests and Woodlands Policy (revised September 2005), 45, 145; partnership with the ANU Fenner School, 139-42; WildCountry Science Council, 16, 283, 285; WildCountry Vision, 16-17, 140, 285
Williams, Professor Jann, 286
'Wise use movement', 374-6
Wilsons Promontory, 20

Wood, Graeme, 101-2, 166, 272-3; and Triabunna Investments, 272

woodchipping, 38-41, 66, 140; Bell Bay chipping and loading facility, northern Tasmania, 273; declining demand for native forest chips, 11; Eden woodchip mill, 51; Triabunna woodchip mill, 101-2, 164-5, 272-3

Wooley, Charles, 105-6

World Heritage Convention, 308-9

World Wildlife Fund, 16, 35-6, 45, 48, 64, 70, 372

Wye River bushfire 2015-16, 196-9, 346; decision not to conduct a coronial inquiry, 198; Inspector-General for Emergency Management (IEGM), 197-8

Ximenes et al 2016, 146

Yale University, 291

Young, Virginia, 163, 286-7

Zero Carbon Australia 2020, 165; *Land Use: Agriculture and Forestry Discussion Paper*, 2014, 165-7

Zero Emissions Network, 143

www.ingramcontent.com/pod-product-compliance
Ingram Content Group UK Ltd.
Pitfield, Milton Keynes, MK11 3LW, UK
UKHW041415180426
11947UKWH00007B/143